Care in Technology

To Bernadette Bensaude-Vincent, who taught me a lot
In testimony of friendship

Innovation and Responsibility Set

coordinated by
Robert Gianni and Bernard Reber

Volume 7

Care in Technology

Xavier Guchet

iSTE

WILEY

First published 2021 in Great Britain and the United States by ISTE Ltd and John Wiley & Sons, Inc.

ISTE Ltd
27-37 St George's Road
London SW19 4EU
UK

www.iste.co.uk

John Wiley & Sons, Inc.
111 River Street
Hoboken, NJ 07030
USA

www.wiley.com

Library of Congress Control Number: 2020951011

British Library Cataloguing-in-Publication Data
A CIP record for this book is available from the British Library
ISBN 978-1-78630-559-6

Contents

Acknowledgments

Several colleagues were kind enough to re-read all or part of this book and give me suggestions for improving it. I thank them very warmly: Bernard Reber first of all, for his very careful reading and his advice, as well as Emanuele Clarizio whose comments are always invaluable. Charles Lenay and Pierre Steiner then, who enlightened me on the TAC thesis and its philosophical sources. Dimitri El Murr who allowed me to see more clearly in certain terminological problems linked to ancient Greek.

My thanks also go to Rionne, for his support and encouragement.

Foreword

A good number of the works that combine innovation and responsibility[1], science and society, or even ethics and new, emerging or controversial technologies do not really take the technologies themselves seriously, nor their relationships to humans and the environment. To be more precise and less unfair, they still base themselves upon dualisms: nature/technology, technology/life, technology/humans, humans/nature, subject/object. What they are desperately lacking is an informed, balanced, and plausible philosophy of technology. However, this well-documented work by Xavier Guchet, who leads an interdisciplinary team at the University of Technology of Compiègne, member of the Sorbonne University Alliance, offers not only a strong thesis, but is backed up by a panorama as broad as it is assembled to good use. I emphasize the fact that his immersion in a university of engineers gives even more credit and plausibility to his work. This book is not only addressed to philosophers, but to engineers and any person who takes account of our technological environments[2], sometimes as threats, sometimes as sources of revelation and solutions, but also with a role to play in our definition as humans.

First of all, we have here a state of the art of the most interesting approaches to the philosophy of technology, ranging from Greek philosophy, especially that of Aristotle, to contemporary work. Configurations of the debate which we have inherited from ancient times have not lost their relevance. Against the Sophists who enthusiastically welcomed the development of technology considered from the perspective of an infinite progress, Plato's metaphysics already indicated this danger

1 For a general overview see Sophie Pellé and Bernard Reber, *From Ethical Review to Responsible Research and Innovation*, ISTE and Wiley, 2016.

2 One approach that strives to combine technology and ethics is that of Armen Khatchatourov, with the collaboration of Pierre-Antoine Chardel, Andrew Feenberg and Gabriel Périès, *Digital Identities in Tension, Between Autonomy and Control*, ISTE and Wiley, 2019.

of unlimited variation of forms in the *polis*, by separating Being and becoming, and submitting the latter to the unchangeable order of the former. Even more, Aristotle had already domesticated technology in a quadruple principle of limitation that prevents a collapse into excess: by form (formal cause), by the "implicit forms" of the material (material cause), by the ultimate purpose (final cause) and by the corporeal possibilities of the living being (efficient cause).

Xavier Guchet therefore presents the broad progress of the biological philosophy of technology, whose connecting thread is the concept of the externalization of life in technology, with great names as Ernst Kapp, Alfred Espinas, Henri Bergson, Edouard Le Roy, André Leroi-Gourhan, Georges Canguilhem, and some lesser known such as William Morris, John Dewey, Lewis Mumford, Laszlo Moholy-Nagy, Gilbert Simondon, whose common feature is that they have considered technology from the point of view of its links with life, with the theme of projection of the organs or of externalization. This also concerns René Descartes, Karl Marx, Arnold Gehlen, Helmuth Plessner, Jose Ortega y Gasset, Paul Alsberg, Francisco Varela, Pietro Montani and Bernard Stiegler. It is gratifying that the *Innovation and Responsibility* set of books devotes several pages to the French philosopher of technology Gilbert Simondon, quite unique in this field and little known in the English language, at least in a form which is not solely exegetic.

One of the originalities of this discussion is the presentation of the Kantian conception of technology as a decisive philosophical reading of the divorce between technology and life, as it was emerging in the 18th century. Kant's genius is to have endorsed this divorce, while indicating, as if in negative space, the conditions for it to be surpassed. Far from continuing to conceive technology, as had been done before him, as *organon*, he proceeded to a split in the very concept of technology in parallel with another, that between organic life and the life of any singular rational and moral person. Technology points toward the representation of goals, but it also sinks into the depths of the body to the point of lodging in the automatism of the living machine. It is somewhat mechanical and somewhat finalized, between causation and purpose. We can understand this "somewhat" as "at the same time".

Companionship with Kant allows Xavier Guchet to discuss the work of Dewey on the aesthetic experience as a perception of the relationship between my action and its effects in the world, between the things I manufacture and the relations that these things will forge with other beings in the world, or recent research in bio-aesthetics. This is the case for example with Montani who sketches a third way, that of the concept of institution, between the heteronomous constitution of the subject and the forms of its experience through technology and the absolute self-constitution of the subject. The concept of institution allows us to say that the subject is not self-created and what brings it into existence is not a set of structures imposed upon it by a foreign authority. The subject does not belong to itself without experiencing a sort of heteronomy.

Similarly, Xavier Guchet's book casts a critical eye over the project of the *Encyclopédie,* following the historical philosopher Michel Foucault in his course at the College of France (1976), but which Kant might also have shared: far from being only a political or ideological opposition to the monarchy and to a form of intransigent Catholicism, the *Encyclopédie* is a political and economic operation of homogenization of technological knowledge according to four mechanisms for the disciplining of knowledge and technical know-how: a selection separating knowledge into legitimate and non-legitimate; standardization and hierarchization of dispersed knowledges and a pyramidal centralization which allows control over them. If the Encyclopedists cannot be accused of having opened the way to industrial technology or even to the scientific organization of labor, Diderot and his partners had already initiated, in some respects, the expulsion of the craftsman from his know-how of experience, by a formalism that drew up an organization of knowledge and technological practices entrusted to the care of a new category of experts, a certain type of logicians, as Diderot writes.

These conceptions are only traces of history; projects such as following the traces of Amazon employees, which Guchet describes and aptly denounces, are of the same vein.

In the relationships between ethics and technology, he is very interesting in the reconfigurations to which he invites us, through his assiduous and precise examination of philosophers of technology, including the "Constitutive/Constituent Anthropological Technology" current, very well represented in his laboratory at Compiègne, as well as the four currents of the biological philosophy of technology which he discusses and distinguishes.

One of the main theses is as follows. It is often and too readily admitted that there is, on the one hand, a positive technology oriented to care and, on the other, a negative technology, animated by the aim of mastery slipping into predation toward all forms of biological as well as social and subjective life. However, according to him, a thought of care in technology includes two fundamental requirements: understanding the technological relationship of human beings to the world according to the category of *insertion* and not that of mastery, and never forgetting the capacity of individuals for autonormativity[3]. His intuitions have a long history since these are the Greek analyses of medicine, with their "square of care" which is designed to impose itself as the whole set of guiding principles for the practice of engineers and

3 In the series *Responsible Research and Innovation* see the book by Marc Maesschalck, *Reflexive Governance for Research and Innovative Knowledge*, ISTE and Wiley, 2017.

designers today. This square of care in the design of technology, which he will detail with the support of a large portion of the traditions of the philosophy of technology, is based on the following principles:

Care results from a double movement of incorporation and insertion of natural processes into the operations of human action.

The human being fulfils its nature as a living being in making a technician of itself. Technology repays its "debt to life", to quote Simondon, by letting itself be guided by the goal of a fulfilment of life, that is to say of value-creating activity once we recognize the value of non-human living beings. Here he is particularly indebted to Dagognet, Simondon and Dewey.

The technological object gains an aesthetic dimension when it is no longer a detachable tool which is transportable everywhere, but a reality inserted into both the natural and human worlds, and when it gives rise to a sense of technological beauty.

The imperative of knowledge.

These practical principles do not allow us to determine in advance what is entailed by care in each situation. They are only regulatory, to employ Kant's terminology. In effect a thought of care in technology is a matter of reflective judgement, and not of the faculty of determinant judgement.

He convincingly shows a large limitation in the ethics of care[4]. Intending to "care for nature" without taking the fight onto the ground of the design of technologies and of the technical organization of chains of production, is to fatally leave intact the distribution of the sensible which defines the social and political order of the moment; even worse, it is to let carelessness flourish, where production process is kept invisible.

Against very superficial conceptions of the idea of responsible innovation which do not question the ultimate goal of these policies and merely encourage and accelerate the passage of fundamental research toward industrial applications while ensuring in advance their "social acceptability"[5], this book goes much further. If responsible innovation conceived in such a fuzzy manner does not guard against the

4 In a complementary way, see Sophie Pellé's book, *Business, Innovation and Responsibility*, in which she interrogates conceptions of responsibility, available in the *Responsible Research and Innovation* set of books.

5 For a more economic critique of this failing see Nikolova Blagovesta, *The RRI Challenge. Responsibilization in a State of Tension with Market Regulation*, ISTE and Wiley, 2019.

creation of new vulnerabilities, and if the human remains outside nature, one may understand that some advocate surpassing this conception through an innovation which is oriented by care. Yet his critique, inspired by Dewey, in some way takes the opposite tack to this critique: the human is an instrument for the improvement of the natural world. Care thus consists of acting technologically in such a way that nature is fulfilled, improved, and optimized according to its own potentialities, and not by reference to the interests of a subjectivity cut off from the world.

Bernard REBER
December 2020

Introduction

> Humanity by itself creates much devastation. A highly technologized humanity, intelligently technologized through a network, which has a geographical sense, is much less dangerous for nature than humanity by itself.
>
> Gilbert SIMONDON

A nagging question haunts our era: can we yet avoid the major ecological disruptions that more than two centuries of industrialization based on fossil energies seem to have rendered inevitable? It is in fact more and more widely recognized that these disruptions, which there is every reason to believe will have disastrous consequences for all of life on Earth (including humans), will occur no matter what we do today to prevent them. We may, at best, limit their magnitude somewhat, provided however that the major industrialized countries take drastic measures without delay – which seems unfortunately not to be the most probable option as evidenced by the tepidity of the (at best, modest) advances made by the international conferences on the environment that have been held for many years.

Should we consider the question as definitively closed, and think that our technological choices have completely committed us on an irreversible path to planetary ecological disaster? Must we even go as far as to consider, with François Jullien for example (Jullien 2004), that the technological relationship with the world as it has been conceived and developed in the western world, and whose model we have inherited from the ancient Greeks, is essentially the expression of a will to mastery of nature, the outcome of which, to the extent that our technological activity acquires more and more powerful means of action, could only be the predicted disaster which we must face today?

This book aims to answer these questions in the negative, showing on the contrary that the western conception of technology is historically articulated with a genuine requirement of care for things and beings. It is true that with the Industrial Revolution, this requirement seems to have deserted our technological action. The latter seems since then to have been controlled only by the power of calculation (*metron*) and seems no longer subject to what should have yet remained the guiding principle of all technological activity: the faculty of *metrion*, that is to say of fair measure, of suitability and of care – which faculty precisely cannot be reduced to the rationality of calculation. How have we come to so radically dissociate technology from care, *metron* from *metrion* – and especially, how to reintroduce *metrion*, correctness and care, into our technological activity? Can we reforge the frayed links between technology and care?

I.1. Care for nature

The need to develop technologies which are more "respectful of nature" is very widely shared among the general public but also among engineers and designers, who consider themselves more and more concerned by environmental issues (especially in the case of students who, notably in France, have over the last few years developed a growing awareness of their present and future responsibility in the matter[1]). The effort to introduce the aim of protecting nature into the activities of engineering and technological design has in fact a long history. The concepts of *green* design, eco-design or *sustainable* design also show a significant change in the openness of technological design to ecological issues in the 1980s and 1990s, reflecting more exactly the fact that this openness has gone hand in hand with an increasingly radical and comprehensive understanding of the responsibility of engineers and designers in ecological matters (Madge 1997).

The links between technological design and ecological thinking are older still. Thus, some of the great figures of the Bauhaus such as Walter Gropius (its founder) or László Moholy-Nagy in the 1930s had already forged contacts with representatives of the ecological sciences in London, where they had emigrated following the closure of their school in Germany by the Nazis (Anker 2010). The reflection on design integrated ecological questions very early on.

The current context is marked by the success of a new mobilizing watchword, which is currently bringing together a growing number of initiatives intended to get the industrialized countries to adopt a less destructive relationship to nature. This watchword is expressed in terms of a fairly simple imperative: *we must take care of*

1 See for example the "Student Manifesto for an Ecological Awakening" in French, https://pour-un-reveil-ecologique.org/fr/ (accessed February 4, 2020).

nature. The expression "take care of nature and of people [may indeed appear as] a new watchword of late modernity. It responds to a social crisis [as well as] an environmental crisis" without precedent (Pierron 2013b). We would thus be in the process of moving from a society whose priority is the search for well-being to a care society (Rodotà 1995, p. 103). Businesses themselves compete in order to win the Grand Prize for Caring, as far as the most trivial acts of consumption – buying recycled toilet paper, for example (Puig de la Bellacasa 2012) – which now seems to relate to taking care of nature[2].

Beyond the excitement that it inspires in advertising executives, care for nature has found important theoretical developments in moral philosophy, more precisely in the so-called ethics of care (Laugier 2012). In opposition to a long tradition of moral philosophy for which the subject cannot be moral unless recognized to be rational, autonomous and free, the ethics of care argue that humans constitute themselves as moral subjects in their vulnerability and in the relations of dependence which flow from it. Vulnerability and dependence are therefore not only accidents which may occur at certain critical moments of life (early childhood, old age or disease); they are our lot. Our moral life is based primarily on our ability to attract attention to, and to pay attention to, everything that allows the ordinary of our lives and which, literally, nourishes us (Pelluchon 2019a). The ethics of care thus put the emphasis on what we do not generally see, on what is rendered invisible by denial of attention, that is to say on everything that makes our lives possible and supports them on a daily basis. However, as noted by Sandra Laugier, indifference, or even negligence and recklessness "in relation to the consequences of our daily actions on the surrounding environment" is expressed more and more today in terms of carelessness (Laugier 2012). In making its irruption into environmental ethics, care does not however limit itself to adding, to the ethical issues centering on the "intrinsic value"[3] of nature or on rational calculation of the consequences of our actions upon it, a generous and caring attitude. Environmental care expresses more essentially the recognition of our dependence on nature – a dependence which, for

---

2 Incidentally, it is not altogether new that human relationships to nature, in industrial societies, should be envisaged by the standard of care, as evidenced for example by the curricula of education on nature in the elementary schools of Western Europe, which have combined at least two ways to teach nature to children throughout the past century: although the scientific and instrumental conception of nature has massively infused the curricula of secondary education, in contrast the primary schools have never really abandoned leading children, by observation, to create an experience and a conception of nature oriented by caring concerns (Postma 2006, p. 3).

3 Catherine Larrère thus emphasizes the existence of an incompatibility between, on the one side, the environmental ethics which proclaim the concepts of intrinsic value and wild nature (wilderness), still dominant on the North American continent, and on the other side the ethics of care for nature (Larrère 2012).

most authors, primarily has a contextual[4] meaning which, moreover, is in no way a characteristic specific to the ethics of *care*.

For all that, the meaning and even the relevance of the injunction to take care of nature are debated (Gaille 2013; Hess 2013). Thus, can we unify care for humans and care for nature in one and the same concept of care, or are there several kinds of care (Pierron 2013b)? Can we export "concepts explored in relations with humans to relations with nature" (Pierron 2013a), by so doing deriving concern for nature and for the world, in its widest extension, from care as it is deployed in inter-human relations, and not the reverse[5]? There is no unambiguous and consensual response to these issues. We must state in fact that extending care beyond individual interactions, toward a care for the world in its entirety, does not necessarily lead to care for nature. Thus, in a book devoted to examining the conditions of care for the world, Elena Pulcini (2013) explains that only a transformation in our concept of the subject can allow us to overcome the "pathologies" of the globalized world (exaggerated individualism, communalist temptations). We must do away with the self-referential and sovereign subject of modernity, dominated by Promethean *hubris* and narcissism, to conceive a subject that is relational, vulnerable, and dependent on others, but also assuming its responsibility with respect to the vulnerability of others. However, Pulcini says nothing with regard to how this relational subject relates with non-human natural beings, and with nature in general. Care for the world certainly implies, as a basic condition, the preservation of life. However, the object of care is not, according to Pulcini, life as such, and even less nature, but the world of human intersubjectivity.

Based on these difficulties some argue that even the idea of care for nature is suspicious and that upon consideration, it is based on very weak philosophical grounds. French philosopher Aliénor Bertrand (2013a, 2013b) thus points to the distortion of principle which exists between the "type of philosophical anthropology" that underlies care and the "naturalism" which guides the relationship with the nature of industrial societies. By naturalism, Bertrand here understands naturalist ontology as defined by anthropologist Philippe Descola (2005): it is a mode of identification of the existing beings which rests on the attribution to all of the same "physicality" (meaning that all beings in the world, living or not, are held to be dependent on the same physico-chemical laws, their behavior can always be explained with reference to these laws) but on the attribution of different "interiorities" between all the existing beings: thus, only humans are assigned a soul, a consciousness or a spirit, all the other existing beings deprived of this. The very

4 "The theme of care allows a pragmatist and particularist treatment of environmental issues, beyond the major principles and incentives to moralise or to (un)assign guilt" (Pelluchon 2019a, p. 14).

5 Pierron J.-P. 2019, pp. 104–105, citing F. Worms.

concept of "nature" is, in fact, unknown to all human groups which distribute ontological attributes among existing beings according to other identification schemes. The concept of nature therefore has only meaning in reference to a particular ontology, naturalism, suspected of having served as a foundation for so-called modern science and technology, and consequently of having paved the way to a mode of development based on the depredation of the environment. How could the concept of "nature" in these conditions become the reference term for an attitude to care with regard to the living and the environment? Naturalist ontology has led us, we Westerners of the industrial era, to conceive in a very impoverished manner our possible relationships with nature; it is inevitable that the concept of care for nature inherits this poverty and in no way allows us to reorient our attitude towards it. In sum, "the problems raised by natural resources are subsidiary in relation to the issue of values … they only arise, in fact, from within a generally anthropocentric conception of the relationship of humanity to nature[6]". "To enact all the changes that would put us on the good path, whether on the level of the fight against global warming or as regards animal welfare… it is important to change our representations of animals and of the place of the human in nature…" (Pelluchon 2019b). The anthropocentric conception classically situates the human outside nature, and assigns to the latter only an instrumental value, which tends to justify the behavior of unlimited exploitation[7]. For us to show more care in regards to nature, we would first of all have to deal with this conception of humanity isolated and cut off from nature, that we replace the human being within the entirety of relations that it has with its environment, and that we cease to consider nature simply as a "background in front of which we perform our activities" (Pierron 2013a).

To assume that human beings are not cut off from nature is equivalent to remembering that this being, who likes to define itself above all by its reason, its intelligence and its freedom, does not yet cease to be a living creature among living creatures. If societies that have chosen industrialization have ransacked nature, it is because the humans in these societies have ceased to think of themselves as living beings and have made of themselves, in the words of André Leroi-Gourhan, a "dematerialized image". They have imagined themselves as separated from the community of beings that they form with all living creatures on Earth, and have assumed the right to exercise over these, and over nature in general, unlimited domination. As Claude Lévi-Strauss also said, "We started by cutting man off from nature and establishing him in an absolute reign. We believed ourselves to have thus erased his most unassailable characteristic: that he is first a living being. Remaining

6 Hess cited by Pierron (Pierron 2013c).

7 It should be noted however that anthropocentrism does not necessarily imply a predatory attitude in respect to nature. "Weak" anthropocentrism may only assign to nature an instrumental value, without justifying its exploitation: this value may be aesthetic, cognitive, etc., which requires a protective rather than destructive attitude.

blind to this common property, we gave frein rein to all excesses" (Lévi-Strauss 1976, p. 41). To reconnect with a conception of the human being as a living being among other living beings should in sum lead us, in a way logically, to put care in our relationships with nature.

The question of how to get out of naturalistic ontology, if that is indeed what it is all about, while retaining the term that it has posed in relation to humans, namely "nature", remains at this point. The question of how it is possible, concretely, to "change anthropology", as by decree, also remains entirely open.

However, apart from these thorny questions, would this planned change in our conception of the human being suffice to make the very notion of a care for nature? Nothing is less certain.

I.2. The two-layered model of care

That there is a close link between, on the one hand, the fact that the human is a living being – which makes it dependent on various conditions to meet its vital needs – and, on the other hand, the fact that it has an essential relationship to the care applied to the maintenance and provision of these conditions, has been widely stressed by the ethics of care[8]. This link is highlighted, incidentally, in a fable which comes to us from a Latin author named Hygin, which Heidegger relates in paragraph 42 of *Being and Time* (Heidegger 1962, p. 242):

> Once when 'Concern' [who is also care, *Cura*] was crossing a river, she saw some clay: she thoughtfully took up a piece and began to shape it. While she was meditating on what she had made, Jupiter (Jovis) came by. 'Concern' asked him to give it spirit, and this he gladly granted. But when she wanted her name to be bestowed upon it, he forbade this, and demanded that it be given his name instead. While 'Concern' and Jupiter were disputing, Earth (Tellus) arose and desired that her own name be conferred on the creature, since she had furnished it with part of her body. They asked Saturn [that is, Time] to be their arbiter, and he made the following decision, which seemed a just one: 'Since you, Jupiter, have given its spirit, you shall receive that spirit at its death; and since you, Earth, have given its body, you shall receive its body. But since 'Concern' first shaped this creature, she shall possess it as long as it lives. And because there is now a dispute among you as to its name, let it be called *homo*, for it is made out of *humus* (earth).

8 And not only by them, by the way; many other philosophies have established this link, including those of the ancient Greeks, as we will see.

Commenting briefly on this fable, Heidegger notes that Concern or Care is not envisaged only as that to which the human *Dasein* remains committed during its entire life and as that which constitutes it, but that "this priority of 'care' emerges in connection with the familiar way of taking man as compounded of body (earth) and spirit" (Heidegger 1962, p. 243).

The meaning of the fable is that the human being will only be a composite of body and spirit at its death, and at this time only can the sharing between Jovis and Tellus be made; before this, what defines the human is the fact that it lives, that it is a living being. As a living being, the human is one, not split, care being precisely that which prevents the split between its natural and not-natural parts. Care, in terms of the human being, means above all else refusing any form of split between the body and the spirit: there is not, on one side, care in the sense of purely technological caregiving, with an objectivized body as its area of intervention; and on the other side care in the sense of taking care, attentive to the human being as spirit and free will. Care cannot be two cares, it is a unity or nothing at all[9].

This requirement of unity in care, an indivisibility of caregiving and taking care, has been particularly pronounced in the field of medical ethics, except however that these two aspects of care, although not separate, are not so much merged into a truly unitary care. There are two layers of care and they are clearly hierarchical. It has in effect become a commonplace to recall that the therapeutic act in the technical sense of the term, that is to say caregiving (cure), is devoid of any moral value if it is not inserted into a more essential relationship of care which envelopes it. The greatest risk is that, while thus conceived according to a two-layer model, care becomes precisely two quite distinct and unrelated cares – and even worse, that care within the meaning of caregiving will eventually prevail over care within the meaning of taking care. As psychoanalyst Donald W. Winnicott deplored at the beginning of the 1970s, "Cure, in the sense of remedy, successful eradication of disease and its cause, tends today to overlay cure as care" (Winnicott 2017, p. 192). However, as stressed by the preface writers of an Italian collective work (Donghi and Preta 1995), the risk of splitting care and giving priority, as Winnicott deplored, to cure as remedy over care, results specifically in an inability to properly pose the problem of the relationships between body and spirit, and their unity. In other words, a relationship of reciprocal implication is forged here between, on one hand, the

9 The reading that Warren T. Reich proposes of Hygin's fable certainly emphasizes the existence of a tension in care – tension between care as a burden and care as solicitude. By its physical and telluric nature, care is a load, a burden; by its spiritual nature, it is solicitude. This reading, however, does not affect the unity of care; it does not mean that there is not one but two cares; there is a single care. However, any care implies this internal tension (Reich 1995).

conception of the human as a living being – but in the sense of a being composed of body and spirit unified by care – and on the other hand, a two-layer conception of care.

However, this two-layer conception of care, well documented in the field of medical ethics, does not confine itself there: it is also found in the invocation of care for nature and it leads to a blurring of its meaning. Not, incidentally, that any idea of care for nature should invariably find its only possible model in medical care (Pierron 2019). Thus, van Rensselaer Potter, one of the theoreticians of the concept of bioethics, considered that medical bioethics, far from being possibly the basis for a global concept of care by extension of the care due to humans to the care due to nature, was on the contrary only a shrinkage of a broader concept of bioethics, linking human health and health of the environment (Gaille 2013). That being said, the need to establish a hierarchical relationship between cure and care now extends to the whole set of activities where care is required. The penetration of the ethics of care into environmental ethics illustrates this phenomenon of extension: we are required not only to interact in a prudent and careful manner with nature, but also to assume that these interactions are of a moral scope and that, therefore, we are morally obliged to comply with the values of prudence and fair measure.

In its most general sense, the idea of care for nature therefore points towards the same two-layer conception of care as that of medical ethics. It is still a matter, here as there, of conceiving care as a single care; however, this care does indeed have two layers in a hierarchical relationship. On the one hand, there is the properly technical level of care, which is the care that we must take of natural things if we are to succeed in our undertakings, in order to avoid the misfortunes of the man of Song. Mengzi, a Chinese philosopher of the 4th century B.C.E., a follower of Confucian thought, tells us that "a man of Song[10], distressed to not see his plants growing fast enough, had the idea of pulling on them from above. Returning home in haste, he said to his people: "I am tired today, I have been helping the corn to grow long" (Mencius, 1970, p. 190). On hearing this, his son rushed to go to see, but the shoots had already dried. Here is an unwise technical act, harmful due to lack of attention and care brought to the specific needs of the plants; but there is also, on the other hand, the moral layer of care, which must provide technical acts of care with their guiding values. The two-layer conception of care extends to nature particularly when it is assigned an intrinsic and not only instrumental value[11]: moral respect for nature must regulate the technical acts that are deployed.

10 One of the Chinese states of the so-called Spring and Autumn period.

11 The concept of the intrinsic value of nature, as well as its relationship to care for nature, will be discussed later.

I.3. The intellectualist conception of technology

There is moreover a conception of technology implicit in the two-layer model of care. This model is in fact underpinned by three assumptions affecting the relations between human life, technology, and care. The first assumption is that human life can be summed up in two dimensions which are difficult to definitively articulate: life is understood on the one hand according to the dimension of the being-in-life, and on the other hand, as a life of the mind. We are "first of all" living beings according to Lévi-Strauss, which means that we are "subsequently", also, something else than living beings. We are not cut off from nature, certainly, but that does not mean that we have absolutely no exteriority in relation to it.

The other two presuppositions of the two-layer model of care relate to the very concept of technology. On the one hand (and thus the second presupposition), technology is defined as the whole set of means available to us to be able to intervene in nature, both inert and living, from a point of view exterior to it. This conception of technology is intellectualist. Technology is the means by which intelligence intervenes upon nature as upon an external, receptive material.

On the other hand (third presupposition), technology as such, or at least a certain type of technology, is considered to be incompatible with care. The two-layer model of care, affirming the unity of all care, is certainly unequivocal: no relationship of care cannot be conceived without technology. A negligent action, performed without precaution and devoid of any real involvement of the one who performs it, is not in the service of care. A gesture that is not careful, that is to say which is not executed with the required technique (which is learned and requires training) cannot be caregiving – such is the message of the parable of the man of Song. Caregiving requires competence. That being said, when technology refers no longer to modestly equipped know-how but to the innovations of chemical and mechanical industry, for example, its relationship to care appears less obvious, more external, and more conflicted. In sum, there are two conceptions of the relationship between care and technology: a relationship of necessary implication on one hand – there is no care without the security of a controlled and properly equipped action, whether it is care for humans or care for nature; on the other hand, a relationship of exteriority in principle, making problematic the possibility of caregiving when the concept of technology signifies, to continue the example, the devices of mechanical industry and chemical products.

The first assumption, affecting both contradictory meanings of the concept of life, is absolutely essential and commands the other two. Indeed, it is because life, in humans, is regarded as dual, consisting of a subject pole (intelligence, the spirit) and an object pole (the body, the being-in-life), that technology must inevitably be understood according to the same division: it is an emanation of the subject pole and it applies to the object pole from a point of view external to it. An intellectualist concept of technology is the result of a bifurcated conception of life in humans. In addition, to the extent that its intervention targets the object pole, that is nature as the theatre and support of the vital processes, technology has an ambivalent relationship to care: certainly there is no care without technology, but the risk of a dissociation between performing cure and taking care is widening as technologies become technosciences.

I.4. From design of nature to care for "ordinary nature"

These three presuppositions thus constitute the unthought-of dimension of the two-layer model of care. However, in the light of the foregoing considerations, it appears more and more doubtful that this model can provide a satisfactory theoretical base for the idea of care for nature. The example of permaculture illustrates quite well the difficulty of transposing this model of care onto nature. Permaculturists[12], however, explicitly put care for nature and for humans at the heart of their system of values. They claim to have a holistic conception of the human being, refusing to separate "the body, the affective and emotional sphere, the intellect, the spiritual[13]". Therefore, they seem to exceed any bifurcated vision of care: care for the body is inseparable from care for the spirit, and care for human beings from care for the earth. "*Care for the earth. Care for people. Fair share*": the motto of permaculture gestures toward a unitary vision of care, and toward a non-dual conception of human life. However, the continuum of care seems to be contradicted here by the prioritization of the concept of design. Permaculturists speak in fact of a real change of paradigm. Their analysis is as follows: historically, human groups have grown up in highly diverse habitats. This accumulated

12 Permaculture is a method of agriculture theorized in the 1970s by the Australians Bill Molisson and David Holmgren. The initial motivation of Molisson and Holmgren was to propose an alternative to intensive agriculture, in which the search for higher yields by the massive artificialization of environments has resulted in the exhaustion of microbiological life present in the soil and, eventually, the sometimes dramatic decline of their fertility. The services that are normally rendered by these microorganisms (degradation of the bedrock to create humus, input material nitrogen, etc.) have thus had to be substituted by other means, including by the provision of chemical inputs. In contrast to these artificializing industrialist approaches, permaculture proposes to return to bio-inspired agricultural practices, capable of taking advantage of natural functions. Nature has in effect given birth over the long term to ecosystems that are extremely complex and robust, based on the existence of multi-level relations of interdependence between many elements, mineral as well as organic: a permacultural farm aims to reproduce this complex network of elements in interaction at all orders of magnitude. See e.g. (Hervé-Gruyer 2018).

13 See the website of the Bec Hellouin permacultural farm – https://www.fermedubec.com/la-permaculture/notre-philosophie/ (accessed November 25, 2018).

experience is a wealth of which permaculture intends to take advantage, drawing on traditional knowledges and know-how. On the whole, however, the growth of the human population on Earth was more like a fumbling adventure than a conscious plan. In place of this globally blind human expansion, the permaculturists intend to substitute a planned habitation, supported by specific knowledge relating to the environment and to the operation of living beings. It is now a matter of thinking of habitation from the viewpoint of design – to propose a "permacultural design". However, if we can understand what permacultural design entails at the level of the farm or even of the territory, what becomes of this vocation of designer when the scale of intervention becomes the Earth itself (care for the Earth)? Must we admit that the human must become the chief designer of global nature? This would be a strange way to defend the idea of care for nature. It is hard to see in any case how we would escape the ambition of becoming masters and owners of nature. "How can one set oneself up as a designer of the world while acknowledging that one is subject to it?" asked Vilèm Flusser (Flusser 1999).

A dual conception of the human; an intellectualist conception of technology; the idea that some technologies are in essence contradictory to care (permaculturists ban mechanization as well as chemical inputs from their farms): permaculture remains well and truly dependent on the two-layer model of care and its assumptions, which leads to a difficulty in rendering robust the goal of care for nature or, in the terms of the permacultural philosophy, care for the earth.

In sum, Aliénor Bertrand sees clearly when she stresses how much the idea of taking care of nature risks surreptitiously returning to a dualistic anthropology – even that one which, in seeing humans as beings outside of nature due to their reason or their spirit, has led to the most predatory behaviors in respect of the natural world. The sincerity of permaculturalists is of course not in question here, and neither are the truly caregiving modalities of their relationships with the cultivated land. That being said, they seem to continue to conceive of the human according to a duality of perspectives, as living beings belonging to nature in their organism, but not in their technological intelligence. Humans belong to nature, but they remain designers of nature. They must simply create a better design for it.

How to get out of this impasse in order to better understand the conditions of a more caring technological action? For what should this action be caring, if not for nature taken globally? A possible answer comes from one current of environmental ethics, the ethics of so-called "ordinary" nature, whose diagnosis is almost that which we have just established: considering all of nature as an object of care, especially by giving it an intrinsic value, in no way leads to escaping from the dualism of the human being and nature; on the contrary, it leads to a hardening of this dualism because the human, despite the assertion that it remains a living being among living beings, appears engaged in a confrontation with nature which has nothing to do with the dialogue of the living being and its milieu. As emphasized by French philosopher Rémi Beau (2013), the shift of environmental ethics toward the concept of care for ordinary nature (i.e. to care, not for nature, but for natural things in our daily environment), could then allow us to dispense with the double conception of a human being which despite everything remains outside of nature, and of a nature without humans.

I.5. Technology, life and care

Let us summarize. The concept of the human as composed of body and spirit unified by care became closely linked to the two-layer model of care, as well as to an intellectualist conception of technology. In this theoretical framework, technological intelligence has a vocation to reconfigure the world – to design it. The only way in this case to guard against the excesses and ravages of technology is to impose upon it a limiting framework, but from outside: ethics seems to have today the function of providing this. Technology is in itself foreign to the values of caution and modesty inherent in care: the imperative of care must necessarily be imposed by an external and overhanging instance.

What however of the relationship between technology and care, once abandoned not only the dualistic conception of human life, but also the two-layer conception of care and the intellectualist conception of technology? As a matter of fact, the philosophical literature on care is almost completely silent on this subject. It still very often proceeds from an attitude of foreclosure in respect of technologies: here these are, so to speak, never questioned in their possible relationship to care, even if some philosophers of care, as we will see, are moving in some respects toward this questioning.

The intellectualist conception of technology seems to act, even today, as the common basis for debates on the relationship between care and technology – even though the non-intellectualist, rather vitalist, conception of technology as *Organon*, that is to say as an extension of a living body acting on its surrounding environment, remained unchanged overall over a very long historical period, ranging from the ancient Greeks (who had rigorously established this definition) until the century of the Enlightenment. In what conditions exactly, and why, did this transition occur? What have been the implications of this for the thought of technology, and for the conception we have of the relationship between humans and nature? We will attempt in this book to make a chronology, exactly, of the historical mutations that have led us to make technology a kind of "unthought-of" of care.

We will thus clarify what conception of technology, breaking with the intellectualist framework within which it still very often remains defined, can orient toward a more careful and caring attitude in respect of nature. What does it mean to think of technology otherwise than based upon intelligence, but as we will see, based upon life? Does this not lead to a counter-intuitive, even disturbing, conception of technology? Does a vitalist conception not mean in effect defending a purely and simply anti-rationalist position? We will show that this is not the case, and conclude by explaining the guiding principles, which are perfectly rational and capable of guiding the work of engineers, for a conception of technology alternative to its intellectualist definition.

This book is therefore a book of philosophy of technology. More precisely, it intends to demonstrate that the problem of knowing how to orient technological design towards the care of all existing beings and of the world cannot be solved only by choosing materials and processes that are more "respectful of nature". This problem engages a global philosophical reflection on technology, which French philosopher Jean-Yves Goffi emphasized at the end of the 1980s (Goffi 1988) that it had always gathered, since Plato, three inseparable dimensions: a phenomenological or ontological dimension, consisting of a description of technical reality through its various manifestations (tools, instruments, machines, gestures, practices, operations, etc.); an anthropological dimension signaling towards the idea that every technique finds its true raison d'être and its ultimate meaning in the necessity of satisfying vital needs; and finally an evaluative dimension. What characterizes contemporary philosophies, Goffi wrote at the time, is a relative fragmentation of this approach. Evaluative, anthropological and ontological approaches are often dissociated: few are the unitary philosophies of technology (Goffi 1988, 52-53). According to Goffi, this situation could be explained by the gradual rise in importance of the evaluative perspective, which has pushed the other two dimensions into the background. The current philosophy of technical artefacts, to which we shall return later, undeniably strives to overcome this imbalance and to regain a unitary conception of the technical fact, by integrating the ontological and evaluative dimensions to a greater extent. However, the anthropological dimension remains to this day the poor relation of philosophical analyses of contemporary technology. The present work is intended to fill this gap. In sum, this book is a book of philosophy of technology, the objective of which is to respond to the following questions: in what sense can we speak of "care in technology"? For there to be care, is it not necessary for there to be a subject to dispense care? According to what concept of care then can we speak of technologies able to perform care by virtue of their very design, if that is indeed what we mean? What meaning can be given to the concept that seems to emerge here, and which seems at best self-contradictory, at worst devoid of meaning: that of an objectified care, reified in technology – that is to say of a care which no longer operates through the activity of a subject who provides care?

One might object to the great naivety of this exercise, or even to its profound ignorance of the real mechanisms that have historically led to the depredation of nature. Can we really believe, we might say for example, that it is due to the persistence of a dualistic anthropology and a two-layer model of care that plastic waste accumulates in hundreds of thousands of tons in the oceans (Eriksen *et al.* 2014), thus forming what we might call a new continent – a continent of waste? Behind this phenomenon, and many others of the same kind, lie not anthropological or philosophical postulates, but quite certainly actors, including industrial ones, who have taken decisions based on their interests. French philosopher Grégoire Chamayou (2019) has recently highlighted the role of brewery and soft-drink industrialists of as early as the 1930s, in the United States, in promoting disposable cans and bottles which came to replace the old system of recoverable and reusable containers (a system based on deposits). The industrial lobbies have worked very well and they certainly have a role in the explosion of astronomical quantities of waste which must now, in fact, be recycled. Let us cease to beat about the bush and point the finger directly at the problem – not in the heaven of philosophical ideas about humanity and about care, but in the strategies of the major industrial players.

This objection is unstoppable, and it must be clear: no miracle recipe will come out of these analyses and, of course, it is above all through struggles and the efforts of regulation that things will change. Bergson warned us: a more frugal mode of life, one therefore less detrimental to nature, would have to go through a very deep moral reform. However, such a reform being apparently unattainable, "we must be content with shifts and submit to more and more numerous and vexatious regulations" (Bergson 1935, p. 274).

There is always an aspect of our daily lives by which, whatever our goodwill and our sincerity, we produce a defect of care. Manufacturing and using a smartphone or a computer, taking a plane trip, turning on electricity inevitably involves a defect of care, since these technological objects or gestures depend upon industrial clusters which are deleterious to natural environments and humans that are sometimes very distant from us. There is no loophole, even if otherwise we promote *low tech* philosophy or agriculture without chemical inputs. We have built our habitation of the world upon the (de)predation of nature; we can without doubt be less (de)predatory, but we can hardly stop being so altogether. But this thought must not discourage us. It in no way prevents us from defining the theoretical conditions and practical principles of caring technological action, to implement as much of it as possible in the aim of reducing the (de)predation exerted upon nature and upon living beings in general. This book has no other ambition than to indicate what these conditions and principles are.

1

Care and Technology: An Anthropological Question

1.1. From mastery to care

1.1.1. *Making good use of technology, anticipating its potential risks: two possible examples of care in technology?*

1.1.1.1. *Care and technology in ancient Greece*

No more limited to medical practices, the moral imperative of care is now required in multiple areas of technological activity: agriculture, livestock farming, as well as all sectors of industry, which, at one point or another in their value chain (or even at all points, from the extraction of raw materials up to the end of life of the products that they dump on the markets), are the cause of massive pollution, as well as devastating effects on human groups who have the misfortune of living where the giants of the industry have (dis)located their productive activities.

The need to problematize technological activities in the light of the requirements of care is certainly not a recent discovery. While the nexus between technology and care has become a problem for us, it seemed self-evident in Ancient Greece. The term "care" there indeed covered a large number of attitudes and practices, that may be reduced to three major themes[1]: firstly, care referred to the set of activities reflecting concern, caring for something or someone. In particular, in this broad field of care as concern, we find the care due to the gods or to parents, but also the care owed to oneself (*épiméléia héautou*); secondly, the term also had a rather more restricted meaning, linked to the development of rational medicine in classical

1 Dimitri El Murr, professor of ancient philosophy at the École Normale Supérieure rue d'Ulm, helped me to see more clearly in these difficult questions of terminology. I give him my warm thanks.

Greece: it designated acts of caring for the sick, of taking them in charge and of the administration of treatment; thirdly, care referred to the governance of a house or a city, or even the administration of business[2]. Economy, which designated precisely this administration in the private sphere (*oikos-nomos*, the good order of the house), was thus a matter of care.

It is to be noted that the Greeks had two terms and not a single one to designate care: thus *épiméléia*, but also *thérapéia*. These two terms were indeed quite close, in particular in Plato who generally used them interchangeably. In both cases, reference was made to caring and to concern, however each of the two terms had its specificities: on the side of *thérapéia*, there was treatment in the medical sense, but also religious care (service to the gods, including the maintenance and repair of the temples) and respect for parents, as well as the daily maintenance of plants and animals. *Thérapéia* also designated services rendered with a view to obtain favors; *épiméléia* predominately bore the sense of "being concerned", of concern in itself, but also of the administration of things and of public charges. *Thérapéia* therefore retained a strong link with this vast field of care, which in archaic Greece was dedicated to the gods, to ancestors and to things, although the term became specialized and eventually came to preferentially designate medical practices (Lombard 2006). Accordingly, for the Greeks it could not be a question of defining all care by extension of the medical model of care. The contrary seems true: the medical sense of care branched out on the basis of a more comprehensive understanding of care. It was understood that *thérapéia*, in the limited sense of the term, should never lose its ties with *épiméléia*. Moreover, *thérapéia* as well as *épiméléia* concerned a multiplicity of beings: humans, gods, non-human living things, non-living things – but not properly speaking of *Nature* taken as a whole.

Admittedly, activities of a technical nature, the *technai*, had different relationships to care. Medicine, agriculture, the activity of shepherds, the political arts (Schaer 2013): all these were activities that involved actions upon living beings and that the Greeks took to be related to *épiméléia* by an essential, constitutive link. In contrast, the arts of manufacture, such as pottery or carpentry, were only considered as arts of care in a derivative and secondary sense. It was of course obvious that the craftsman should be careful with his materials and his tools, and Plato himself explains that the different arts of manufacture which support the care we have for ourselves are arts of care (Plato 1997, pp. 585, 586). Thus, concern for the feet (therefore of the body) entails concern for shoes. The art of shoemaking is therefore in a sense an art of care, as are all the arts that deal with the body.

2 Note that in Latin also, there are two terms: *curare*, which has the greatest extension and covers the different meanings of care in the Greek sense of the term (with a predominance of the sense of worry about the administration of business), and *medicare* which means more specifically to administer treatment in view of healing.

However, stresses Plato, a distinction should be made between the care we bring to our belongings, and the care that we bring to ourselves as regards the body (*épiméléia héautou*). If the arts of garments or footwear are indeed in some way arts of care, they focus on things related to the body and not on the body itself – whose specific art is gymnastics. The care that we need in dealing with our body, in concerning ourselves with it, is reflected in some way in these particular arts which aims at providing our body with whatever it needs. It is not as arts that these arts have a relationship with care, but as arts of the body. The object of care is the body, and only indirectly the shoe or clothing.

In addition to the fact that they did not put the arts of manufacture on quite the same level as the arts with an essential relationship with care, the ancient Greeks also made a distinction within the arts of care themselves: if care in medicine had a moral significance, it was not exactly the same for the culture of the soil or the rearing of herds.

However, on these two points, the current extension of the field of care introduces a clear break with the Greek concept of *épiméléia*. On the one hand, care now also concerns the art of manufacture, in this case the most diverse varieties of industrial production; on the other hand, care for natural things, ecosystems and non-human animals (domesticated or not) now has an eminently moral meeting. In our relationships with nature and the living world as a whole, our constitution as moral subjects is at stake – which would have seemed devoid of meaning to a Greek of antiquity. Acting technologically meant for the Greeks to act with prudence, applying care, but this care passing through *technè* had no moral significance – except without doubt in the medical *technè*. The decoupling of technological activity and care has led to the depredation of nature, although it is widely recognized today that reconnecting with technological activity which takes care cannot consist of returning purely and simply to the ancient attitude of prudence, in the contemporary form of good management of natural resources. We must now not only be cautious and circumspect, but also take care of nature in granting moral significance to its care.

1.1.1.2. *For a moral evaluation of technologies independent of the uses made of them*

What are the technologies deemed responsible for the widespread mismanagement to which the industrialized societies seem to be shamelessly devoted? Not all are concerned, of course. It would for example not come to anyone's mind to incriminate the so-called traditional know-how: on the contrary, these competences are generally seen as custodians of a form of wisdom and a fine knowledge of local environments, likely to help us to change direction onto the path of a less predatory social and economic life, more imbued with the requirements of care. All art implies a form of care, at least in the sense of a cautious and attentive state of mind in respect of what the activity involves. However, technological developments related to industrialization for more than two centuries seem to have forgotten this. The latter seem to subscribe to a reductive idea of technology, in that they seem to ignore the elementary rules of caution and modesty to which, according to the Greeks, any true *technè* should bow.

The technologies which stand accused of negligence are also not readily listed, and the field that they describe is difficult to mark out clearly: extractive industries, industries based on fossil energy, chemical industries, new technologies generating strong uncertainties about the environmental consequences of their products (bio- and nanotechnology, synthetic biology for example). Also indicted are all the activities which have massively relied on the sectors of innovation and production mentioned above – for example, intensive agriculture dependent on chemistry, mechanization and biotechnologies, or industrial farming that uses pharmaceutical products or even genetic engineering, so that animals can withstand the conditions of life which are imposed upon them in contempt of their most basic needs.

The accusation consists most often of lifting the veil on the way in which all of these areas of technological innovation and development lead to objects or processes whose "mode of existence" has a causal link with the degradation of natural environments. This expression of mode of existence of technologies, coined by French philosopher Gilbert Simondon (2017), means in effect that an object or even a technological process must not be considered only according to how it is used. This point is essential: the lack of care, that is to say mismanagement (*in-curia*), is not explained mainly by accidental reasons, that is to say, by adverse circumstances or by bad use of technology. It is more essentially implied by the structural or operational characteristics of the objects or of the technological processes that we adopt – which moreover is not always easy to admit, to the extent that the so-called thesis of the neutrality of technology, illustrated by the famous argument of the knife, seems to be the most widely shared thing in the world: the knife is neither good nor bad in itself, everything depends on the use made of it. I can use it to prepare my meal, or to assassinate my neighbor. As American pro-gun lobbyists like to say, guns don't kill people, people kill people. This corny argument of the "morally-neutral-in-itself" knife (or gun), assumed to apply to all technologies, is often placed at the service of an evaluation of technologies centered on the value of the intentions themselves: the moral value of technological action involves the choices that we make to use our technologies for one use and not for another. Technology does not in itself bear a choice of values; it can be put to the service of good or bad intentions.

However, although this thesis seems convincing at first sight, it faces a formidable difficulty shown by the example, among many others, of the smartphone: the extreme diversity of uses that it is capable of, and therefore the underlying intentions of the users (and designers), does nothing to change the fact that the whole of its manufacturing process is proving to be ecologically and humanly harmful[3]. Mismanagement lies not only or even primarily in what we are doing with our technologies, but in their conditions of production and working (or functioning).

The imperative of care therefore appears to require us in the first place to reconsider the very concept of technology – to abandon its definition as a simple neutral means at the service of external and pre-established goals (only these purposes of use are then assumed to be subject to moral evaluation), in favor of a conception which emphasizes the way in which, in its properties of structure and its patterns of operation, a technological object or process may give rise to a kind of moral evaluation about its effects on the natural world and/or on the human world – that is to say on areas of reality that are either structured according to moral values (such as the human world) or conceived in such a way that our relationships with them now have a moral meaning (such as the natural world).

Far from being a means that is clearly limited to its known properties, something in which there is nothing more than the intention of the designer and/or manufacturer, the technological object or process always exceeds the list of its specifications and is defined above all by the relations it is led to establish with beings and processes on all levels (natural or human). These relations are most often invisible: in the smartphone that I use, I do not see the soils devastated by the extraction of metals necessary for their operation, or the impressive quantity of water used for its manufacture, let alone the forms of exploitation including slavery that this manufacturing generates, at least in some of the countries where the production is (de)localized. With this smartphone, it is possible to send electronic messages: how many users know that this innocent gesture, without apparent environmental consequences, in reality bears a very high environmental price? These messages are in fact stored on servers located in gigantic data centers, that is to say warehouses or "data farms". These emit a lot of heat and require a lot of electricity, not only for operation but also for cooling. Sending a simple electronic message is reflected finally by the issuance of a quantity of CO_2 equivalent to that which a tree can absorb in a day. According to a study published in 2015 by the

3 See for example the article in the French newspaper *Le Monde*, dated July 19, 2012, on the activities of the Chinese state-owned enterprise Baogang. About 97% of rare earths necessary for the manufacturing of electronic devices are extracted and processed by this company in the city of Baotou. In this region, vegetables no longer grow, and herds are decimated, due to the quantity of toxic substances dumped in the soil and groundwater.

Global e-Sustainability Initiative, data centers are responsible for 2% of the total emission of CO_2 in the world, and the situation is expected to worsen.

Our means of communication therefore link us to natural and human processes which, on the one hand, are not immediately visible at the order of magnitude and in the limited field of our existence and our perception and which, on the other hand, cannot always be located in the strict sense in space and in time: the questions of knowing exactly where an electronic message will generate CO_2, as well as determining when exactly its effect will be felt, are indeed devoid of meaning.

Broad areas of technological research and innovation today seem to give birth to strange beings who carry with them, in the relations they have with other beings and with the natural environments, their own topology and their own chronology. Space and time are no longer on the side of the transcendental subject and its operations, as Kant wished, but on the side of the objects themselves. It is likely that a whole aesthetic, that is to say a theory of sensory responsiveness, must be reformulated, such that the subject (individual or collective) is no longer considered as the absolute origin of categories in which objects are seen and known: learning to see what we have not learned to see, namely the constituent process of our technological objects of current use (from the extraction of resources used in their manufacture, up to their scrapping, passing through the multiple ways in which they can be used), involves first admitting that these technological beings themselves define the spatio-temporality and the relevant categories that allow us to apprehend, perceptively and cognitively, everything that their mode of existence implies.

1.1.1.3. *The impossibility of being quite sure that our technologies avoid a defect of care*

For all that, the simple fact of going beyond moral evaluation of technology focused on the kind of use, good or bad, that we make of it – with the aim of giving preference to a moral evaluation consisting rather of establishing a comprehensive review of what our technologies, by virtue of their mode of existence, really do to the beings of the natural world – does not lead de facto to an attitude of care in respect of these beings; at least, it does not lead there automatically. It so happens in fact that the consequences resulting from the implementation of technology cannot at all be identified and known in advance, which diminishes the possibility of absolute certainty that in acting technologically as we choose to do, we are moving effectively and indisputably in the direction of care, and not in the opposite direction. As well, bio- or nanotechnological laboratories bring into existence entities which are intended to come and take their place in the natural world, and thus to enter into relationships with other entities or existing processes. The actual (and possibly toxic) behavior of these entities will depend on these relationships.

However, it is not possible to establish an exhaustive list of these in advance[4]. The invisibility of relationships is not only an aesthetic question, but also an ontological one: it proves to be inherent in these technological objects of a new kind that are less, according to the distinction drawn by Hannah Arendt (1958), "works" intended to occupy a well-defined and stable place in the human world, than entities capable of triggering unprecedented natural processes whose characteristics as well as future effects it is impossible to predict. Uncertainty renders ultimately quite vain the hope of accounting in advance for everything that our technologies, at all stages of their lifecycle, may do to natural beings, and of proposing on this basis a satisfactory, that is complete, cost-benefit analysis. In technological matters, is a morality of care not inevitably, in sum, a provisional morality, always likely to be disturbed by the discovery that in the end, what we thought to be a careful and care-giving technological activity reveals itself finally contaminated by a defect of care?

These findings may lead to pessimism and inaction, by making the efforts that we may take to reduce our ecological footprint appear very limited. Permaculture offers testimony to this. To bring about a caring and careful attitude in respect of nature, permaculturists opt for self-manufacture of their agricultural tools. They thus give themselves the possibility of not using the technologies of intensive agriculture. This being said, as explained on the Inriality site[5], urban agriculture applying the principles of permaculture can derive great benefit from communicational electronics, by the use of "small electronic probes and [of] control stations to optimize plant growth through the successive stages of observation, analysis and experimentation with parameters. The data can be collected on smartphone or shared with other users on the Internet[6]." In addition, "for the future of some areas of urban agriculture (…) exchange of knowledge via the Internet is a crucial issue." Followers of these innovative agricultural methods can therefore reject some technologies of industrial and mechanized agriculture, refuse ploughing to preserve the structure and the microbial life of the soil, take advantage of synergies between plants: despite all this they remain dependent, to develop and disseminate their projects, on technological clusters which are at the antipodes of the values that they claim – that is, on industrial sectors involved in the manufacture of electronic

---

4 Approaches to regulation of nanotechnologies through anticipation of risks, with a view to establishing a cost/benefit balance, have been the subject of very many criticisms. For a summary of these criticisms, see e.g. (Guchet 2014).

5 Media on the uses and the challenges of the digital, attached to the INRIA, *Institut national de recherché en sciences du numérique* – the French National Institute for Research in Digital Sciences.

6 Project entitled *Brains for your garden.* [Online]. Available at: https://www.inriality.fr/agriculture/agriculture-urbaine/consommation-collaborative/lagriculture-urbaine-fertilisee-par/ [accessed November 10, 2019].

devices, in many respects contradictory to the watchword of "earth care, people care, fair share" proclaimed by permaculturists.

The followers of de-growth on the basis of low technologies are without doubt right when they argue that no "technological solutionism" will allow us to overcome the major crises with which human societies, on a global scale, are now confronted. Expecting geo-engineering, artificial intelligence, or nano-bio-technologies to provide solutions for global warming, the depletion of resources and the massive extinction of biodiversity is evidence of ignorance and of great naivety. It is better to assume that our needs for resources and energy must decline drastically, and that we should opt for technological innovations less dependent on extractive industries, on the plastics industry and energy production.

However, what is it precisely that Philippe Bihouix, a French engineer and spokesperson for low technology, says? Let us quote him at length: "We should be (…) rethinking our objects in depth, designing them as simple, robust and user-friendly, repairable and reusable, standardized, modular, based on simple materials, easy to dismantle and using scarce resources sparingly. We should also limit their electronic content, at the cost of accepting a less demanding design brief, lesser or intermittent performance, a loss of efficiency and even of usability. This would allow us to considerably increase their lifespan and to optimize the recycling rate of resources" (Bihouix 2019, 104). However, smartphones with "limited" electronic content, less powerful and running intermittently, with a long lifespan and therefore produced in smaller quantities, are nonetheless technological devices incorporating rare earths in their components, which will in no way change the ecological and human situation in the territories where these metals are extracted, in China or elsewhere. "Our work," says a short video from *Low-Tech Lab*[7], "consists of sourcing low tech everywhere in the world and encouraging documentation in the form of tutorials via our collaborative online platform". The low-tech philosophy thus aims to lead to the emergence of a community of practices and to disseminate technical know-how – impossible, in the end, to consider this movement without the Internet and digital technology. Being low tech and de-growthing here therefore risks exacting a very high price at the other end of the world, among those who, precisely, continue to supply the world market with rare metals, or other substances involved in the composition of the devices necessary for the dissemination of the "low tech philosophy".

In an interview in April 2018, Antonio Negri explains in the same vein but in another register (the profound transformations of capitalism in the digital era) that as an intellectual worker, it is within the commodities of this capitalism that he is

7 [Online]. Available at: https://www.youtube.com/watch?v=SfGORMt0nGQ [accessed June 30, 2019].

himself, now, led to live and work – in particular the computer tools (Negri 2018, 65). The problem here, says Negri – is that to theorize the struggles of the present against capitalism, and to organize these struggles, we cannot do otherwise than to accept, at least up to a certain point, the goods produced by this same capitalism. In sum, as Negri says, to produce care here, we must never lose sight of the fact that we may always produce, elsewhere, the opposite of care.

1.1.1.4. *The limits of a purely contextual evaluation of technologies*

It is not certain, moreover, that this situation, which is only schizophrenic in appearance, can be surpassed. Does it not in effect suggest the existence of a tension in any care, at least in all cases where this care passes through technological devices – that is to say, when one thinks about it, almost all the time?

The ethics of care agree in saying that care can only be contextual – we cannot know *a priori* what makes care in a given situation; we must have an empirical and pragmatic approach to the question, we must listen. Contextuality and locality of care, therefore, which implies means of care that are contextualized in themselves. However, by our technological devices, we produce effects in the world well beyond the perimeter of our action. The lifecycle analysis of devices allows us to apprehend this phenomenon: the technological device can contain materials whose extraction, at the other end of the world, is reflected in the devastation of ecological and human resources; its use and scrapping can generate serious toxicities for human health and the environment, much later and without these effects being easily locatable. In other words, we have to face the problem of the spatio-temporal delocalization of the effects of our technological activity, including when this activity has a local purpose of care. In limiting the device only to its purpose of use and in the context of its implementation *hic et nunc*, that is to say, creating what is called a "black box", this phenomenon of delocalization remains unapparent. The device is conceived as an undifferentiated unit. What it produces, that is to say that which makes it the subject of ethical analysis, is generally held to be equivalent to what it is possible to do with it, or to make it do. In reality, this is not so. Any technological device is multi-layered and operates at multiple levels. A caregiver uses it in a context of care, in the hospital or in the fields, in virtue of the features it has, but it is also made of different materials that may relate to the realities of the world seemingly external to this context, in spatio-temporal orders of magnitude very different to that where care is provided. As it is manufactured, the technological device is also connected to the industry which manufactures it and which may, if it should disappear or change its strategy of development, cease to manufacture it: what will then become of the caregivers and cared-for, once care has become dependent on the device?

The ethics of care, flourishing and plural, insist on the need to hear the different voices (in French, *voix*) of care, those of the caregivers and cared-for, which are often reduced to silence. These ethical issues pay attention to what has been rendered inaudible. It is perhaps appropriate to complement this perspective by recognizing that care, or its opposite, may also pass through multiple channels (in French, *voies*) – through the features of the technological device, but also through its materials, or its production process. It is here less what has been rendered inaudible that deserves attention, than the invisible or rather the invisibilized, as is also recognized by the ethics of care: in this device that I use, I do not see the lack of care which is embodied in the rare metals that it contains, or in some of its features likely to break down professional practices and to cause caregivers to lose the ability to heal. How in these conditions can I be sure that in abandoning a tool, I do not remain dependent on other tools that produce a defect of care elsewhere in the world?

The situation is therefore particularly difficult: we must be able to assess and morally judge technologies by calling them "good" or "bad" in the light of what they really do to the beings of the world that they affect; surely, not all technologies are created equal with regard to the desire to perform care and we must be make an account of this – in a situation, however, where this evaluation inevitably has, for reasons already mentioned, a provisional and indefinitely revisable nature.

It so happens that the two major approaches to moral evaluation of technologies mentioned above, by the choice of their use or by the anticipation of their actual effects on the world, very widely dominate the ethical debates that accompany major programs of research and innovation today. It is indeed invariably demanded that research and innovation be "responsible" in the sense that, on the one hand, they are expected to be oriented towards socially desirable and ethically acceptable applications, in compliance with certain general principles and "societal values", but also in the sense where, on the other hand, it is expected that their forecast benefits be maximized and that the risks or the damage they are likely to incur be on the contrary minimized. All this ultimately summarizes fairly well the manner in which ethics has been involved these last 20 years in the development of genomics, nanotechnology or synthetic biology.

As a consequence of the previous findings, these two traditional approaches to moral evaluation of technology are not completely satisfactory when it comes to proposing concrete solutions for regulating or even guiding technological development. However, what they both have in common is envisaging problems in terms of mastery. The question is invariably the following: how to master our technologies, either by *a priori* respect for unconditional principles guiding the choice of uses that we make, or by anticipation of the consequences and a calculation of the costs and benefits which are supposed to arise from them? Are not the difficulties faced by these two approaches to evaluation explained, therefore, by the limits inherent in their shared premise – a premise consisting in sum of making an equivalence between, on the one hand, the goal of greater care in our technological relationships with nature, and on the other hand, the search for "mastery of our technological mastery"?

1.1.2. *Do we need to learn to master our technological mastery?*

1.1.2.1. *"We must re-bind Prometheus"*

In order to introduce the requirement of greater care into our technologies, must we not yet, indeed, begin by (re)-learning how to master them? It is not clear how the answer to this question could be negative. The theme of technological *hybris*, that is to say the escape of technology from any possible control, has in fact been a commonplace of philosophical questioning in the past century, and in some respects it still is currently. What was then called the industrial Machinery was almost systematically, among philosophers anyway, the object of quite accusatory evaluations on the basis of a diagnosis formulated in almost constant terms: in less than 300 years, we have accomplished the dream of modernity in succeeding, according to the famous words of Descartes, in making ourselves "masters and possessors of nature". Unfortunately, this search for greater mastery over our conditions of existence through technology has degenerated. Prometheus is unbound: we must urgently rebind him, that is, learn to master the power of our means of action. We must in summary learn to master our technological mastery, building up safeguards against the pressure of technology which has become uncontrollable. Doubled in this way, the word "master" here takes two different meanings: on one side, it signals the intensification of our power to act on the natural world, and to transform it in our image; on the other side, it designates an imperative of vigilance and prudence in respect of the same power.

"To master our mastery": the idea, if not the expression, is often associated with the name of Hans Jonas (Larrère 2006, Larrère and Larrère 2012), whose *The Imperative of Responsibility* called in effect for a new ethics able to restore the leash to Prometheus unbound, became uncontrollable (Jonas 1985). The expression itself appears under the pen of Michel Serres, in *Hermès III. La Traduction* (Serres 1974, p. 93): "The question, now, is to master mastery, and no longer nature", and then in an article in the daily newspaper *Le Monde* published on April 15, 1979: "Now, perhaps we should think no more of being masters of the world, but of mastering our mastery" (Robert 2005, p. 150). Serres resumed this theme in *The Natural Contract* (Serres 1995a, p. 34) as well as in his *Conversations with Bruno Latour* (Serres 1995b, p. 171).

It is also found in an article by Olivier Godard, published in 1997: "it is now time to master mastery[8]" (Godard 1997).

"Prometheus unbound": this cliché which was often repeated in the last century, beyond the rhetorical effects which were dear to an era but little to our taste today[9], showed the very common belief among philosophers that to overcome the difficulties of the present, it was necessary to return to the wisdom of the ancient Greeks. Exercising *phronèsis*, that is to say prudence and wisdom in the development of technology: only a moral and spiritual strength can oppose the *hybris* of industrialization and its devastating machine technologies, denying any form of authentic human life. Did Bergson not ask that the great increase in technological power be counterbalanced by "a bigger soul", that is to say by a leap in the moral and spiritual domain? (Bergson 1977, p. 310). As for French sociologist and philosopher Georges Friedmann, he invited us to counterbalance "power" (understood as technological) with "wisdom" (Friedmann 1970).

Drawing inspiration from the Greek example in order to establish a framework limiting technological development might at first seem relevant: in fact, although they in no way associated technology with the risk of depredation of nature, and not even with the risk of seeing technology become dehumanizing and alienating, the Greeks took the need to contain the productive activity of artisans within defined limits very seriously. Did this imperative of limitation, however, have the same meaning in ancient Greek thought as the desire to master technological mastery today?

1.1.2.2. *Hybris and technè: the Greek model of the limitation of technology*

The Greeks in fact had different, even opposite, positions with respect to technology. The Sophists, for example, who were in a way enthusiasts for progress, were the antipodes of Plato and the Stoics. The position of Plato is particularly interesting and allows us to understand why artisanal activity might be seen as potentially dangerous, and how it should be contained. Although in several passages of his dialogues Plato undeniably shows a form of distrust of artisans[10], he also considers them in a positive light due to the fact that they are creators. Artisans are

8 Cited by (Larrère 2006).

9 Although there are other recurrences, see e.g. (Zarka 2013, p. 4).

10 In *Phaedrus* for example, Socrates establishes a hierarchy of souls, from the more heavenly to the most vile: In the first rank is "the lover of wisdom or of beauty, or who will be cultivated in the arts and prone to erotic love"; in the ninth and last rank is located the "the tyrant." The soul of "the manual laborer or farmer" is located in the seventh row, just in front of "the sophist or demagogue" (Plato 1997, p. 526). In *Alcibiades*, farmers and all the people of trades are inferiorized to the extent that they spend their existence on the affairs of the body, instead of dealing with their soul and the knowledge of themselves (Plato 1997, pp. 589, 590).

indeed causes of passage from non-existence to existence of what they manufacture, insofar as all operations which are of the domain of the arts are creations, and that all the workers of these operations are creators (Plato 1997, p. 488). As stressed by Richard Sennet (2008), Plato is still close to the spirit of the Homeric hymn to Hephaestus which celebrates artisans as *poietai*, in a sense of the term which does not yet distinguish tradespeople from those who make verses and who sing – poets and lyric poets. Today, regrets Plato, only poets are called creators: this is because the original unity of creative activity has been forgotten, creation is fragmented.

However, it is precisely because they are creative that activities falling within the scope of *poiesis* must be rigorously limited. The artisan is indeed the one who imposes forms on matter and, if he is not limited, nothing prevents him *a priori* from varying the forms according to his whim. He can always make something into something else, losing that which served to define it. The artisan is therefore the one who risks delivering the human world into permanent instability, even while the main concern of human beings is to ensure the stability and sustainability of the habitat that they have built against nature's power of erosion (Arendt 1958).

Against the Sophists who welcomed with enthusiasm the development of technology and considered it in the perspective of an infinite progress (Vergez 1956), Plato's metaphysics can thus be understood as a response to this danger of unlimited variation of forms in the polis: by separating Being and becoming, and submitting the later to the unchangeable order of the former, Plato has put a yoke on the unlimited power of invention of forms by the artisan. The fact there is an *eidos* of the chair, that is to say an eternal essence fixing very precisely what makes a chair a chair, allows it to be detached from the illimitation of technological *hybris*.

For Plato, the good artisan is not primarily one who has know-how acquired from long experience, a skillful hand or a good way of working: it is above all one who is ruled by the *eidos* of what he manufactures. The human artisan must confine himself to imitating the forms created by the divine Demiurge (Beaune 1994, p. 48), and transmitted by tradition (*technè poiétiké* is in no way, in fact, a matter of inventive genius, Sennet 2008). He must work according to a principle of limitation imposed by the essential nature of what he produces. His directing precept must be to act with reason as Plato says, that is to always be able to justify what he is doing by knowledge of the *eidos* and of the cause of that which his activity concerns (Cambiano 2013).

Technological activity must therefore be contained within the limits of the model. In Book II of the *Physics*, Aristotle contends with the materialists: they consider that the true nature of a thing is its matter; like Antiphon, they take it for evidence that if one buries a bed in the earth, and the wood sprouts, that which emerges will not be a bed but wood. The form of the bed is therefore accidental, contingent, it is the result of a certain arrangement of the subject matter according to the rules of art. The material is, on the contrary, the real substance of things, it is that which remains throughout change. This metaphysical position concerning the nature of things is also a political position: if the material is the substance of things, one may thus rigorously conclude that these have no definite essential form. This may lead to a variability of appearances of things in the polis, affecting what the French philosopher Jacques Rancière calls the distribution of the sensible (Rancière 2004). The distribution of the sensible refers to the way in which sensory experience is configured in a society at a given time. It operates through a system of *a priori* forms determining what may be sensed. By defining the true nature of things, natural as well as artificial, by their form and not by their material, Aristotle also imposes, like Plato, a principle of limitation of productive activities.

This dangerousness of manufactured things, as they can always disturb the distribution of the sensible which organizes a definite distribution of roles and places in society, is not a concern of the ancient Greeks alone. According to the expression of Walter Benjamin, things are bearers of revolutionary energies (Benjamin 2016); the design of things has from the outset a political impact. When the Bauhaus designer Marcel Breuer manufactured his famous B32 tubular chair, in 1925, he introduced to the world a chair which no longer has feet and whose base is flexible – a chair that no longer corresponds really to what is a chair, which escapes from the unanimously accepted definition of a chair – from the *eidos* of the chair, in sum. It reshapes definitions, it destabilizes the world – we understand that the Bauhaus, as explained by the philosopher Pierre-Damien Huyghe, was the *bête noire* of the Nazis (Huyghe 2015). Containing these revolutionary energies, or even purely and simply extinguishing them, was thus able to appear as a priority for totalitarianism – the Bauhaus was closed in 1933.

In addition, for the Greeks and in particular Aristotle, matter introduced into artisanal activities a second principle of limitation. Far in fact from being the pure principle of indeterminacy as is wrongly believed, matter always has implicit forms which destine it to receive certain forms rather than others. Matter, or rather material, is inhabited according to Aristotle by a desire for form (Schiefsky 2007). We cannot do whatever we want and not everything is possible: the artisan must act with accuracy in respect of a material or of a situation whose complexity cannot be mastered by theory.

A third principle of limitation circumscribes the field of productive activities in Greek thought. In *Nicomachean Ethics*, Aristotle defines *phronèsis*, practical wisdom or caution, as the disposition, accompanied by true rule, which is capable of action in the sphere of what is good or bad for a human being (Aristotle 2009, p. 106). He also defines art (that is to say, the whole set of productive activities) as a state of capacity to make, involving true reasoning (Aristotle, 2009, p. 105). While *phronèsis* relates to action (*praxis*), *technè* relates to production (*poièsis*). These two concepts distinguish two major types of activities which Aristotle prioritizes: activities that have no other purpose than their own exercise, such as virtuous conduct, and activities that have their end outside themselves, such as the manufacture of an object. Activities falling within the scope of *technè poiétiké* have their purpose outside themselves: the manufacturer deals with the means and the immediate goals, but not with the ultimate purposes. Technologies of manufacture are therefore maintained in dependence on a purpose which is of a political nature above all, which it is not in the power of the technician to fix. Any activity aims at a particular good, although these limited aims of practices only find their true meaning and their justification by exceeding themselves in the aim of the good life (Ricoeur 2001).

We should also add that the intermediate ends themselves regulate the *technè poiétiké*: the artisan is not the judge of what he produces, his activity is finalized by the requirements that it must meet, that is to say by the appreciation of the user. The best judge of the house, said Aristotle, is not the architect, but the one who lives in it. It is ultimately he, the user, which is the sole judge: his vital needs are used as terms of ultimate reference.

We may add here an additional consideration, raised in particular by Spanish philosopher Jose Ortega y Gasset, namely that for the ancient Greeks, the concept of *technè* does not really exist as a generic concept: *technè*, it is always a particular *technè* and an indissoluble link binds the *technè* to the one who performs it (Ortega y Gasset, 1941, pp. 146–147): the cobbler cannot become a carpenter, he possesses shoemaking as a gift which is fixed and given once and for all.

Of course, the Greeks know very well that the *technai* are not natural, they are acquired, but the one who acquires a *technè* owns it as a quality now inherent in his own nature, a treasure defined and without opportunity for substantial extension Ortega y Gasset says. The artisan can do only what is most appropriate to him, that is to say only one thing. He is either a shoemaker, a blacksmith, or something else, but he must not be several things at once. He has a nature which destines him to such or such a well-defined activity and he must comply with it.

Finally, the Greeks conceived *technè* as *organon* (Vernant 1985, p. 305), that is as an application of muscular strength, human or animal, through a variety of instruments, with a view to bring something into existence. The activity of the artisan is limited by his bodily, biomechanical and muscular potentials. A man, through living, fits into the rigid system of the movements of his body, just as in the fixed system of arts (Ortega y Gasset, 1941, p. 147).

*Techn*è is therefore subject to a quadruple principle of limitation that prevents it from collapsing into excess: limitation by the *eidos* or the form (formal cause), by the implicit forms of the material (material cause), by the ultimate purpose fixed by others than the manufacturer (final cause) and by the power originating from muscular and motive strength (efficient cause). The famous four Aristotelian causes also circumscribe the broad field of poietic activities, maintaining them within well-established limits[11].

The Greeks ultimately had a deep sense of the cautious attitude demanded by any *technè*. Any *technè* aims before all at the interest of the thing being created, whether it is a table, a beautiful and abundant harvest or a navigation. *Technè* is a form of knowledge that always deals with the singular, with the uniqueness of this thing. Aristotle insists on it: each thing is the bearer of its own good, its own excellence (*ergon*). It is this excellence that *technè* has in view in the act of manufacturing or in the activity of piloting (health in this body, a generous harvest here and now, or even the security of navigation in these special conditions). In this way, technological activity cannot consist in a mechanical application of rules: it is a correct assessment, a practical judgement, an assessment of the situation or of the thing to be done, with regard to its own characteristics. Acting by *technè*, for the Greeks, meant above all else to allow oneself to be instructed by the singular situation itself. This dictates the way to apply general rules to good use, according to its own requirements.

Thus, even in the *technai* of manufacture, which seem to indicate most clearly an authoritarian application of the rule by the subject – as with the potter who imposes on the clay a form to be determined – technological activity seems to consist rather in ensuring that the matter itself generates the form. The famous *Prigioni* of Michelangelo, at the Museum of the Accademia in Florence, perfectly illustrate this conception of *technè* as letting it happen[12], including in artisanal manufacture. The *eidos* of the thing is not the general rule that the artisan applies authoritatively to formless matter; the *eidos* is that which penetrates the *technè poiétiké*, which imposed itself on it by providing it with its *telos*, its particular goal.

Concerning the artisan, Jean-Pierre Vernant thus explains, "The poimea, which is the goal of his work, is beyond him… Whether it was a house, shoe, flute or shield, the product answered a specific natural need. It was not seen as something artificial,

11 "The theory of the four causes of poiesis strengthens in a way this conceptual framework of industrious activity: it is rendered dependent as a quadruple effect of the idea (eidos) of the thing to be realised (the house for the builder), the characteristics of the material to be worked with (marble for the sculptor), the control of the prescriber (the request of the client) and the energy and drive of the body of the craftsman" (Schwartz 2008).

12 See also (Heidegger 1958).

in the true sense of the term. It was an eidos, a Form laid down in advance, as was the case for natural objects. The artisan did not invent it, nor could he alter it… Form is superior to the workman and to his technè, and it directs and controls the work that brings it into being. It gives this work its goal, fixes its limits, defines its framework and its means" (Vernant 1983, p. 316). As also stressed by the anthropologist Tim Ingold, in referring to Vernant, manufacturing is not for the Greeks an activity of intelligence imposing a design upon the material[13]: manufacturing is letting things come into existence according what it must be by nature.

This being said, the Greeks were above all keen to limit technology as a creative power likely to destabilize the polis and the distribution of the sensible that defines it. At no time however would they have had the idea of considering technology as a power potentially destabilizing nature (*physis*). Contemporary human technological activity completely reverses the situation because now it can change nature globally and no longer only locally.

In addition, the ancient Greeks certainly recognized the need to contain the illimitation which was attached in principle to productive activities, but they also understood to what extent limiting principles could only expect to have any efficacy on the condition of being inherent in these activities, intrinsic and not extrinsic to their exercise. That *technè poiétiké* is limited by the essential nature of what is manufactured, by the materials, by the biomechanical possibilities of the productive body, by the attachment of *technè* to the very nature of the technician, by life and its needs. All of this involves limitations which are not imposed from the outside on productive practices, but which are intrinsic and which serve to define it. It is not in our power to decide what the *eidos* of the table is, and no more to shift the limits of the possibilities of the human body at work, no more than to make it so the shoemaker can become a carpenter.

The position of the Greeks is decisive on this point, and we must acknowledge this: according to them, to prevent any danger of technological *hybris*, there should be nothing less than a detour through general philosophy, defining what technology is in its relationship with the essence of things (an ontology), explaining also the way that it is articulated among human beings, both as living beings (an anthropology) and as users (an evaluation). On this last point, note it is not the use that is considered, good or bad; it is the user who judges.

---

13 "The forms which the artisan realized in his material issued not from the human mind, as constructs of a rational intelligence, but were themselves inscribed in the order of nature. Thus, the idea of making as an imposition of rational design upon raw material would have been entirely alien to Greek though" (Ingold 2000, p. 81).

1.1.2.3. *Regulating by ethics: an externalist vision of the mastery of technology*

However, in the past century and still today, "to master mastery" means above all imposing external limits, of an ethical, legal or political character. The Greeks were themselves undoubtedly aware of this form of limitation. Thus, in Plato's *Phaedrus*, the king of Egypt Thamus rebuffs God Theuth, the inventor of writing, who meets him to offer him his invention. The inventor is congratulated and thanked, but his invention will not be accepted due to the fact that it deeply risks transforming the way in which collective memory has always been transmitted up to now, that is to say orally – a development deemed dangerous by Thamus. The Greeks conceived therefore that *technè* could be politically prevented.

However, the limitations that Plato and Aristotle imposed on *technè poiétiké* were above all of a metaphysical (the *eidos*, the essential form) and anthropological (the possibilities of the body, the needs of life) nature. In the 20th century and still today, "mastering mastery" means giving technological development a thoughtful and chosen direction, which can only come from deliberate actions and human decisions. Ethics and politics are called to impose from outside a framework on technology.

This externalist vision of the mastery of technology continues to guide the accompanying discourse for major programs of technological development today. Indeed, in order for these programs to come to fruition, defenders argue as we have seen that it is necessary to ensure, on the one hand that these programs will not infringe upon fundamental values, and on the other hand, that their potentially damaging impacts, on nature as well as on society, will have been seriously anticipated and, if necessary, that provisions will have been taken to prevent them. The terms of the debate are perhaps less grandiloquent; it is no longer a matter of "re-binding Prometheus" as we might have demanded a trifle emphatically in the past century. However, it involves attaching to technical progress, if not limits in the strict sense, at least a certain regulatory framework.

This externalist position seems at first sight fully justified by the characteristics of contemporary technological development, which seems to be an unlimited process. The range of possibilities that it deploys is manifestly infinite. As was already stressed by the German philosopher Arnold Gehlen at the end of the 1950s (Gehlen 1980), a major reversal has occurred in the relationship between technology and its purpose: while technological activity had always meant reflection on the best ways to resolve a problem, it now means exploring all the possibilities offered by a given process or method, to see what emerges from it. "Feasibility", rather than the answer to a problem which is well defined at the outset, has become the master word for technology. There is no doubt that Gehlen is correct, inasmuch as techno-scientific research seems oriented today towards "proofs of concept" – that is to say, towards demonstrations that it is possible to do something, rather than to the development of technological devices or processes to meet specific needs. Modernity, Peter Sloterdijk emphasizes, does not lie in the discovery of virgin spaces that were waiting to be surveyed by human explorers and conquerors; rather, it is defined by the opening up of spaces of expanded possibilities thanks to the implementation of new operational routines. The watchword of modernity is "experimenting" (Sloterdijk 2012).

In this context, we must realize that the growth of technology no longer seems to be restrained by at least three of the four limiting principles defined by the ancient Greeks, to such a point that only obligations external to its own dynamics, of a legal or regulatory form, now seem to be able to impose limits.

The grand program of technological convergence, known under the acronym NBIC, seems to confirm this observation, illustrating it in a particularly symbolic way. Nanotechnology, as well as biotechnology, integrated into this grand vision of technological convergence, appear to be completely free of any constraints with regard to form and material, at least in their declared ambitions. *Shaping the world atom by atom*: this slogan, which has served to define an ideal horizon for nanotechnologies, means that ideally, we should be able to build anything at the nanoscale, without any kind of limitation imposed by essential forms. In research laboratories, it is now less a matter of forging hypotheses which may be falsified by experimentation, than of opening a domain of unlimited possible technological interventions (Nordmann 2012).

As for matter, it must allow itself to be docilely arranged according to the goals of the engineer, and thus correspond to an ideal of absolute passivity and receptivity. In the so-called "bottom-up" conception of nanotechnology, it is supposed to offer no resistance, it possesses no "implicit forms". The same goes for biotechnology: there, living beings seem to be regarded as nothing more than matter to be reconfigured at will, especially in regard to microorganisms. It now seems possible (and even permissible) to make the latter, after reprogramming, do almost everything and anything. In this domain, ingenuity appears to have no limit. Synthetic biology is a good illustration of this apparent lack of limit to technological tinkering with living beings, aiming for the construction, by genetic engineering, of novel metabolic pathways, the complete reprogramming of bacteria, or even the construction of organisms *de novo* from a catalog of "standardized BioBricks" (Bensaude-Vincent and Benoit-Browaeys 2011).

As for the efficient cause, that is to say the living corporeality of the artisan defined by a limited field of biomechanical and motor possibilities, the processes of automation which have intensified with the industrial revolution have had a major impact in marginalizing these.

What then of the final cause? The need to submit technological choices to collectively discussed values, and therefore to ultimate purposes, is unanimously affirmed. Promoting "responsible" science by anticipating its "ethical and societal impacts" is today a requirement imposed at the highest level, in institutions of scientific and technological policy. The ends appear quite disassociated from the means – that is, politics from technology – and the first must take priority over the second. If the *eidos* of the thing produced (formal cause), the "implicit form" of materials (material cause), and reference to the possibilities of the body (efficient cause) now seem to have no capacity to inherently limit technological development, the conclusion is obvious: this limitation can only be performed from outside through discussion of purposes and values (final cause). We thus understand the importance taken by ethics in accompanying convergent technologies since the 1990s.

It should therefore not surprise us much that a rather externalist ethical position appears today to be the only possible way to regulate technological development. This finding is reinforced by the fact that now, the activity of engineers in many areas seems governed not by the ideal of mastery of the real, but on the contrary by a search, by the engineers themselves, for a form of non-mastery – changing the genetics of living beings so that they accomplish human goals through their metabolism; delegating to physico-chemical processes the self-assembly of functional structures; changing nature so that it expresses novel possibilities which are inaccessible to the classic approaches to manufacturing. Many areas of research and technological innovation are today inspired by the strategy of "making it do", in place of conventional engineering approaches consisting of arranging materials based on a predetermined plan.

Technological activity, in a distant past, might certainly have consisted of steering natural processes, excluding perfect mastery of the operations in play: the farmer, the rancher, the ship's pilot had to rely on processes which operated without them and which they did not master. Their activity thus consisted in taking advantage of these processes at best, or even seeking to guide them, to steer them in sum, in order to achieve human goals. That may be, but such technological activities through actuation of natural processes (Larrère and Larrère 2015) were distinguished precisely from the technological activities of manufacturing which, on the contrary, essentially consisted in following a predetermined plan. Manufacturing involved human mastery; the artisan was not to depart from the model (*eidos*, form) of the thing to be manufactured. Today, however, the activities of manufacturing themselves seem to have also been devolved to natural processes which the scientist or engineer does not strictly speaking control, and which at most they can seek to guide in a better way. Rodney Brooks, the former director of the MIT Artificial Intelligence Laboratory, predicted in 2002 that within 30 years, we would have acquired such control over the genetics of living beings that instead of waiting patiently until a tree grows to be able to cut it down and chop it into boards in order to be able to make a table, we would be able to "grow" the table – that is to say, to genetically program microorganisms to self-assemble the table at the molecular or even atomic level. Mastery of processes is therefore no longer necessarily the directing value for the technological activity of fabricating objects for use, even as perfectly ordinary ones as tables. In the 2000s, a CNRS team had envisaged exploiting the mechanisms of self-assembly of living beings in order to reconstruct, in the artificial context of the laboratory, the very complex structure of the flagellar motor of bacteria. Until not so long ago, a motor was manufactured from a plan; now, it seems conceivable to "grow" a motor on a biomimetic surface capable of hosting proteins which, if only we provide them with the appropriate microenvironment, are supposed to self-assemble and, in some way, "self-manufacture" the flagellar motor. The distinction between manufacturing and steering of natural processes has been scrambled.

Non-mastery of manufacturing operations, according to some[14], has thus become established as a desirable goal in some lines of research and innovation in technology. Add to this another suspicion of non-mastery, related to the uncontrollable and unpredictable character of the behavior, or even the evolution, of some of the beings to which science gives birth, in particular in biotechnology laboratories.

Indeed, technological activity, as Arendt stressed, consists more and more of triggering natural processes whose effects it is impossible to anticipate. Nature itself eventually appears as a major source of uncertainty, unable to provide safeguards against technological *hubris* – we are thus far from the belief of the Greeks that *technè*, to succeed, had to comply with the nature of things[15] and was therefore limited by it. In this context, no one would dispute the importance of putting in place drastic safety standards, as an exact consequence of the very little control that we have over the processes that we launch.

This also means that the contemporary engineer, as Gehlen already suspected, moves towards seeking to systematically test everything which the technological means at their disposal enable them to do, in the aim of seeing what happens. Only ethics, conceived as external to nature as well as to the activity of the engineers, still appears able to provide a limiting framework for technological development.

In these conditions we understand the trend towards externalist modes of regulation of technology, dominated by the identification of a few general principles to be respected and by a calculation of risks, little inclined to seek limiting conditions within the technologies themselves. This is in any case the analysis at the basis of the concept of a search for mastery of our technological mastery.

---

14 See e.g. (Dupuy 2004).

15 Of natural as well as man-made things, the *eidos* or the Form of things.

1.1.3. *The limits of the externalist approach to technological regulation*

However, this purely externalist approach to the regulation of technologies raises two considerations that make it questionable.

1.1.3.1. *First consideration: the ambiguity of the means employed to master technologies*

The first is that the very goal of mastering technologies carries ambiguity and may lead to serious misunderstandings. Indeed, this goal, understandable in its most general formulation, leaves absolutely undefined the nature of the means which would accomplish this. Thus, to master the foreseen disastrous consequences of climate warming, some today advocate very seriously for the deployment on a large scale – that is to say at the level of the whole Earth – of more or less fanciful technologies of geoengineering, in order to contain the rise in temperatures. Geoengineering, as its name indicates, means implementation of engineering of the Earth itself, by making its major biogeochemical cycles a domain of intervention within the reach of human technological action. This would mean, for example – limiting ourselves to some of the solutions considered the most "realistic" – spraying millions of tons of sulfur dioxide into the stratosphere each year. Given their strong reflective power, these sulfuric particles would reflect an increased fraction of the solar radiation reaching the Earth back into space and, in so doing, compensate for the greenhouse effect with a lower level of radiation[16].

Another idea, which also has its adherents, would be to "fertilize" the oceans by emptying into them astronomical quantities of iron particles, in order to stimulate the growth of phytoplankton whose activity of photosynthesis already provides a good portion of the recycling of carbon dioxide in the atmosphere.

We already knew that human beings in industrial societies had ceased to be living beings among others – daring certainly, but finally insignificant in the face of the terrestrial elements – to become great disruptors of the basic conditions of life for most living beings on Earth. Now we find that they have also ceased to be simply the engineers of their own artificial habitat and that they now aspire, moreover, to become chief engineers of the Earth-system – not to mention that this engineering appears very adventurous and devoid of any possibility of control. Non-mastery is inherent to it. We can therefore see, from this extreme example, how much the goal of mastery over our technological mastery may be contradictory to an approach based on prudence, modesty, and the goal of greater care in our relations to nature.

16 The British Royal Society thus considers the spraying of sulfate aerosols the "most promising" of the possible ways to decrease solar radiation.

1.1.3.2. *Second consideration: uncooperative matter*

The second consideration which applies to the externalist approach to technological ethics refers to the belief upon which it is based: that no regulating or even limiting principle for technological development is thought to be able to come from the technology itself, or from technicians – that is, today, scientists or engineers. Thus, this regulation can only come from an institution external to science and engineering themselves. However, this conviction is far too abstract and ignores in many respects the reality of technological practices. This idea of a technology devoid of any internal limitations on its deployment would indeed puzzle many scientists and engineers who, in their laboratories or their research centers, on a daily basis, are faced precisely with difficulties which are strictly internal to technological practice. Their work is primarily to deploy the treasures of patience, to multiply the most ingenious tactics, to spend their time in subtle negotiations with materials and recalcitrant living beings which never do exactly what is expected of them. Can we pass silently over this reality of material technological practices, refusing to see that in their daily exercise, they gesture towards other dynamics than that of *hubris*, of uncontrolled and ravenous rage?

Let us look a little more closely at the actual work done in one of the most emblematic areas of this so-called contemporary technological *hubris*: nanotechnology. Are the scientists and engineers in this field today doing what Plato feared from artisans – namely, making the world of ordinary things very unstable, arrogating to themselves the power to vary forms without limit? Although the discourses accompanying nanotechnologies have undoubtedly pressed upon this sensitive spot, we must acknowledge that achievements in this domain are much more conservative than predicted. There are countless cases in the literature of analogies between nano-objects developed in the laboratories and the most common, banal objects of our daily existence: molecular switches, molecular scissors, molecular wheelbarrows, molecular shuttles, molecular ball bearings or even nano-cars. When it comes to infinite variation of forms, we can probably do better. It seems that in order to be able to penetrate into the common world, and make sense to all of us, nano-objects must necessarily fit into forms which are perfectly recognizable, well identified and in their place (Schummer 2006): the "nanotechnology revolution" trumpeted in communications appears in reality far from having released the "revolutionary energy" of things, as Benjamin said.

Undoubtedly, technological research and innovation may result in surprising cross-fertilizations, leading us to suspect a crossing of boundaries which we might have believed to be definitive. Genetically modifying a goat to make it produce, in its milk, the protein that is part of the composition of the silk of the spider's web, as was attempted in Canada at the beginning of the 2000s, is clearly a form of technological *hubris*. The matter as well as the essential form of beings, including living ones, no longer appears to be able to pose any kind of limit to our desire to technologically change the real. However, this does not mean that ventures of this type are systematically successful; far from it. In this case, the genetically modified goats secreted a protein which proved to be unsatisfactory for manufacturing a silk thread identical in its mechanical properties to that of a spider's web: the metabolism of the spider clearly does what the metabolism of the goat cannot do. The tinkerers with living beings here ran up against limits coming from life itself, against which it does not seem so easy to make anything do anything.

1.1.3.3. *Are scientists and engineers the sorcerer's apprentices?*

Matter, including living matter, therefore offers limitations to the hopes that scientists and engineers may have to be able to reconfigure it according to their will. Incidentally, what exactly is this will? Do scientists and engineers most often behave like the sorcerer's apprentice, ready to run all kinds of risks to society in order to satisfy their drive for knowledge and experimentation? This does not seem to really correspond to reality, and to be precise, the reverse seems to be true: the scientist as well as the engineer never cease to claim that they have no intention of exceeding the boundaries, which they claim to be very narrow in any case, of their mission. This mission can be summarized, on the one hand, as advancing so-called pure knowledge, which is not supposed to pose a risk to anyone since it does not involve, at this stage, producing effects outside of the laboratory (barring accidents, which are always possible). On the other hand, as seeking to produce these effects while limiting themselves strictly to responding to specific needs, through development of technological solutions for which there is a well identified and circumscribed social usefulness, in the field of health or in transport, energy, or communication, for example. As explained by French sociologist Christelle Didier (2008), the engineers often consider themselves not to be competent to say what ends their work should serve, and which applications it is preferable to develop as a priority: it is for society to do so, they will say; only society is competent in the field of values, and not the engineer. Effacement of the engineer, therefore, at least in terms of the choice of orientations for technological applications of their devices.

It is worth noting however that this assertion of modesty opens itself up to two criticisms.

On the one hand, it gestures towards a conception of science which has been considerably tarnished by several decades of work in the social sciences: there is indeed no such thing as a pure-and-disinterested science, in the sense of a science floating in space and enclosed in its ivory tower. Thus, a geneticist may very well, with good reason, claim to be conducting pure and fundamental science, still far from application; but in doing this, they may be guided by the prior epistemological choices consisting, as was the case in the past, of giving priority to research on the genome rather than on other levels of complexity of living beings, and may be encouraged in this direction particularly by the level of financing granted to genomics compared to other thematic areas[17]. The heavy investments from which genomics has benefited, and continues to benefit, may indeed prove to be very attractive to our geneticist claiming to perform so-called pure-and-disinterested science. In turn, the work they carry out is supposed to allow, by raising hopes of successful spin-offs in the clinic, validation of the relevance of genomic approaches for helping patients. Even if they devote their work to the study of a signaling pathway in a fundamental manner, our geneticist is led to produce powerful effects upon the course of care.

Another reason, on the other hand, to add a note of caution to the affirmation of modesty, this time rather on the side of the engineer, is that it appears to refer to a conception of the ethics of technology which is questionable, and inadequate in many respects: that consisting of the claim, as seen previously, that technology is value-neutral and that moral evaluation should only focus on its applications and the uses which are made of it, in society.

These two criticisms seem to provide arguments in favor of the necessity of an externalist approach to the ethical regulation of technology: to continue the same example, in the study of the genome, we might suspect choices relating at the same time to issues of biological philosophy (the primacy granted to the genome in explaining the pathophysiological process) and to issues of health policy (organization of the course of care). In these conditions, it seems legitimate to submit these choices to public discussion, which implies, as some claim like Philip Kitcher (2004), that scientists should not be the only ones to decide on their thematic areas of research. Science, we could say in parodying a well-known formula, is too serious a matter to be left only to scientists.

With regard to engineers, they often retreat behind the argument of the value-neutrality of technology, but again this argument has been seriously depleted by several decades of the social study of technology[18]. Technologies, by their structural properties or their functionalities, are likely, as we have seen, to have effects well beyond the uses for which they are originally designed. Here also therefore, an externalist framing of technological development, establishing what is desirable or not to do given the high uncertainty attached to its future consequences (for nature

17 We indeed see this today with the launch of the Plan France Médecine Génomique 2025, intended to equip France with a set of genomics platforms with a very high throughput, spread over the whole of the territory, whose objective is to accelerate the translation of research towards the clinic.

18 A presentation of different ways of understanding the non-neutrality of technology will be made a little later in the book.

as well as for society), also establishing the risks that a community agrees to take or not, seems to show itself to be both legitimate and necessary.

1.1.3.4. *Evaluation (appraisal) versus valuation (prizing)*

These findings are decisive. However, the meaning of the expression "externalist regulation of technology" has surreptitiously drifted with regard to our point of departure. We began from externalist approaches consisting either of putting forward a few general principles, for example in the form of "societal values" to be taken into account, or proceeding to a calculation of expected costs and benefits, with all the difficulties that we have mentioned. The Pragmatist philosopher John Dewey defined *appraisal* as the type of evaluation consisting of rationally estimating the means adequate to achieve an end (Dewey 1939). Establishing guidelines for good practice, charters, or codes of conduct which set out general principles in the aim of framing the work of scientists and engineers, or proceeding with the calculation of the risks and benefits expected from their respective activities: all of this comes under *appraisal*. Dewey pointed out, however, the need to proceed beforehand to another type of assessment to establish what we really value, that is to say what we prize – *prizing*, or "valuation" (Dewey 1939). As *prizing* and no longer *appraisal*, "externalist regulation" now means that the choices of direction that underlie science and technology policies are now to be subject to public debate in pluralistic arenas: it is a matter of asking collectively in what kind of world we want to live, and what our conception of the good life is. This Aristotelian question here raises the same difficulty that Aristotle indicated: how to ensure that the diversity of individual points of view on what is good, what is right, what is happiness, can be translated into a unified position and lead to decisions? If pluralism of values prevents any *a priori* identification of universally accepted principles, and if proceeding to a comprehensive calculation of techno-scientific risks seems to be impractical, the expression "externalist regulation" of technology then means inventing procedures allowing us, in spite of everything, to orient ourselves in contexts of high uncertainty.

Can we moreover continue to speak, here, of "externalist" regulation? We might doubt this for two reasons. Firstly, because these preliminary discussions on the values and aims of research and innovation are assumed to translate into choices of orientation in the process of research and innovation itself, as well as in the design of technological devices – that is to say, in sum, to be "internalized" into the process of science and technology "in action", as Latour would say. Secondly, because scientists themselves (engineers less so, no doubt) are very active in the pluralistic institutions of the ethical assessment of science and technology. They are for example very present in the ethics committees of the major research institutions in France (CNRS, INSERM, INRIA etc.); they have an important role in public debate as well as in the hearings of the Parliamentary Office for the assessment of scientific and technological choices (OPECST). Not only therefore is ethics increasingly internalized in the processes of research and innovation, particularly by the obligation for leaders of research projects to testify to real consciousness, throughout a given project, of the "ethical and social issues" that the research is likely to raise (otherwise the projects in question are more and more likely to be denied funding); but still, actors in science and technology are very involved in the pluralistic arenas where choices of values are discussed. The idea of an exteriority in principle between science and engineering on one side, and ethics on the other, does not stand up to scrutiny.

1.2. In what sense can technologies be "inherently" caregiving?

1.2.1. *Can there be an intrinsic morality of technology?*

1.2.1.1. *The empirical turn in the philosophy of technology: towards internalist approaches to the ethical regulation of technologies*

The empirical and pragmatist turn taken by the philosophy of technology these past decades has in fact led to a widespread abandonment, in the academic literature, of this theme of *hubris* and its corollary, the search for ways of "mastering technological mastery" from an externalist point of view. What currently predominates is the very pragmatic question of learning how to guide technological research and innovation in a direction considered to be morally desirable, in finding a way to translate choices of values (*prizing*) into characteristics of structure and into features of technological devices themselves – according to a perspective that we could therefore qualify, in the first analysis at least, as "internalist".

Dutch philosopher Peter-Paul Verbeek thus assigns philosophers the task of performing technology accompaniment (Verbeek 2010); they are invited to adopt an interventionist approach in order to assist in the design of "good technological mediations". This expression is certainly very problematic, especially in the context of pluralism of values: it requires the justification of the basis on which, and for whom, some given technologies might be said to be "good" mediations. The thorny question of who has jurisdiction to decide what is a "good" technological mediation in any case echoes the slippage, highlighted above, from a normative meaning of externalist regulation (identification of general principles to be enforced and/or calculation of risk) towards a more pragmatist and reflexive meaning (expression of pluralism of values and collective discussion on what we want). There is thus every reason to believe that a more internalist approach to regulation of technologies, in a sense which will be specified later, will have to combine with an externalist approach in this second sense of the term.

Leaving aside at this stage this thorny question of "good" technological mediations, the expression "technological accompaniment" shows an interest in putting the emphasis on the design of technology itself, rather than only on its uses. Since technologies are said to possess their own "agency", in the sense that they actively contribute to configuring human situations[19] – since, in sum, the good is now embodied in things (Coccia 2019) and not only in human interactions – it appears highly desirable to seek to ensure that this power of agency realizes what we want to promote, that it operates in the way that we have decided collectively. The values to be promoted, such as for example respect for the private life of individuals in the use they make of digital technologies, or concern for the environment, must be literally embedded within the properties of structures, in the features, and in the technological functions themselves.

Efforts are currently made to closely intertwine the philosophy of technology and the philosophy of design, in precisely the idea that the philosopher can help the conception of technologies that integrate moral requirements in their features and their operations (i.e. internally and not only externally). The idea that there is an "intrinsic morality" of technology (Verbeek 2011) is experiencing huge success today, which certainly covers multiple meanings and situations[20], but indicates an overall approach to the regulation of technology according to which it must be internalized in the technological processes and operations themselves. Thus, there are countless approaches to design centered on values, among which the concept of *Value-Sensitive Design* set the tone over the 1990s, in the field of informatics.

The context is therefore very favorable to clarify the idea of care "in" technology, that is to say, internalization of the requirements for care in the design of technological objects and processes.

1.2.1.2. *Can there be care in technology?*

That being said, and even before having clarified what exactly this internalization of care "in" technology may be, it is worthwhile to dwell on the expression itself, which raises difficulties. One possible understanding of care (the only possible, some would say) is indeed to see it as the need for a frankly deflationary attitude in the field of technology: promoting the most careful and caring relationships with human beings, but also with the environment and the non-human living beings, seems inevitably go hand in hand with practices that are less dependent on technology. Does it not mean returning to more "natural" ways of

19 See e.g. (Verbeek 2005). For a critique of this concept of the agentivity of technologies, see in particular (Lemonnier 2012); see also (Quéré 2015) as well as (Ingold 2007).

20 It may in particular involve conceiving of artificial (explicitly or implicitly) moral agents (Moor 2006).

doing things, and does this not imply reducing the number and the complexity of our technological mediations?

Thus, in the medical field, care as implementation of increasingly sophisticated knowledge and technology in order to treat and cure patients on the one hand, and care as a relationship oriented by values of attention to the other, listening and kindness on the other hand, are often presented as falling into two conflicting registers. "When care is emptied of its initial intention – to take care of the other – to redefine itself in an operational and technological version where it is reduced to a gesture of repair, in the most mechanical sense of the term" (Marin 2013), there is inevitably danger.

In a famous article on robotics in care, Robert and Linda Sparrow (Sparrow and Sparrow 2007) thus underline how unethical is the desire to substitute robots for human caregivers: regardless of the fact that we are far, according to the authors, from being able to develop robots able to perform all acts of care as well as or better than a human being would, the perspective itself is morally suspect. Indeed, considering economic pressures, there is a safe bet that the choice of developing *carebots* – a contraction of *care* and *robot* – if it were enacted, would result in a reduction of the number of human caregivers assigned to the care of the elderly. However, a robot will never be able, according to the authors, to give care in the sense that caregivers give care. In an activity of care, the caregiver performs certain well-defined tasks, but, in so doing, does much more: listen, speak, touch. With a human caregiver, care in its technical dimension, that is to say in the actions performed (even the simplest), already carries an ethic of attention to the other. Transferring care to a robot would be tantamount to splitting care, to giving precedence to care in its technical and prescribed dimension, forgetting the other, relational, and moral dimension of care, which cannot be prescribed. Activities of care are all the more caring the less technological they are.

This situation may be found in other areas, such as agriculture. On this point, the analyses of Heidegger are famous and continue to serve as terms of reference. In the conference entitled "The question of technology" (Heidegger 1977), he opposes without ambiguity the work of the traditional peasant, for whom cultivation meant encompassing with hedges and encompassing with care, to agriculture which has become a motorized industry of food supply. It consists, at the polar opposite to any care, of challenges against the materials and energies of nature. The alternative to large ultra-technologized farms thus seems to necessarily reside in cropping practices consisting of intervening as little as possible in natural processes – letting do and making do, rather than commanding. For example, by maintaining trees in the middle of the fields, the followers of agroforestry expect that the combination of mycelial and microbial life that flourishes in the vicinity of roots will provide valuable services (aeration of the soil or inputs of nitrogen). Careful and caring relationships seem at first sight to require a withdrawal of the technologies accused of carelessness, and not a vain effort to force the requirements of care into them.

It must be admitted that the manner in which the question is formulated – can there be care "in" technology? – may leave us perplexed or even give rise to misunderstanding, as it seems self-contradictory. A common feature of the ethics of care is in fact maintaining, on the one hand, that care is of a relational nature, and on the other hand that it is constitutive of the beings in relationship: care is not a relationship between previously constituted subjects; it constitutes both the subjects who dispense it, and those who receive it. However, to speak of care "in" technology risks raising the belief that there might exist somewhere, in the functional and even material specifications of technologies, a care prior to any effective relationship of care – a care materialized prior to care performed, which appears inconsistent with the definition of relational care. If care is relationship, how can there be care "in" anything, prior to the relationship?

A first way to respond would be to emphasize that care is not only relationship, but also disposition. It is however a falsely illuminating response, inasmuch a caring disposition only owes its reality to the practices which actualize it. As Carol Gilligan also stresses, care would only be watchfulness if conceived as pure disposition without actualization. It would be a false care. Again, therefore, how to understand that care "in" technology might exist, if care as well as technology have the meaning of a relationship and/or an activity? How could care constitute an inherent property of a material being, when it constitutively refers to an action? What does "in" technology mean? In fact, we are not completely incapable of resolving this difficulty.

1.2.1.3. *A plurality of possible approaches to the thesis of the non-value-neutrality of technology*

Actually, the question of how technology can, "in" its very design, embed or embody value choices, has already received many responses.

A first response was given at the beginning of the 1980s by the political scientist Langdon Winner (1980) who, in a high-impact article, did not hesitate to speak, not of an ethics but of a "politics of artifacts" – in the subjective meaning of the possessive. According to Winner, technological design may be guided by political choices in the sense that it can, by its most material aspects, contribute to achieving an explicit policy goal. How can a policy, so to speak, materialize itself in a choice of technological design? Winner gives an example which has become famous, that of the walkways constructed in the 1930s to allow pedestrians to cross a freeway in New York[21], but this case has been challenged from a historical perspective. So, let

21 This freeway linked the city to Long Island and Jones Beach, a resort for New Yorkers. However, the builder Robert Moses intended to implement his own segregationist ideas, in

us illustrate this point by another example, concerning the design of digital platforms aimed to facilitate public debates. French political specialist Clément Mabi studied a number of these (Mabi 2014), including those deployed in the context of the CIGEO debate on the project of deep geological disposal of long-life nuclear waste, in France. Mabi explains that the choice of technological features selected for the digital platform aimed, as is normally done so that debates can move forward and lead to proposals, to channel the issues likely to be discussed; this being said, it shows that one purpose of this channeling of exchanges was also to exclude their most dissident participants: the format of contributions did not allow the expression of the most virulent positions.

Winner gives a second meaning to the notion of "a politics of artifacts". He moreover judges this second meaning to be even more important than the first: it no longer entails uncovering a conscious but not explicit political program, made effective by technological choices, but recognizing that in itself, a technology can by virtue of its design require a given socio-political organization. Winner supports himself with an analysis by Friedrich Engels concerning authority. To anarchists who rejoice in a society in which any form of authority will have been abolished, Engels responded that in the context of industrial society, authority is necessary. Large-scale industry cannot exist without authority. Steam power fixes its pace, and it requires rigorous coordination of tasks and an iron discipline exerted upon the workers, whatever the regime for the appropriation of the means of production. The mechanization of industry, explains Engels, is ultimately much more despotic than small capitalist manufacturing.

We can no more, to take another example, imagine a decentralized and local production of nuclear energy: this type of energy implies a strong and centralized State, a political organization incompatible with decentralization and local ownership, by communities, of energy production.

A third response has been given by Canadian philosopher Andrew Feenberg (2010): technology materializes an unequal ideology or social order in the form of a "technical code", making the problems raised by this ideology or this social order appear to be purely technical problems, calling for solutions which are also purely technical – which, in fact, immunizes them against criticism. Feenberg gives the example of child labor in the 19th century: putting children to work implies designing workstations adapted to their morphology and to their strength, obviously

---

the absence of rules explicitly prohibiting African-Americans access to Jones Beach. The very singular characteristic of these walkways designed by Moses, according to Winner, is that they were very low. By opting for this choice of construction, he in fact made the beach accessible to motor vehicles only, but not to the bus. Thus, only rich whites and owners of automobiles would now have the possibility to go to the beach, but not poor blacks who were forced to move by public transport.

different from those of adults. Formulated like this, the question appears purely technical, anticipating a response in the same register – which contributes to legitimizing the very existence of a deeply iniquitous social order which allows the labor of children, and which of course had to be challenged.

A fourth answer may be formulated as follows: a technology, when implemented, may redefine in depth the fundamental coordinates of a moral situation. Verbeek gives the example of ultrasonography (Verbeek 2011): the generalization of this imaging technique in gynecological practice has contributed to changing the moral status of the unborn child. Through these images, future parents have had their perspective on the fetus changed by giving it the sense of being a fully-formed person, well before birth. This being said, it is now also possible, at the same time, to identify *in utero* severe malformations, leading future parents to take sometimes painful decisions (medical interruptions of pregnancy). Verbeek thus speaks of "intrinsic morality" of technological artifacts, as opposed to a conception of the morality of technology relating only to the uses which are made of it. Here, the problem is not whether ultrasonography is used for "good" or "evil", at the moral level; the fact is that this technology, due to its features that open up new possibilities in terms of medical imaging, has greatly contributed to an in-depth reconfiguration of a situation with a moral character.

This idea had incidentally already been put forward by Dick Willems with regard to a portable medical device, the mini-spirometer, designed for monitoring asthma and detecting crises. The device, explains Willems, bears a "moral function" to the extent that the requirements of the device have become rules of conduct which are no longer inscribed in the small machine, but in the body itself (Willems 1995, cited by Soler and Trompette 2010).

A fifth response is that of Latour: the designer of a technology can seek, by the choice of a particular design, to encourage, guide or even compel the behavior of users in a direction considered desirable at the moral level. The expression may seem curious at first: does moral action not presuppose freedom, that is also the ability to not act morally? If I am "forced" to behave in the way prescribed by the technological devices that I use, am I still a moral being?

Latour gives many examples to support his conception of human morality, such as the Berlin key (Latour 2000). This key is designed in such a way that it obliges the inhabitants of a building in Berlin to lock the entry door, each time they enter and leave the building – which they obviously do not do spontaneously, that is to say, by pure moral respect for the caretaker of the building who, because of the carelessness of the inhabitants, must leave his lodge all day long to close this door which is always open. Latour also gives the example of the speedbump, which requires motorists to slow down in front of a school – which, again, they do not do spontaneously, that is thinking simply of the safety of children crossing. Technologies therefore have their own "moral agency": they require us to adopt certain behaviors, judged morally desirable, that humans do not adopt spontaneously.

In these conditions, however, can we still talk about morality? Should we not rather recognize that humans are inherently deprived of morality, and that to obtain from them the same results as if they really behaved morally, it is necessary to multiply constraining technological artifacts? A Kantian deontologist would respond negatively to the first question, and affirmatively to the second. They would conclude without any doubt that technology, far from contributing actively to the moral situation (an absurdity for Kant), only conceals the deep immorality of human behavior. If the people of Berlin or automobile drivers are actually doing what it is morally expected of them, this is simply an appearance: they act in reality out of well-understood self-interest – unless they lock the door behind them, the Berlin key remains stuck in the lock; avoiding having to use the vehicle's shock absorbers – and not out of moral concern for others. According to the famous Kantian distinction, neither Berliners nor motorists operate *out of* duty, that is to say, by pure respect for the moral law: they act *in accordance with* duty, natural inclination here being the true subjective principle of the action, and not respect for the law.

In contrast, to consider technologies as implicit moral agents is to situate the morality of an action, not in the human agents alone, but in the multi-agent context, populated by humans and by non-humans, where the action is deployed. Latour's proposal is thus to assume that human agents are never moral on their own: they are moral only with their technologies. Where a Kantian would say that there is no morality here, Latour argues that the morality proceeds from the situation and not the way in which the will decides to act.

A sixth response is to say that technology presupposes, in its design, underlying representations which are often not made explicit and which by their nature direct the positions that will be taken, epistemic ones for example. Thus, as historian Hallam Stevens explains (Stevens 2016), the algorithms used today in genomics have an historical kinship with algorithms for textual analysis developed in the 1950s. Consequently, the metaphor of the genome as a text to be read and deciphered, although it has been exhausted at the epistemological level[22], has found itself operationalized through the dissemination of these algorithms. This metaphor, although false and unfounded, therefore weighs on the orientation of research programs and on the type of questions that are asked to living creatures.

---

22 Nature is not a book written with four letters, that is to say the four nucleotides which enter into the composition of DNA!

A seventh response focuses on the materials used for the manufacture of technological devices, at a somewhat infra-functional level. No need to insist on this point which has already been discussed above, using the example of electronic devices and their dependence on rare earths, the extraction of which is very destructive at both environmental and human levels.

An eighth response, an extension of the previous one, is located at a supra-functional level: the technological devices that we produce can also be evaluated, from a moral point of view, over the whole of their lifecycle. Aiming for completeness, this response in a sense aims to sum up all the others: it combines the different considerations that each of the others have highlighted. Let us note, however, that life-cycle analyses are currently far from covering all these dimensions of the morality of technology.

None of these eight possible conceptions of the non-value-neutrality of technology can be disqualified at the outset, and none of them should be *a priori* excluded as non-relevant, if our goal is to think of the possibility of an "inherently" caring technology. Caregiving in technological design may indeed require the engineer to make explicit their representations and their choice of values; or to consider to what extent the technology currently being designed, when it deploys all its effects, will not contribute to strengthening an unfair social and political order; or again, to be conscious of the structuring values of the environments in which the technology in development will be used, and which it will be likely to reconfigure.

1.2.1.4. *Is ethics in technology a categorical, dispositional, relational, or systemic property?*

To clarify these different ways of understanding how technology can embody values "in" its very design, it may be useful to refer to the concepts by which French philosopher and biologist Lucie Laplane (Laplane and Solary 2017) proposes to define what stem cells are. The "stem property" can, according to her, fall under four different concepts: it can be either a categorical property (a property of which an entity disposes due to its intrinsic characteristics and not by virtue of its interactions with the world), a dispositional property (an intrinsic property but whose expression depends on extrinsic factors), a relational property (a property that emerges from a relationship between two entities), or a systemic property (a property regulated and determined by the system).

The transposition of these four concepts to the thesis of non-value-neutrality of technology discussed above provides interesting results: firstly, none of the eight meanings of this thesis falls under the conception of a categorical property. If it is possible to say that a technology "embodies" choices of values, in virtue of its features, the materials that comprise it, or even of its overall design, it is certainly not in the sense that these features, these materials and this design in themselves possess these values.

Neodymium – a metal belonging to the class of rare earths and featuring in the composition of electronic devices – is not a factor of carelessness in itself; however, the systemic conditions in which it is extracted (for the most part of it in China, in close proximity to the city of Baotou), as well the relations that the residue generated by this extractive activity creates with the surroundings, are environmentally and humanly devastating[23]. A workstation adapted to children, to return to the example of Feenberg, is not in itself an anti-value: it becomes so in a system which legalizes the work of children, and when it is put into relationship with the bodies of children. Verbeek's ultrasonography is not moral in itself: it manifests its "moral agency" in a specific situation, it involves a dispositional property. Latour's speedbump is only involved in human morality in relationship to the characteristics of vehicles (suspensions): it involves a relational property. With respect to the operationalization of the textual metaphor in the algorithms of genomics, this seems to be rather a dispositional property.

Talking about care "in" technology, or an "inherently caregiving" technology, therefore calls for two comments. The first is that by "in", or "intrinsic", we should not understand a "categorical property", that is to say one inherent in the substantial sense. This remark removes the perplexity of a care "in" technology which would precede any effective care, an idea actually void of significance. The second remark is that the terms "in" or "intrinsic" are opposed here to "instrumental" (care and its opposite are decided only in usage, according to an instrumentalist conception of technology), which moreover does not in any way prejudge whether this care "in" technology can be defined as a property involving a strong external component – relational or systemic, to return to Laplane's terms.

1.2.2. *Technology and care: a difficult articulation*

1.2.2.1. *Reconsidering the two-layer model of care: the state of the art*

The idea of a care "in" technology is therefore not as insoluble and even contradictory as it might seem at first analysis. Several works, especially issuing from the Science and Technology Studies (STS) current, thus seek to overcome the supposed opposition between technology and care[24]. In refusing to essentialize concepts such as "nature", "society" or "care", these approaches have indeed

23 See on this subject the newspaper article in *Le Monde* dated July 9, 2012.

24 For an overview of the literature on this point, see (Bates *et al.* 2017).

produced valuable studies on the processes, always contextual and multi-agential, according to which care is specifically configured by technological devices[25].

Unsurprisingly, the criticism levelled against the assumption of an opposition in principle, supposedly unsurpassable, between technology and care, mainly concerns the medical field. Dutch anthropologist Annemarie Mol has thus unpacked a particularly insistent premise in conceptions of care, namely, that care above all requires taking patient choice into account (Mol 2008). This "logic of choice", which is obvious at first sight (who would want care in which the patient does not have a say?), fits despite everything within the two-layer model of care. Indeed, it gestures towards a dualistic conception opposing, on the one hand, a human subject considered as a free and rational will, author of its own choices, and on the other hand, a body belonging to the same subject over which he/she has full control, in authorizing interventions upon it, or not. However, stresses Mol, it is not at all obvious that this logic of choice inevitably leads to care. Care has its own logic (logic of care). The latter consists less of referring only to patient choice, and therefore to the value of personal autonomy, than looking for the best way to combine the heterogeneous elements that make up the situation of care, including technologies.

As Jeannette Pols also stresses (Pols 2012), the supposed opposition between *cold technologies* and *warm care* is concretely surpassed in practice: caregivers are always negotiating the possibilities of technology, directing them towards the purposes of care (Mol *et al.* 2010).

The testimony of caregivers, incidentally, corroborates the existence of these situations which are more nuanced, more complex, than the coarse opposition of *cold technologies* and *warm care* would suggest. Thus, Elisabeth Hermes, a dialysis unit nurse, describes her experience of the very heavy care that people with terminal renal insufficiency must undergo (Hermes and Sifer 2005). At first sight, dialysis centers are highly technologized universes, where we might therefore expect to see *care* pass into the background behind *cure*. Is the situation of care in such a place not entirely configured by the dialysis generator? However, explains Hermes, in hospital dialysis units, the line between caregivers and patients is less marked than in other services. Not only does the dialysis generator – high-tech and very unfriendly machine though it be – not ruin the relationship of care, but as counter-intuitive as it may seem, it may be conducive to bringing caregivers and patients together. The opposition of technology and care does not stand up to investigation, including areas where it might seem *a priori* to obviously apply.

25 See the introduction of (Schillmeier and Domènech 2010).

Such contextual and pragmatic "tinkering" also involves, incidentally, the designers of technological devices. Engineers are stakeholders in care. This is what emerges in particular from the study of French sociologists Julien Soler and Pascale Trompette (Soler and Trompette 2010), on the design and development of an algorithm for home detection of epileptic seizures in children, that the authors themselves describe as a technology of taking care.

Soler and Trompette have examined the close interactions between computer engineers and parents all throughout the development process of the algorithm. Starting from the premise that the signs of epilepsy are not the same in all children, and also that many contingencies may interfere with the algorithm and lead to false alarms, the whole challenge was that the algorithm should incorporate interaction with the caregiver (Soler and Trompette 2010, p. 347). Based on day-to-day experience gained over a long period, the caregiving parent has very accurate knowledge of the characteristics of outbreaks of epileptic seizures in their child, which is indispensable for customizing the detection algorithm.

The intertwining of engineering and care may also exist in contexts where, however, the engineers are very distant from care practices and have no direct contact with the patients. The example of tissue engineering (a booming field of bioengineering, aimed at the reconstruction of tissues or organs from cells) illustrates this phenomenon. Philosopher and ethicist Mechteld-Hannah Gertrug Derksen (2008) has shown how engineers in this area establish a link, even a tenuous one, with the universe of care, though they do not have to deal directly with the caregivers nor with the patients. Derksen's initial observation is very banal: for engineers, human bodies are objects of study and intervention which can be described in strictly functional terms. The parts of the body are pieces of a machine. Engineers work with "biological materials" (cells) whose link with real people remains very abstract and distant. It is exceptional that the engineer refers to the body as a *lived body* and not as a *living body* or organism. So far, nothing very new; this argument is also made by anthropologist Lesley Sharp (2013).

Yet, notes Derksen, in their day-to-day activity, engineers are not as foreign to care as they would seem to be at first sight. This irruption of care into engineering work proceeds precisely from the use of cells. Derksen summarizes a problem confronting engineers: in the field of tissue reconstruction, the results differ depending on the cells used. For example, the cells of a diabetic patient do not give the same results as the cells of an individual without a particular pathological condition. In other words, the engineer is forced to understand the singularity of the individuals from whom the cells were harvested, and the fact that cells are not indifferent and generic "biological materials". Each cell comes from a single person, whose characteristics are vital to understand. This is doubtless not yet quite an integration of the requirement of taking care of individuals into technological design

– we are still at the cellular level, and not at the point where engineers take into account what Derksen called the *lived body* – but we have already left behind this mechanistic and homogeneous representation of the body which is supposed to be part of the standard baggage of the engineer.

The question of the links between technology and care also raises a growing interest beyond the strictly medical domain. In the field of caring for vulnerable persons (at home or in an institution), *carebots* have already given rise to numerous reflections on this point. *Carebots* are robots who assist the personal caregiver in their tasks (lifting, feeding, and toileting in particular), or who assume these tasks by themselves. The general question of a "caregiving" technology here takes on a particular coloring. Can we say of a robot that it is truly a caregiver, that is to say, that it can actually provide care for vulnerable persons, and if yes, under what conditions?

The two authors of an article (Meacham and Studley 2015) have answered this question in the affirmative, in criticizing the presupposition that according to them underlies the positions that deny the possibility of a robot caregiver. This presupposition is linking care to the possession of internal emotional states – which of course robots lack. However, maintain the authors of the article, the possibility of care is in no way dependent on a prerequisite of internalist character: there is no need to identify, in the caregiver, the possession of such a mental state that they can actually give care, and be recognized as a caregiver. Care is not "in" the heads of carers, but in their behaviors and in the environment where they are deployed. This externalist conception of care, which the authors call the "environmental hypothesis," translates into issues of design which can be formulated in behaviorist terms: how to design a robot that can display movements which demonstrate appropriate attention and responses to perceiving expressions or signs of vulnerability, and movements which make us able to reasonably rely on the permanence of this attention and this response (Meacham and Studley 2015, p. 179)? Although externalist as regards the definition of what makes care, the position of the authors clearly shows in consequence the need for reflection on what is care "in" the very design of such robots, which guides the analysis towards the work of the engineers.

In the same vein, Aimee Van Wynsberghe (2013) intends to propose a conceptual framework and a general methodology allowing engineers who design carebots to understand what exactly the requirements of care are, at least in the particular context where their robots will have to intervene. The aim is to make them able to take account of these requirements in the very design of carebots, which implies undertaking a quasi-ethnographic process of inquiry among caregivers.

That being said, this type of approach to carebots suffers from a serious limit: it actually endorses the duality of technologies of cure and technologies of care, in the sense that, for example, surgical robots do not fall within the perimeter of the analysis – Wynsberghe excludes them from the outset. The term carebots designates the class of robots who assist nurses in their activities, and not the whole set of robots used in medicine. The dichotomy of technology and care is not itself questioned; it is pushed further away. Thus, in an article devoted, precisely, to robots used in tele-surgery, Wynsberghe and Chris Gastmans do not address the designers of these robots (Wynsberghe and Gastmans 2008): they draw up a balance sheet of robotized surgery and examine to what extent it meets, or not, the ethical values that constitute the patient/doctor relationship. It is therefore no longer a question here of embodying care "in" technology, but of assessing available technologies in view of medical care and the moral requirements inherent to it. Not, therefore, care "in" technology, but technology "in" care. The two-layer model of care does not appear to have really been surpassed.

This is, substantially, the finding of French philosopher Frédéric Worms and his colleagues, who in turn proposes a definition of care which aims precisely to go further in surpassing the reciprocal exteriority of, on one side, technology in its functional and material dimension and, on the other, care in its relational dimension. "The philosophy of care [...] seeks to think of ways in which technology can, including in its materiality, become integrated into the goals of care. It does not intend to add from outside a caregiving dimension to existing technological medicine, but to think of care at the heart of technology and of medicine" (Benaroyo *et al.* 2010). The technology referred to here has no limited boundary; it is any technology, not only the supposed "good technologies" of craft or gesture. The two concepts of care that Worms identifies – the technological concept of "caring for a body" and the relational concept of "taking care of a being, of a person" – are not separable; they must be kept together at risk of losing sight of the very requirement of care. Care, insists Worms, cannot be conceived as a dimension that is added from the outside to technological practices. In medicine, as elsewhere, he stresses, "How can we conceive of care that does not involve technological invention?" (Worms 2013). It remains however for philosophers of care to develop this fruitful viewpoint, including increasing the number of empirical studies which are still too few. What does it mean concretely, ultimately, to "think of care in the very heart of technology"?

1.2.2.2. *Nature, technology and care*

The area of caring for persons is not the only one concerned by this questioning of the opposition between so-called *cold technologies* and *warm care*. The study conducted by sociologist Maria Puig de la Bellacasa on the "science of soils"

(Puig de la Bellacasa 2015) well illustrates the effort to introduce care into the human-nature relationship, yet without this effort translating into a pure and simple rejection of technosciences. These may also contribute to taking care.

In a productivist optic, the soil is only worth what it can produce. This conception of the soil, explains Puig de la Bellacasa, is dependent on philosophical presuppositions which are deeply contradictory to "soil care". Firstly, human beings do not belong to the soil, they are outside of it and must treat it as an object; secondly, time is an arrow and, in this linear conception of time, the link between present and future boils down to an imperative of short-term profitability – we must extract economic value from the soil; thirdly, the soil must be controlled, mastered, in order to avoid its depletion and to preserve its maximum productivity; fourthly, we are in an emergency, and we must act quickly if we want to extract what is needed from the soil to feed almost ten billion human beings. These assumptions affect the type of technologies used in the exploitation of the soil (mechanization, fertilizer chemistry).

Puig de la Bellacasa then analyzes a significant evolution in the way of conceiving soils, moving from productivism to an attitude of care and, based on this, calling for a transformation in technological mediations. The soil is now seen no longer as a resource, a simple support for agricultural activities subject to the imperative of performance, but as a complex system consisting of the intertwining of biophysical processes and the metabolic activity of microorganisms, fungi and worms, without of course forgetting that of plants. The soil mixes multiple temporalities, including that of human activities, which cannot be reduced to the linear temporality of progress. Human beings are a part of this "community of the soil"; their relationship to the soil is not limited to an activity of exploitation alone, also stemming from activities of maintenance, life support and even repair – all activities, importantly, which must mobilize very advanced knowledges and technologies.

Puig de la Bellacasa in no way argues, in fact, that this evolution in our relationship with the soil has gone hand in hand with an elimination of science and technology; quite the contrary. The most advanced science and technology are not antithetical to care, they are even required for the goal of care to be now able to guide the relationship between humans and the soil. This goal particularly requires extensive scientific knowledge on soil structures, as well as on the network of relationships involving the diversity of living beings inhabiting them. The criticism of technoscience, which projects onto any reality and onto the future itself the imperative of producing economic value, does not yet imply a massive rejection of contemporary technology.

1.2.2.3. *Care for engineers*

Can we however speak of "caregiving" technologies if the engineers who design these technologies, and the agents who maintain them, are themselves the object of a lack of consideration and attention – that is a lack of care? That is the question asked by Puig de la Bellacasa in discussing Latour's position (Puig de la Bellacasa 2011). Latour, she noted, already introduced the problematic of care into the sociology of technology. The book entitled *Aramis, or the Love of Technology* (Latour 1996) illustrates this point well. In this work, Latour follows the traces of the fate of an automated Metro project which was to serve the "Petite Ceinture" south of Paris in the 1970s, in seeking to understand why this project was finally aborted. Latour's interpretation is that Aramis – the acronym of this famous metro system – failed because it was not subject to attentive care: it was neglected in the sense that the definition that the designers originally had for it did not evolve in response to the controversies, the displacements, and the translations imposed by the different actors involved over the years. It did not vary one iota; everything happened as if Aramis from the very outset had been considered as a *matter of fact* (an established, stabilized, frozen fact) and not as a *matter of concern*, that is to say the object of shared development, and therefore a common concern. Aramis remained a project stored at the bottom of a box, due to never having been loved, says Latour – that is to say, in the absence of having been an object of care.

However, stresses Puig de la Bellacasa, nourishing a "concern", in regard to technologies, is to put oneself in a position to consider their fate, to ensure that these technologies embed values in their material characteristics. She evokes on this point the position of Latour in respect of SUVs, those elevated vehicles which are very greedy for fuel, and therefore pollutants. It serves no purpose to demonize them, disqualify them as "Frankenstein" technologies, says Latour; better to also make these *matters of concern*, that is to say, take care to consider their fate (for example, to introduce more consideration for the environment).

The problem, says Puig de la Bellacasa, is that this methodological position demands a high price: to reject alleged radical, ferocious and dogmatic critiques, on the grounds that they prevent us from understanding the process of negotiation and translation to which technologies are subject, is to take a serious risk of not seeing that a given socio-technological assemblage, made up of translations and displacements though it may be, may contribute to strengthening an asymmetry of power between the different actors concerned. Puig de la Bellacasa refers to a study by Lucy Suchman concerning the development of so-called "smart" interfaces which ratify, so to say by design, the effacement of artifactuality itself: designing interfaces which are as "natural" as possible means making invisible all the underlying work of engineering and maintenance. Introducing care into social studies of technology, in sum, means becoming attentive to those who are rendered invisible or devalued – including the engineers. As Mumford says, as formidable (or desirable, should we say today) as the automatism of the machine appears to us, the machine always conceals a human being who governs, corrects, and takes care of it (Mumford 2015). For technology to be caring, it is essential that he (or she) who designs it also be the subject of care.

1.2.2.4. *Care for technology*

Here it is necessary to add that technological devices, in order to contribute to "taking care", must themselves be objects of care. "Caregiving" technologies are indeed above all technologies which are taken care of. This point has been emphasized for example by Mianna Meskus (2018) concerning the use of induced pluripotent stem cells (IPSs) in biomedical research. These biological tools are generally presented from the point of view of their extraordinary potential. However, they are above all very fragile, precarious, and unstable elements, which must be taken care of. One might object that this conception of care gestures only towards the technical and instrumental dimension of care. It would entail that form of care whose breach is sanctioned by failure, as opposed to a care of a higher nature, of a moral order. Bioengineers are like the man of Song: if they do not want their cells to die, they must take care of them in an appropriate manner. However, Meskus defends a more radical position: based on Dewey's theory of valuation, she argues that there is no distinction between the instrumental and moral perspectives of care (between the ends and the means, in sum). The two form a continuum of care and, when the bioengineer bestows upon their cells a lot of attention, thoughtfulness and even emotion, they are already within the moral dimension of care.

Outside the biomedical field, the study of French sociologists Jérôme Denis and David Pontille on signage in the Paris Metro (Pontille and Denis 2015) also promotes the dimension of care inherent to the activity of maintenance staff. Paris's public transport authority, RATP, committed at the beginning of the 1990s to a reflection on the complete overhaul of signage over the whole of the network, with a concern for harmonization and standardization (signage was quite different on the metro, regional rail and buses). This consequently involved bringing order to this plural space of Parisian transport – stabilizing this space by giving it a homogeneous structure, via signage which was pervasive and identical everywhere. However, as Denis and Pontille show, production of order comes not only from the adoption of shared norms and standards, but also from sometimes tiny activities performed in the name of caring for things. The maintenance agents of RATP are actors of care, attentive to maintaining the objects of signage (the different display panels) which they perceive first of all as fragile, vulnerable things, subject to all kinds of hazards.

It therefore appears that the goal of "care in technology" is in reality a triple goal. Guiding technology towards care goes hand in hand with the requirement of care for the technology itself, "taking care" of our technologies – of a *care of things*, say Denis and Pontille – but also of care for those who design, produce and maintain the technologies, in the sense of attention to be given to them, of attracting attention to them, as Puig de la Bellacasa also stresses.

Nonetheless, these stimulating contributions have not yet really become integrated into the philosophy of technology. On the whole, philosophical studies of technology on one hand, and the ethics of care on the other, have yet been only very timidly articulated. There are undoubtedly very many philosophical studies devoted to examining the "agency" of technologies in the areas of health and personal care. However, the effort made by the ethics of care to propose a more global concept of care, well beyond the problematic of medicine and personal assistance, has not really percolated into the philosophy of technology. Up until now, in sum, the philosophers of technology have not really begun to theorize care.

1.2.2.5. *Design with care*

Are theorists and practitioners of design better placed than philosophers of technology on this point? We might doubt it – this is, in any case, the finding made by the editors of a collective book on urban design (Bates *et al.* 2010, p. 7, 10). It should be noted first of all that while recent years have seen the flourishing of many so-called "by design" approaches to technology – *privacy by design, safety by design, ethics by design* – *care by design* seems to have a very limited meaning. The expression simply refers to a set of programs for organizational transformations in order to improve the care pathway in health institutions (Marshall *et al.* 2015).

Designers are undeniably not indifferent to the problems of care; far from it. The development of living labs for health is testimony to the importance given to these issues in contemporary design. In fact, though *design with care* refers today essentially to a set of commercial proposals in the field of health technologies, at the beginning of the 2000s the expression had a more interesting and ambitious meaning. It then referred to a methodology developed to deal with issues of design in environments of care, whether it be hospitals, residential facilities for dependent elderly people or private homes (Cheverst 2001, Cheverst 2003). Its proponents of the time argued that in the field of care, design is guided primarily by technology, to the detriment of a finer understanding of the needs, values, and life habits of the users. More specifically, design is always controlled by the preconceptions of an able-bodied user; designers struggle to account for the specificity and diversity of the needs of vulnerable people, in a situation of disability. *Design with care* is therefore first of all a method of designing devices for assistance and care oriented towards use (*user-centric design*), adopting the point of view of those who will be the recipients of the devices designed.

Thus conceived, *design with care* has however a clearly defined perimeter, care for people, and is very far from covering the whole of today's problematic of care. Interactions between technologies and non-human natural beings do not enter into this equation. In sum, concerning a possible care for natural things, *design with care* has nothing to say.

In any case, these approaches appear to remain heavily dependent on the two-layer model of care. In a voluminous report on design in medicine, French designer Roxane Andrès thus clearly distinguishes design oriented towards cure from design oriented towards care – to which she adds a third dimension, design oriented toward the safe, that is to say toward a better understanding of situations of risk (Andrès 2013)

It is worth noting however that the situation seems to have changed. *Design with care* now refers to a more ambitious and comprehensive approach to the issues of care. At the end of the 1990s, two French industrial designers, Antoine Fenoglio and Frédéric Lecourt, founded a design agency called Sismo. Their intention was to upset the codes of design (hence their name, "Sismo", meaning earthquake), surpassing the opposition between design as creation and design as a response to a brand's order[26]. Today, 20 years later, Fenoglio engaged in a reflection with French philosopher Cynthia Fleury around what they call, precisely, *design with care*, associated with a methodology called *proof of care* (Fleury and Fenoglio 2019). This new concept of *design with care* aims to extend well beyond the issues of health and personal care. It aims to "connect design and philosophy from the point of view of care", which definitively has little relation with *design with care* as it was understood at the beginning of the 2000s. An essential point of the design with care promoted by Fleury and Fenoglio is the priority it gives to empowering individuals, especially the most vulnerable. Indeed, by designing relevant solutions to well-identified problems, the aim is to make individuals more empowered, with the ultimate goal being that they continue to be the "subjects of their own norms" of life according to Canguilhem's words (Canguilhem 1947).

Urban design is without doubt today one of the areas where reflection on the articulation between design and care is most advanced. The design of urban spaces and habitats is indeed underpinned, more and more, by two guiding ideas (Imrie and Kullman 2017): firstly, the objective of urban design is not to seek to predefine, normatively, what the behavior of the inhabitants must be. On the contrary, it means

26 See an article in *Le Service économique hebdomadaire du design* dated September 24, 2012, p. 7.

to think of spaces in such a way that individuals can invent interactions and ways of being together for themselves. In particular, rather than designing standard spaces and habitats with a preconception of what individual bodies and needs are, urban design must recognize the diversity of bodies and aspirations. Urban design guided by care thus avoids predetermining, in a coercive manner, the way in which urban forms will operate. It also avoids essentializing the categories of "body" and "human collective"; in other words, it does not normatively impose what a "good" collective is. On the contrary, it seeks to ensure that spaces will be sufficiently flexible to enable individuals to appropriate, negotiate, and reconfigure them according to their wishes and their needs.

This initial prerequisite for urban design guided by care firstly echoes the need to give up the desire to control. Secondly, and correlatively in a sense, urban design can only integrate care when it focuses on the materiality of the spaces, habitats and objects that serve at the same time as frameworks and as possible media for human relations, rather than on these relationships themselves – in opposition to the conceptions of design which conversely give priority to experiences and to relations rather than to objects and their materiality[27]. The designer is not called to design the relationships themselves, but that which makes them possible without yet predetermining them: their material foundations (Bates *et al.* 2017).

These two premises imply that care, in terms of urban planning, is not only a matter of concern for designers: it is also a matter of concern for the inhabitants of the city, who must take care of their habitat and the things that inhabit it, in order to invent their own way of living together. A design which performs care is therefore not a design in which designers are the masters and the agents of care, according to a unilateral and top-down conception of what is "good" care; a design which performs care is a design that promotes care, that encourages those whom it targets to take care, to create *configurations of care*. A design which performs care is a design in which care is not predetermined as to its content: it is maintained in a metastable state, charged with the potential to become, as French philosopher Gilbert Simondon would say. Care is not contained in the result produced by the design, it is a kind of "continuous designer", in the sense Descartes spoke of continuous creation.

27 Vincent Beaubois explains very well that in reaction to the functionalism that was dominant until the 1950s (of which the *Streamline* movement is a symbolic illustration), but also faced with criticism directed at industrial design (the latter was accused of serving the interests of capitalism, encouraging mass consumption), designers have responded by giving a new definition of design: its purpose is not to design objects reduced to only their functional utility; it aims above all to create sensorimotor and affective-emotional experiences, as well as relationships. See (Beaubois 2019).

In this sense, as Michael Schillmeier insists commenting on Heidegger (Schillmeier 2017), care is indeed the fundamental, temporal structure of our relationship to the world and to others. To exist is to take care. Schillmeier here goes very far in his reflection on care, implicitly replicating Heidegger's argument in the fable of *Cura*, that is to say concern within the meaning of care, of taking care: conceiving the human according to a duality of registers – namely the body and the spirit, and therefore assuming that there is on one hand a care turned towards bodily, natural and worldly things, and on the other hand, a care turned to the spirit and human interactions – inevitably misses the profound nature of care. There is no mutual exteriority between taking care of the things of the world, be they natural or artificial, and taking care of self and others in view of living together.

The problem is that to inhabit the world eventually means for any of us to adopt habits, routines and even automatisms, which makes us lose sight of the work of care that we must continuously accomplish so that we can, precisely, live. Care means before all else the struggle against forgetfulness of care – that is to say never taking for "normal" and "objective" the existing configurations of our habitat, which are always based, in the end, on mechanisms of invisibilization and a distribution of the sensible which obscure the lack of care where it thrives. These mechanisms are used to circumscribe what can be made visible in the city, what falls within the perimeter of society, casting into invisibility and outside any possible care the non-social, the radical otherness – which may however be another social, the social of other humans or non-human living beings. Children in the Democratic Republic of the Congo, forced to work 12 hours per day in the mines to extract tantalum, or coltan – a metal essential to the manufacture of certain types of capacitors that feature in our electronic devices – are thus our own non-social, we who live in the European industrialized spaces; they are rendered invisible and excluded from any care.

There are indubitably, in these reflections on design, real advances in regard to the alignment between philosophy of technology and philosophy of care. In particular, the idea has arisen that a design oriented towards care requires investigation of what it means to "inhabit" the world and of the distribution of the sensible that lies at the basis of our categorizations, in the aim of revealing the mechanisms of invisibilization that lead us to forget everything upon which we depend.

1.2.2.6. *Ethics of care and of technology: the missed connection*

Does the literature on the ethics of care contribute in its turn to the alignment between the conceptualization of care and the conceptualization of technology? It does not seem so, in the first analysis at least. The ethicists of care show rather little interest in the fact that care is, in most cases, inextricably linked to the implementation of technology. Certainly, technology is not completely obscured, deprived of analysis, but it must be admitted that the place it occupies in the examination of problems is generally very discreet, and for a good reason: the question seems determined in advance. Contemporary technologies are from the outset seen as foreign to the values and the requirements of care. If ethicists of care are generally in agreement that care is not only a disposition but also, and inseparably, an activity, they do not seem to consider the fact that, necessarily, this activity takes place at one time or another through technological operations or by the use of technological devices. Or rather, it is often held to go without saying that care can only be gained beyond, if not against, the technological features of the activities it involves – at least once these technological features, again, exceed the skillful and cautious gesture, to become high-tech machinery and large systems.

If we must illustrate this observation by references or quotes, there is an embarrassment of choice and we have already given a number, but let us continue: in the medical field, Pierron (2010) for example considers care to be "the repressed of technological medicine" that is "buried under techno-medical devices, which sometimes make it unrecognizable". Thinking of care, according to him, is above all to "no longer be naïve in the face of the drunkenness of technological power"; it is "thinking at the bedside of a medicine which is sick from its hyper-technology" and victim of the "technological ideology which reduces health to maintenance of the body".

Beyond the medical field, Pierron stresses the general dissociation of knowledge and technology: "Our knowledge knows us to be members in symbiosis with nature, but our technology still maintains a relationship of power and domination of nature of nature" (Pierron 2010, p. 65).

In her book *Care monde* (Molinier 2018), French sociologist Pascale Molinier relies on a very interesting field study that she and her colleagues conducted in a nanomedicine laboratory, in the aim of examining how the researchers working there dealt, concretely and in daily life, with moral issues. Here we see workers in science who maintain an ambivalent relationship to these questions: as scientists, they retain the conviction that the two areas of morality and knowledge must be kept carefully separated. At stake is the value of the objectivity of science, which in their eyes requires this separation of science from morality, facts from values. However, when it comes to their daily tasks which may lead them, for example, to manipulate stem cells from patients, these same scientists testify to very strong moral concerns, having regard in particular to what should be said or not said to these patients.

However, this whole analysis is conducted without ever mentioning the nanotechnologies themselves. Let us quote Molinier at length, who explains herself on this point: "We were asked to respect professional secrecy, which is why I will not address any of the concrete products of the physicists we encountered, nor will I mention their names. As will be seen, *this is not necessary to enter into the dynamics of their moral questioning* (emphasis added). But it deprives us however of the opportunity to comment on one very rich aspect of their activity which consists in the creation of a metaphoric language, with an ability to create images, that is to situate nano-objects in a field which, rendering them perceptible and transmissible, moves them closer to ordinary life" (Molinier 2018, p. 47).

In other words, the impossibility of involving nano-object design in reflection on the ethics of scientists is of little consequence: the relationship between technology and morality is quite inessential. In addition, we understand that what is nonetheless lost still concerns subjective productions, and not the objects: not to be able to speak of objects is above all not to be able to analyze the imaginary constructions of scientists. It could not be said more clearly that an ethics of care has not much to do with technology, even while the concept of "relational responsibility" that Molinier promotes, and which Tronto borrows from her, seems to require including technology in moral reflection. Indeed, this concept stipulates that responsibility stems not only from the fact that we are moral agents responsible for our actions, but also from the fact that we participate in networks of relationships which are more or less extended, tangled and complex. Yet, as a matter of fact, our relationships move through our technologies.

Most often, however, the authors mentioned above explain that genuine care in no way implies completely banning technology. Pierron insists on this, speaking of care in medicine: "a philosophy of care does not condemn medical technology" (Pierron 2010, p. 13); we need "a definition of health that takes into account the importance of technology" (Pierron 2010, p. 52).

French philosopher Michel Dupuis summarizes the situation: "never less biomedicine or technology, but that they should be always more subject to human projects, at the service of human beings, their desires and dignity" (Dupuis 2013, p. 50). In sum, the only way to overcome the opposition of technology and care appears to be to bring the first into dependence on the second. The idea of an exteriority in principle of technology and care is not in question. Technology can only support the requirements of care by being constrained by it, from the outside, by appropriate regulations and in demonstrating the greatest vigilance. A matter of mastery, again and always.

On the whole, the ethics of care therefore do not seem to be able to provide precise indications to enable us to better understand the conditions of a "caring" engineering. Can we not, however, find another reading of the ethics of care, which makes them appear less oblivious to technology? Can we not in particular find, in the famous definition that philosophers Joan Tronto and Berenice Fischer give for care, a few elements of reflection which promote this point of view?

1.3. Taking care of living beings

1.3.1. *Care and technology: from ethics to anthropology*

1.3.1.1. *The definition of care according to Tronto and Fischer, or the denial of the role of technology in care*

Care, write Tronto and Fischer, consists of "a species activity that includes everything that we do to maintain, continue, and repair our 'world' so that we can live in it as well as possible. That world includes our bodies, ourselves, and our environment, all of which we seek to interweave in a complex, life-sustaining web" (Tronto 1993, p. 103).

Three essential elements emerge from this definition and are directly relevant to the issue of what a concept of caring technology might be.

First of all, care is an activity consisting of maintaining, sustaining, or even repairing the world; all these activities clearly require technology. In addition, there is every reason to believe that some technologies are of such a nature that they support these tasks of maintenance and repair, while others, by very reason of their properties of structure and functioning, are not favorable to them. The issue of care is therefore concerned very directly with our technological choices. As we have said, Tronto and Fischer pay no particular attention to technology (totally passed over in silence in Tronto's book). However, this should not convince us that their definition of care necessarily excludes the latter as a question and as a problem.

Secondly, the purpose is to "live as well as possible", which returns us in the final analysis, to the problem raised by Aristotle: what is this good life, or at least this as-good-as-possible life of which Tronto and Fischer speak? We need a term of reference to which to relate our moral assessments – what Aristotle called the Sovereign Good.

Thirdly, what we must take care of is a complex set of interdependent elements, including "our bodies, ourselves and our environment". The reason why we must take care of them is that only under this condition is it possible to support what appears to be the famous term of reference for our moral evaluations: life. The concept of "life" here refers to the life of living beings, the being-in-life, and not only the intellectual and spiritual life of humans.

We must acknowledge that the definition of care proposed by Tronto and Fischer may be understood in two opposing ways. It may indeed, at first sight at least, accommodate two radically different conceptions of the subject which provides care; furthermore, these alternatives also lead to quite different ways of taking account of technology in activities of care.

1.3.1.2. *Two possible conceptions of the subject that provides care*

The first conception would be to consider the subject assumed to provide care as rising above, transcending the realities that must make the object of care. This conception would not in any way question the dominant position of the subject, its ability to impose upon things an order decided by itself. It is this type of relationship of care that French anthropologist André-Georges Haudricourt illustrates by the breeding of sheep (Haudricourt 1987): the herd should be the subject of attentive care, otherwise the animals waste away and eventually die – which of course is not in the interests of the breeder! This type of care consists, for the herder, of directing the existence of the animals under their control and providing them with everything they need. The herder knows what is good for them and decides everything.

However, to this rather vertical and authoritarian relationship of care, Haudricourt opposes another relationship of care which he illustrates by the example of the culture of yams: the yam is a fragile plant, to intervene directly upon it as a herder does to their animals can only lead to failure. The activity of the grower is rather to act upon the environment of the yam, on the conditions surrounding it – the action is indirect and not direct. The grower does not see themselves as a subject which overlooks and dominates that to which their activity applies: they think of themselves rather as a being acting upon a reality which exceeds and encompasses them. The relationship is not vertical but horizontal.

German philosopher Arnold Gehlen had already, in the 1920s, identified this modest and careful position in respect of nature and of living beings, characteristic of what he called the agricultural morality (Gehlen 1980). Gehlen also included in this category the rearing of animals. This morality is characterized by the requirement of care, conceived as a service of reciprocal action. Living beings exist for the human, and the human exists for the living beings of which he or she takes care. This structure of reciprocity which defines care has a major anthropological significance: by it, in it, human beings align themselves with the fundamental structures of their condition as living beings, and the conditions of reproduction of plants and animals.

When human technological activity impacts upon living beings, the guiding concepts of this activity are thus, according to Gehlen, service and duty – concepts which structure a general attitude implying the voluntary renunciation of mastery over things. The agricultural morality therefore goes hand in hand with acceptance, by the human, that he or she is also a creature, a living being among living beings. The passage from the agricultural morality to the industrial morality, that is the passage from technological activity turned towards living beings to technological activity involving inert nature, has deeply transformed, says Gehlen, humanity's conception of itself: it now sees itself as the "master and possessor" of nature, in a position of superiority and absolute domination towards it. "Thus", says French philosopher Pierre Macherey, "we may compare the conception of a transcendent subject, abstracted from the world and assured of an exceptional position of dominance and authority towards it, with that of a subject that is immanent to reality and to the complex process of its internal relations, who, without extracting itself from this process, fulfils towards it a critical function of review, interrogates it about the values that it spontaneously implements, discusses its orientations in a way which is not only theoretical but practical: the first is a metaphysical entity, and the second a living being, an organic subject" (Macherey 2016).

Thinking of care in technology therefore implies not only articulating an ethical requirement and a conception of technology; it also calls for, perhaps first of all, taking an anthropological position.

However, this duality of perspectives, regarding the position of the subject who provides care in respect of the object of this care, also engages two very different conceptions of technology.

1.3.1.3. *Technology and the two meanings of life*

Technology may first be conceived as a set of means at the service of life. "Man, even physical man is not limited to his organism. Having extended his organs by means of tools, man sees in his body only the means to all possible means of action" (Canguilhem 1978, p. 118). Human beings, adds Canguilhem, "want to be the subject of their norms" (Canguilhem 1947), that is to say that they want to be able to choose, based on their system of values, among a plurality of possible paths of life. This being said, under the life that I choose to live according to my own preferences and values, there is life inasmuch as I am a living being. Yet, this life, life as a life of the living being, is also an activity oriented by norms – what Canguilhem called "vital normativity". An essential point of Canguilhem's analysis is that vital norms cannot be reduced to the norms which the subjects want to have the freedom to choose, according to their values – there is on one side the normativity of life as a life of living beings, and then on the other side the norms of living well, the "living-one's-life", of which the subject wants to be sole decision-maker.

Vital normativity exercises itself through the means provided by the structure and the functioning of the organism. Disease is reflected by the unavailability of some of these means, or by their failure, which obliges the living being to reconfigure the range of its possibilities of life – to invent new vital norms, new lifestyles, doubtless narrower ones. In humans, this range of means of living is enriched by technology: human beings, the living beings who invent non-organic means to support their life as living beings. However, technology also supports the capacity of the human as a subject to choose what "living well" means for him or her. It can be understood as the set of means at the service of the norms according to which the subject wants to live his or her life. The aircraft, for example, extends to a large number of individuals the opportunity to live the life of a traveler, discovering the four corners of the world: it is a choice of life within the meaning of "living well", a possibility that bears a value.

However, the norms according to which each person gives meaning to their life can always enter into conflict with the vital norms of the being-in-life. To resume the same example, flying in an airliner, at eleven or twelve thousand meters of altitude, exposes organisms to intense solar radiation. These radiations are very harmful, and, over the long term, they may have destructive effects on the health of those exposed to them.

Technology at the service of the accomplishment of the goals of the subject, "living well", consequently only receives its full meaning in reference to a normative activity of a different type than that, precisely, of the subjects who define for themselves the standards of their "living well". This other normative activity, designated by Canguilhem as vital normativity, gestures towards an activity of constructing norms that is performed, not by the subjects that we want to be, but of the beings-in-life that we do not cease to be.

Technology is therefore indeed a creative activity through which humans give themselves their own norms of "living well". However, through this activity, humans do not yet establish themselves as self-referential beings, self-producers of the entirety of their own reality: they remain beings-in-life, fruits of a long evolutionary process, and as such they are also involved in a normative activity which they do not have the power to reconfigure as they will.

This first conception of technology therefore orients towards a well-defined task: to ensure that the standards of the "good life" that the subjects have chosen, and which they make effective in particular through technology, do not do violence to the normative activity of the being-in-life which they do not cease to be. This is doubtless what Macherey meant: reflecting upon the values that we implement, discussing their orientations in not only a theoretical but a practical way – that is to say: critiquing our choices, those that we have made in the perspective of "living our life well" – is to refuse to look at things from on high, imagining ourselves as pure subjects, but to assume that we are also living beings, immanent to a whole set of natural processes which constitute us. In summary, for subjects, critiquing the choices that they make based on the conception they have of what "living well" means, implies seeing themselves from the point of view of the being-in-life that they also are, and that may have to suffer severely from the consequences of those choices.

1.3.1.4. *Is the conception of the human as a being which produces itself compatible with care?*

However, there is a second way of conceiving of technology, which shows itself to be unconcerned with this task. Far from being conceived as the set of means available to each individual to forge a harmonious, non-destructive dialogue between his or her aspirations as a subject on the one hand and his or her being-in-life on the other hand, technology can indeed be seen as the set of means available to a society to realize its own goals.

Thus, to the vital normativity of the being-in-life and to that which existentially defines the activity of a subject deciding for himself or herself what "living its life well" means, we come to add a third normativity that must be named a social normativity. This may insert itself flexibly into the immemorial dialogue of the living human being and the human subject, coming to support the capacity of the latter for self-determination. However, it may also assert itself as autonomous and superior, self-referential, and negating any normativity which is not itself. Human technological action can then no longer be conceived as always inserted into the possibly conflicting dialogue between the two meanings of life, "living one's life well" and the being-in-life; it is now designed as superior, devoted to enlisting any reality in the service of its full accomplishment. There is no longer any vital normativity, nor even existential normativity, and also no tension and possible dialogue between the two: there are only metabolisms to repair or to tinker with, and psyches to be guided.

The subjects then appear completely cut off from their being-in-life and their vital normativity; they hold themselves in a position of superiority faced with nature towards which, as Gehlen said very well, no ethical attitude is therefore required. The conception of the human which prevails here is based on the duality of the body and the mind: through their body, the human beings belong to nature, but through their mind they dominate nature and can reconfigure it according to principles totally foreign to its own requirements and to its organization. The two-layer model of care, which we have seen to constitute an impasse, is in sum directly implied by forgetting that technology is what creates a tension, but also an uninterrupted dialogue, between the two meanings of life.

1.3.1.5. *In what sense should vital needs be objects of care?*

How to locate Tronto and Fischer's definition of care in light of this alternative between the two conceptions of life and its corollary, the two conceptions of technology?

Tronto analyzes care as an activity of work passing through four necessary steps (Tronto 1993, pp. 106–107). The first, consisting of caring about, is the moment when I recognize a need that must be satisfied: it is the moment of attention. The second, taking care of, implies examining to what extent the need might actually be satisfied, and to design the appropriate means to achieve this: it is the moment where I assume the responsibility to act. The third, care giving, is the activity of care itself, the moment of implementation: it requires skills. The fourth, care receiving, is a moment subsequent to the act of care itself, but absolutely fundamental: it is to ask the person who has benefited from the care whether the latter was quite adequate to its need.

However, this analysis of the stages of care pulls away from decisive consequences to clarify what the expression "care for nature" might indicate.

The recognition of a need to satisfy (the first step) clearly means doing something other than objectively identifying a lack, a deficit to be filled by appropriate actions. This step only proves to have a moral significance if this need is not just the objective input of a problem to be solved. Caring about does not consist of establishing a list of all identifiable or even measurable lacks, but of assigning a moral significance to the lack itself. This first stage is therefore crucial, since it makes me understand that if the being in need requires my "caring about", it is precisely to the extent that it is not reduced to a sum of needs which are identifiable and possibly quantifiable in view of an appropriate technological response: it is the being whose needs are a moral obligation for me.

Confucius (2005) spoke very well about these two meanings of need, one which creates a moral obligation and the other which does not. He said one day to his discipline Ziyou, who questioned him about filial piety: "Filial sons nowadays are people who see to it that their parents get enough to eat. But even dogs and horses are cared for to that extent. If there is no feeling of respect, wherein lies the difference?" (Confucius 2005, p. 89). One's parents' need to be fed is not assimilated to the need to be fed of horses: the first is imbued with a dimension of moral obligation that the second does not have.

The term of reference for "caring about" is what, in the vulnerable being, is not only the determinable need, but through it the moral requirement attached to it.

This requirement has no objective determination, it is not a second-order need, a super-need subsuming all others. It is the non-determinable aspect of need – which is precisely what makes the need of the vulnerable being of a different nature to that of a being who does not call for moral consideration. Contrariwise, a car has also a "need" for an overhaul or a draining, from time to time. And Confucius's horses have the "need" to be fed[28].

However, if I can grasp the moral requirement within need, it is because I have the ability to put myself in the place of the vulnerable being: "caring requires that one start from the standpoint of the one needing care or attention. It requires that we meet the other morally, adopt that person's, or group's, perspective and look at the world in those terms" (Tronto 1993, p. 19). Hence the importance of the fourth phase of care, evaluation from the point of view of the one who has received the activity of care. This ability to occupy the point of view of the other implies, on the one hand, that I do not encounter them based on what defines me or constitutes me, and on the other hand, that I do not encounter them as beings summed up as what defines and constitutes them.

"Starting from the standpoint of the one needing care or attention"; "look at the world in those terms": the being in need must be recognized in its claim to define for itself the norms according to which it wishes to organize its existence, as a function of its own values. However, in accordance with what has been established above, the being must also be recognized as incomplete in itself, belonging to a totality broader than itself, of which it does not control the constructive processes. Tronto and Fischer's definition clearly expresses this: the being in need must doubtless be supported in its ability to establish its own reality, by defining its guiding values. This is the aspect of "ourselves", the reflexivity of this "we" that gestures towards the self-referential dimension of the subject; however, this "ourselves" is inserted into a greater whole which also includes the body and the environment, that is to say the two terms that the very fact of being-in-life brings together.

If the capacity of each person to establish their own norms of life is indeed the purpose of care, this object cannot be targeted directly and in isolation, as if it were a separate reality, self-sustaining, and floating in a vacuum: taking care, rather, means to act indirectly by paying the greatest attention to the body and to the environment, that is to say to this dimension of totality that is broader than each individual existence – a dimension into which the individual subject is inserted and only within which they can be capable of autonormativity.

28 Today we would obviously no longer say that the needs of horses do not oblige us morally, as Confucius thought.

The autonormative subject is indeed that which has the ability to organize, based on the idea they have of "living its life well," an environment which they have not created in all its parts and which constitutes for them an initial given, a problem to be solved. However, in Tronto and Fischer's definition of *care*, this dimension of the real that exceeds the constituted individual is very explicitly identified with life – not life as a set of determinable natural and biological structures, but precisely inasmuch as it constitutes for human beings a reservoir of potentials for normative activities, in the meaning of the capacity that any human being must possess to choose for themselves their own standards of "living life well". What must be the subject of care, according to Tronto, is therefore not so much the already constituted individual subject as the set of conditions that allow it to self-determine within existence, which implies, as we have seen, the ability to keep constantly open, and in tension, the dialogue between the two meanings of life.

Taking care reveals itself, in summary, incompatible with an anthropology according to which the socialized human being is considered to be the producer of the entirety of its own reality. Taking care requires, on the contrary, assuming that the human being is not precisely the producer of itself, or rather that it actually produces itself but by marking out its reality against the background of a broader totality, that it has not made and that surrounds it. We must take care of that without which we would be beings cut off by ourselves, isolated from our complete reality – in a word: alienated.

1.3.2. *Caring about valuating living beings*

1.3.2.1. *Vital valuations and care*

It follows from the previous analyses that extending *care* to our technological relationships with nature and with the whole set of non-human living beings requires recognizing, in them also, needs which bear a moral obligation. What does this mean exactly?

Remember that the concept of vital normativity, within Canguilhem's meaning, means that being-in-life is not a simple state of fact, arising causally from mechanistic interactions between the physical-chemical characteristics of the organism and those of its environment. Being-in-life means, for the living being, organizing its relationship with the environment according to vital values, be they positive or negative; being-in-life means, for the living being, "polarizing" its environment according to assessments that it itself performs, based on the needs it must meet (feeding, breeding, perpetuating itself in existence by fleeing from predators, etc.). Being-in-life in sum means proceeding to "valuations". This neologism, "valuation", translates what Dewey designated by the concept of *prizing*: the activity by which a being gives a price or value to something. Strictly speaking, according to this constructivist position in regard to values, there are no values in the world, there are only "valuative" activities through which living beings create goods and situations to which they assign a meaning and a value. Being-in-life means always proceeding to a differential allocation of meaning and value to the elements of the environment. Von Uexküll's tick is probably not a subject that wants to lead its life according to the norms it chooses, individually; but living, for the tick, means discriminating between physical-chemical signals in its environment, giving some of them a value – it is therefore valuing.

Among some non-human animals, behaviors are attested that reflect, already, a tension between vital normativity, that is to say the valuating activity of the being-in-life, and the desire of individuals to organize their existence according to evaluations which do not derive from pure being-in-life. Thus, the adoption by Japanese macaques of a technique of washing sweet potatoes to eliminate sand has apparently been controversial: some individuals have refused to follow this new method. Macaques, too, can obviously choose according to what standards they want to "live their lives well".

However, the introduction of the concept of valuation can illuminate anew Tronto's definition of *care*. Every living being must be recognized as a valuating being; its existence consists of structuring its living environment according to values, which may be binary (predator/prey etc.) or more complex. The requirement of care as a moral requirement has meaning only in respect of beings who are recognized as valuators. Claiming that only human beings are valuating beings and worthy of moral consideration, and that any other reality can only be granted a simply instrumental value in the narrow sense of its availability for exploitation, means inevitably to empty the requirement of care towards living creatures of any moral content.

Recognizing valuation among all living beings thus requires us to be concerned with knowing what the human acts of valuation or *prizing*, performed through technology, do to these countless activities of valuation or *prizing* of all living beings. The technological activity of human beings is the means for valuations which are no longer reducible only to the valuations of the being-in-life. As Canguilhem says, among human beings, life has flourished in "Technical plasticity" (Canguilhem 1978, p. 118), creating an environment organized according to values which are not those of the being-in-life. However, technology is an organizer of the environment which may prove to be very toxic for all living beings, by narrowing the range of their ability to conduct their own valuations. Canguilhem distinguished "propulsive" ways of life, in which the living being defines for itself its own norms of life, and "repulsive" ways of life in which the capacity for autonormativity of living beings is exhausted (Canguilhem 1978). We must admit that human technological activities massively propel other living beings to adopt repulsive ways of life (such as when the gorilla sees its forest shrinking to nothingness).

The concept of valuation dear to Dewey thus reveals itself promising, in the first analysis at least, if we evaluate technological developments in relation to the requirements of care, well beyond the care due to human beings. It is however possible to point out two difficulties in this connection between care and the concept of valuation.

1.3.2.2. *First difficulty for the valuating subject: a dated conception of the living being?*

Canguilhem's argument has allowed us to introduce a third term between existential values and the objectifying point of view of science, namely vital normativity (Canguilhem 1978). However, Canguilhem himself had clearly grasped how this definition of the living being by vital normativity might be shaken by the growth of molecular biology and medicine in the 1950s. What indeed becomes of vital normativity when life, as well as disease, are understood in terms of entangled maps of molecular interactions and metabolic networks? "Normal and pathological," recognized Canguilhem, "have no meaning on a scale where the biological object is reduced to colloidal equilibria and ionized solutions" (Canguilhem 1978, p. 59) – today we might say, "reduced to intracellular signaling pathways".

Does not so-called "personalized medicine", potentiated by high-throughput technologies for acquiring molecular data, particularly genetic, in large numbers (*big data*), as well as by bioinformatic tools for modelling and interpreting these data, remove from the medical field any possible expression of a vital normativity? It is common to castigate genomic personalized medicine for its molecularist reductionism and for its extreme objectivism, resulting in the inability of this medicine to take into consideration the subjective experiences of patients. *Cure* versus *care*, the opposition seems impassable. What is the "place of the patient" in this new medicine, one often asks. However, the real threat weighs perhaps not so much on the patient as a subject wanting to choose their own norms of life, now in the process of being reduced to their molecules, as on life itself – on vital activity as a normative activity. Would not the real threat be in the occultation of this activity, in favor of approaches to health and disease which are merely content to counterpose genes and the environment, two supposedly objectifiable and even quantifiable areas, maintaining between them relationships of a mechanistic nature? (Guchet 2016). In other words, do not recent developments in biomedicine inexorably reinforce the division between, on the one hand, the subject who intends to "live their life well" by choosing their own directive norms, and on the other hand, a body-object which is fully "actionable[29]" by novel tools of comprehension and intervention? This body-object is certainly no longer the body conceived as a totality of organs, it is decomposed at a lower, molecular, level; nevertheless, it is still a body assumed to be fully transparent to the gaze of objectifying science and its tools of intervention. No place for taking account of the normative activity of the being-in-life in the molecularist approaches of personalized medicine. In these conditions, care must seem to flow back towards the subjective pole, rediscovering the two-layer model.

In summary, when the being-in-life is described today at a molecular level, does its holistic understanding in terms of vital normativity and valuating activity not emerge seriously weakened, unable now to be a relevant term of reference for evaluation of technologies in the light of the requirements of care?

1.3.2.3. *Second difficulty of the valued object: does the concept of care for nature suppose assigning an intrinsic value to nature as a whole?*

When referred back to the three classical conceptions of nature, the aim of care for nature appears in itself problematic, even contradictory.

Unnecessary to dwell at length on the relationship to care of *natura vexata* – that is to say nature as an object of experimental sciences, "tortured" to give access to knowledge. Submitting nature to experimentation certainly demands care, but in the sense of an experimental framework that is carefully designed and arranged – care that is purely technical, and not moral.

Natura naturans – that is to say nature as the generative power of beings, a process of individuation – does *natura naturans* oblige us morally to provide it with care? It seems not. The activity of the forester is without doubt beneficial to the forest itself, in that he or she exercises upon it a regulating role. However, the forest does not cease to be the forest if the forester should disappear: it continues despite everything to produce living beings which are always new, whose generations succeed each other. *Natura naturans* certainly requires, as a minimum, that we not destroy any possibility of procreation, of fertility: a soil saturated with chemical inputs eventually becomes sterile, nothing more grows. One will agree, however, that this is not, strictly speaking, taking care: it is rather being careful, not doing just

29 This concept refers to the ability to inhibit, by appropriate medication, altered elements of the cell (proteins, enzymes, etc.) which are involved in a maladjusted metabolic pathway. In "targeting" them, this drug prevents the activation of the metabolic pathway in question and thus the activation of cellular functions commanded by this metabolic pathway: the cell loses certain functions, for example those of growth and proliferation, which might be interesting for fighting cancers in particular. See (Nelson *et al.* 2013).

anything. However, taking care cannot be summed up by avoiding the devastation of natural environments, unless if we give care a purely negative meaning, consisting only of refraining from doing something – which is not nothing, certainly, but cannot be fully satisfactory if it is to define a positive attitude of care in respect of natural beings.

There remains the conception of nature within the meaning of *natura naturata* – that is to say, nature as a set of individualized natural beings, regardless of the order of magnitude considered (individual living beings, inert things such as rocks or rivers, ecosystems, the Earth as a whole). The fundamental question of environmental ethics, explains John Baird Callicott (1985), is whether nature thus understood possesses an intrinsic value or not. This concept of intrinsic value is however polysemic and must be clarified (O'Neill 1992).

Firstly, in its broadest meaning, intrinsic value is synonymous with non-instrumental value. According to the most commonly chosen definition, a being with instrumental value is one which has value only as a means to achieve a goal considered desirable. On the contrary, a being with intrinsic value is one which has a value in itself. Secondly, intrinsic value refers to the value that an entity possesses only in virtue of its intrinsic properties, in the sense of non-relational properties; the value of the entity persists regardless of the existence or non-existence of other entities, and it can be characterized without reference to other entities. Thirdly, intrinsic value is synonymous with objective value, in the sense of a value possessed by an entity independent of the existence or non-existence of valuating beings, human or non-human, and independently of the valuations which such beings carry out if they exist.

I can thus assign an aesthetic value to a landscape and therefore see in it something other than a means that can be exploited to achieve a goal: it is nevertheless a relational property which moreover depends on the valuations of a valuating being. This value is therefore intrinsic in only the first meaning of the term, that is to say within the meaning of a non-instrumental value, but not within the meanings of a non-relational property or of an objective value respectively. I can also maintain, like Holmes Rolston III (Rolston III 1994), that living beings only attribute an instrumental value to certain resources that are useful to them because they value their own life as their own good, by giving it an intrinsic value. This argument is the teleological reasoning of Aristotle in the *Nicomachean Ethics*: there are clearly things that are means towards something else, for example horses for making war. However, Callicott argues, the existence of means implies the existence of ends. In addition, although a means might exist because of other means, the chain of means must somewhere reach an end, which is not the means to another end – an end in itself – without which the chain of means would be infinite and would lose all anchor. And to the extent that means are assessed in an instrumental fashion and that the end in itself is valued in an intrinsic way, if an end in itself exists – and it must exist for the means to exist, and we know that means exist – then intrinsic value exists (Callicott 1985). Thus understood, intrinsic value has not only the meaning of a non-instrumental value, but also that of a non-relational property. That said, the value that a living being assigns to itself depends on a valuating act which, though not necessarily a conscious act, is no less a subjective act: intrinsic value does not here have the meaning of an objective value. As Callicott strongly emphasizes (Callicott 1985), the more difficult question of environmental ethics becomes in the end: do intrinsic values within the meaning of objective values exist in nature? How does one move from the subjective assignment of intrinsic value to one's own life to the assertion that there are objective values in nature?

The essential point of Callicott's argument is in substance that the gap separating the subjective definition and the objective definition of intrinsic value is impassable, but also that there is in the final analysis only a gap to be crossed if we remain prisoners of the theoretical framework of modern Cartesian thought, Callicott says, which basically consists of a dualism of subjects and objects. As soon as we abandon this framework, the gap disappears in favor of a non-subjective and non-objective conception of value, whose inspiration is provided in particular by quantum physics which, precisely, has severely depleted the modern conceptions of subject and object (Callicott 1985). This new theoretical framework is supposed to allow us to move from a biocentric type of ethics, for which value is always assigned by a valuating entity, a living valuator (a subject, in summary), to objects, to an ecocentric type of environmental ethics for which value is not linked either to the existence of isolated valuating subjects nor to that of isolated valued objects either, but to the biotic community taken as a whole: individual entities, that is to say the subjects and the objects of modern thought, have no value in themselves, regardless of the place [they] occupy in the whole and which assigns to them their value (Larrère 2010, p. 408). If we understand properly, this assignment of value cannot be the act of a new entity that would be external to the biotic community – which would lead us back to the classical dualistic framework, opposing an overseeing and valuating super-subject to a super-object valued from the outside by this super-subject. The assignment of value to the biotic community can only be the act of the community itself, although this evaluation is not the result of an act of attribution of value by a valuating subject who is a member of the said community – which, again, would put such a member in a certain position of exteriority towards the community. Ethics, as Callicott said, is the description of the structure of the community made, from the inside, by its own members (Larrère 2010, p. 408).

However, as C. Larrère notes, an ecocentric ethics, which is basically a consequentialist ethics – the morality of an action is measured by its effects on the biotic community as a whole – falls under the blow of the criticism which can be made of any consequentialist ethics, namely of losing sight of the value of individuals, who can always be sacrificed to the common good (Larrère 2010, p. 410). This point seems to us at a minimum to be very thorny when the action affects, among a large number of natural beings, human beings – that is to say almost always, since our technological activity, when it culminates in damage caused to nature, never fails to also result in damage to human groups. Must we thus subscribe to an environmental ethics that radically rejects taking account of the individuality of living beings in favor of biotic communities, including when these living beings are human beings? If we refuse to go as far as this extreme conclusion, might we answer that in the case of human beings, that is to say in the case of individual living beings who are also persons, it is better to maintain a deontologist point of view affirming the unconditional and absolute value of individuals? Do we not then risk having to live with a two-speed environmental ethics – ecocentric when it comes to living beings who do not have the status of persons, but biocentric when it comes to living beings who have this status and to which an individual objective value must still be assigned, if we are not to fall into the worst difficulties?

In the face of these formidable questions, some philosophers such as Bryan Norton have proposed to revert back to an anthropocentric conception of environmental ethics, and therefore to an instrumentalist conception of nature, in arguing that anthropocentrism is not necessarily synonymous with the depredation of nature: we can exploit nature without mercilessly destroying it, when for example we assign to it an aesthetic value, a value for the progress of scientific knowledge or a value for the ecological services it renders. These interests lead to protecting nature rather than destroying it.

C. Larrère concludes with the need to shift the dividing line between the different environmental ethics. To the classical opposition between anthropocentric and non-anthropocentric ethics, she counterposes a distinction between, on the one hand, environmentalist positions based on respect for otherness, recognizing in nature an Other to which moral consideration must be granted – which leads to the ethics of intrinsic value of non-human beings, living or non-living – and on the other hand pragmatist positions based on the recognition of the importance of the relationships of humans to the environment – positions leading rather to soliciting the expression of conflicts of values in relation to the environment, as well as to seeking a satisfactory compromise through pluralist deliberation and democratic decision-making (Larrère 2010, pp. 410–411). That said, C. Larrère asks, though they do not abandon concern for nature, do the pragmatists not turn away (despite everything) from what can be considered the main teaching of non-anthropocentric ethics: we are not alone in the world, non-humans also matter in themselves? (Larrère 2010, p. 411). In sum, she suggests that an environmental ethics should be able to reconcile, on the one hand, the constructivism of pragmatist approaches (values are not given in nature, they are constructed), and on the other hand the recognition that there exists, in nature, something that matters independently of our own valuations.

This requirement leads incidentally to redefining the individual living being in such a way that the two perspectives, on one hand of axiological self-referentiality – that is the self-evaluation Rolston III spoke about – and on the other hand, of the impossibility of formulating a concept of the individual independent of the relations that constitute it, are not two opposing perspectives but on the contrary inseparable. We find for example in Simondon such a conceptualization of the individual. According to Simondon indeed, the individual is nothing else than the refusal of an unsurpassable incompatibility between, on the one hand, the fact that it is a relational being, i.e, defined by relationships with that which is not itself, and on the other hand the fact that it is defined also by intrinsic characteristics. Simondon goes as far as considering the individual to be the very process of making compatible these two sets of criteria, respectively extrinsic (relational) and intrinsic – this is what he calls an individuation. One may ask why an individual is what it is. One might also ask why an individual is different from all others and cannot be confused with them… In the first sense, individuation is a set of intrinsic characteristics; in the second sense, a set of extrinsic characteristics, of relationships. But how can these two sets of characteristics be connected to each other? In what sense do the intrinsic and the extrinsic form a unit? (Simondon 2020).

Can we therefore propose a possible understanding of care for nature which is based on this necessary conciliation between a constructivist approach to value and the existence of intrinsic values in nature? If this is the case, how should we understand this concept of intrinsic value: in O'Neill's second or third meaning?

Let us summarize. The difficulty raised by bio-centric ethicists is that they enclose value within living individuality, they make it an essential property of living beings taken individually. However, as the proponents of ecocentric ethics rightly point out, living beings are not isolated entities; they are in a relationship with other living beings and with abiotic elements, the whole forming a complex system of interrelationships and interdependencies without which there would be no living beings at all. What has value, the proponents of ecocentrism accordingly argue, is the biotic community and not individual living beings. However, can we not subscribe to the ecocentric position claiming that a living being is never locked within the limits of its organic envelope, that it is partly constituted by the relations it has with its environment, without that yet implying a pure and simple abandonment of any form of biocentrism?

1.3.2.4. *Natura instruens*

Environmental ethics based on respect for otherness lead in a straight line, stresses C. Larrère, to positions which confirm rather than they exceed the duality of human beings and nature (Larrère 2010, p. 412). To avoid this pitfall, it is helpful to describe the situation of non-human living beings in nature according to exactly the same categories as those used to describe the situation of humans in nature. The human, as we said with Canguilhem, is the living being whose life has flourished in Technical plasticity, manufacturing for itself organs of a new kind. In other words, the human is the living being with an extended range of instruments for living – in addition to its native instruments, it also has the instruments which it has manufactured. Then let us say that the non-human living being must also be envisaged as an instrumented being, a being who needs instruments to live and to proceed to its valuative activities. Can instrumentality be interpreted as a general biological category, not separating humans from non-human living beings?

To say that non-human living beings can also be seen in terms of the biological tools they have at their disposal to live is a commonplace, however, this perspective has generally been placed under the banner of a discussion on the alternative of mechanism and finalism in the living world. Does the existence of biological tools make it possible to distinguish between the ateleological and teleological conception of the living? (Tétry 1948, p. 12). The extension of the concepts of tooling and instrumentality to the biological equipment of non-human living beings has therefore fueled a great debate in the philosophy of life, but not a reflection of moral philosophy. These discussions on the biological tools of animals have not led to ethical considerations. However, the question may be asked: cannot making instrumentality a general biological category support an ethical stance towards the living world? On the condition, no doubt, that this concept of biological tool or instrument can be freed from the limited meaning conferred on it by Tétry, that of being an organ with a well-defined function (the claws of crabs, etc.), to give it a more general scope by extending it to nature itself.

The metaphor of nature as a theatrical scene where the valuative activities of all living beings can flourish and intertwine, at multiple levels in space as well as in time, may help to clarify the conception of nature that is emerging here. The theatrical scene is indeed not only a space where actors develop in playing their respective roles – let us understand by this, to refine the metaphor, where living beings evolve in deploying their respective valuative activities; it is also what gives the actors the instruments, the means of their play. If there is a bed, a chair, a table, it is for the players to be able to do something with it: no element of the scenery here is devoid of instrumental value. To continue the metaphor, would nature as a theatre not also be the great provider of instruments for living in the meaning of being-in-life, for all living beings? To define a concept of care for nature, it would be therefore appropriate to enrich the classical conception of nature: in addition to *natura naturata*, *natura naturans* and *natura vexata*, we should add *natura instruens* – not instrumentalizable and instrumentalized nature, offered up to human exploitation, but "instrumenting" nature, that is to say a provider of the instruments which living beings in general and not only humans need to proceed to their valuative activities. Taking care of nature would then mean ensuring that its *instruens* function in respect of living beings in general, including humans, is not threatened. The ravaged Amazonian rainforest is not only the lungs of the planet, that is to say a kind of machine for manufacturing oxygen, which is in danger of no longer being able to operate properly: it is the purveyor of instruments of life for many living beings, human and non-human, which is becoming deteriorated. Nature here then loses the universalist meaning that it assumes in naturalistic ontology: it now designates the entanglement of innumerable fields of instrumentalities, in a sense of the instrument that no longer knows the division between the technical and the biological.

As Corine Pelluchon (Pelluchon 2020) points out, "sentience" (a concept that refers to the capacity of non-human animals to feel pleasure and pain, to experience suffering, to have interests to preserve and individual preferences (i.e., living their lives in the first person – a concept that underpins the idea that animals are moral patients in need of protection) gains from being understood not only in terms of their capabilities, but in terms of both "vulnerability", i.e., dependence on various "foods", and "agility", which implies considering animals as competent beings. By this, Pelluchon intends to bring out concepts that can be applied to humans as well as to non-human animals – notably the phenomenological concept of corporeality – which allows her to lay the foundations of a philosophical anthropology free of the idea of anthropological cut-off, at least when it comes to defining moral duties towards animals (which does not mean that no distinction between humans and non-human animals can be found – a position that Pelluchon unambiguously rejects). This being said, in the absence of a general conceptualization of instrumentality, not limited to humans, there is a great risk that the two concepts of "food" and "agency" will remain both abstract and partial.

Abstract in the sense that the "foods" on which the living depend in turn depend on certain realities that produce and provide them, and which are therefore not "foods" in themselves. Thus, the forest of Achuar dear to Descola is the great provider of all kinds of "foods", both material and symbolic, which the Indians feed themselves. The forest can rightly enter into the great category of instrumentality, if one understands by this, in a general way, what by which "foods" are produced and made accessible to living beings. The Achuar forest is not itself "food"; it is natura instruens.

Partial in the sense that, precisely, what counts above all is the guarantee that "food" will continue to be produced and made available, as well as the possibility for the living being to have access to it – which presupposes that it has the adequate instruments and that these instruments are not deteriorated. "Food" without the instruments that provide access to it is useless. It seems that the sonars used by the military disorient whales and cause them to strand: an instrument that the whale possesses and that is indispensable to it is obviously rendered useless by man-made noise pollution. It has now been proven that bees are also disoriented by the use of certain insecticides: in this case too, the instrument that bees use to access the "food" they need to live is deteriorated. A philosophy of animal "food" and "agility", without a philosophy of instrumentality as a general category of the living, beyond human artifacts, risks remaining incomplete.

1.3.2.5. *Towards a renewal of the instrumentalist conception of technology?*

It is a curious paradox, by the way: while the philosophy of technology was developed as a full-fledged discipline in the unanimous rejection of the instrumentalist definition of technology as a means, this definition can precisely be shown to introduce a better understanding of the possible relationships between technology and care.

In a famous article devoted precisely to castigating the instrumentalist conception of technology, Latour prognosticates indeed the end of means (Latour 2002), that is to say the untenable character of this conception of technology as a means for achieving predefined ends, or as an instrument. As an alternative to this instrumental definition of technology, he proposes a new concept inspired by Deleuze: technology as a fold. Technology according to Latour is in effect a fold in the sense of a folding of time, space, and actors. This hammer I have here, under my hand, is not contemporary with my action of today; it mixes heterogeneous times: it gestures towards the duration that it took the earth to produce iron ore, but also towards the duration that it took for nature to grow the tree from which the wood of the handle was taken, or even towards the shorter time of human industry which produced it. Technology is also a folding of space – the hammer connects the mine which provided the ore, the forest of the Ardennes which provided the wood, the German factory which produced the hammer. Technology is finally a folding of actors, that is it holds together a whole set of beings, human or non-human, who are involved in its operation.

This understanding of technology as a fold seems to introduce to the question of care for nature: to unfold what has been folded, is this not indeed to make visible what has been rendered invisible in technological functioning, that is everything to which the technological object is closely linked – through the industrial clusters involved in its production, through its materiality and through its features? Is not the instrumental conception of technology, on the contrary, oblivious to any possible care?

Let us recognize rather that the instrumental conception of technology may find new life, on condition of extracting it from its intellectualist interpretation. Technology would be an instrument, a means, but not a means at the service of ends defined prior to any activity by a superior intelligence. This rehabilitation of the instrumentalist conception of technology in no way requires us to rebut the criticisms which have been addressed to it: these are definitive; technology is not a simple neutral means at the service of ends laid down in advance. Given this, these criticisms are directed at the whole concept of technology as a "means *of*" in the objective sense of the possessive: a means at the service *of* ends, a means to achieve these external ends, defined by an intelligence external to nature. Understanding technology as a "means *of*" within the subjective meaning of possessive – a means *of* the living being, a means which the living being uses to conduct its valuations – could perhaps allow us to open a new and possibly instructive perspective as to the idea of care for nature passing through technology.

In short, it would be a question of dissociating the instrumentalist perspective on technology from any teleological reference to the very idea of the "end" – that is, of thinking of the means without the end or, to parody Latour's formula by taking the opposite approach, of proclaiming the "end of ends".

Let us return to our two difficulties in order to examine if an instrumentalist approach to nature, in the sense indicated here, can help to resolve them. The first concerned the conception of the living being, inspired by Canguilhem, likely to guide us to a conception of care for nature as care of valuating beings. The conception can be dated, we suggested, to the era of molecular approaches in biology and medicine, which seem to have definitively excluded any possibility of considering the living being as a unitary whole, an active center of valuation. Then let us say, still with Canguilhem, that one possible outcome would be to revise our conception of the molecularized body of techno-scientific medicine, and this precisely in the instrumentalist perspective developed here. A metabolic network can be seen otherwise than as a set of mechanistic interactions between molecules; perhaps it is not impossible to see there also an instrument or a means for the being-in-life, the means of all the means of possible actions as Canguilhem said.

The second difficulty concerned the question of intrinsic value. Between the attribution of value to individual living beings and the attribution of value to the community that they form with other living beings and with abiotic elements, the instrumentalist approach takes a third path. The allocation of intrinsic value to the very act consisting, for the living valuator, of conducting valuations implies giving particular importance to relationships, to the instruments available, which also means as Dewey stresses[30] that one cannot separate values and instruments: without instruments, values do not exist. Instrumentality objectively brings values into existence – which leads rather in the direction of Callicott's ecocentrism, for which the duality of the valuating subjects and the objects of valuation must be surpassed. However, the act of valuation remains the act of a living valuator, which leads rather in the direction of biocentrism. In addition, to say of living beings that they are valuators in the same way as humans is to *de facto* assume a multi-species constructivism of values: human valuations are deployed in a very pluralistic context within which a large number of living beings are also carrying out valuations, although these countless valuative activities can be telescoped. The task of defining what are the good arenas for these conflicts of valuations to be expressed, how to identify the "spokesperson" for non-human living valuators, what are the criteria for prioritizing valuative activities (and if it is necessary to prioritize them), leads to a rather pragmatist approach to environmental problems. The instrumentalist approach, which as we see is not yet a return to anthropocentrism (even a weak one), thus combines the three points of view: individualistic, relational and pragmatist.

An instrumentalist approach to technological action in nature can finally help us to refine the criteria for moral evaluation of this action. Indeed, making a general opposition between intrinsic value and instrumental value seems too coarse to provide these criteria. Thus, consider respectively the activity consisting of producing cheese by means of microorganisms which perform fermentations, and that consisting of genetically modifying microorganisms so that they over-express certain substances in which we are interested (in order to produce drugs or biofuels for example). These two cases undeniably both refer to methods of instrumentalization of living beings; however, the two cases doubtless do not refer either to the same concept of instrumentalization, nor to technological activities whose meanings and moral implications are identical. We will return to this point later.

1.3.2.6. *Should recognizing that living beings are valuators lead to assigning them moral consideration?*

It is clear from the foregoing that no incompatibility in principle exists between, on the one hand, recognition of the fact that any living being is a valuating being, and on the other hand, the fact of assigning to these valuating beings an instrumental value. It is not at all inconsistent to recognize the valuative activity of living beings which, furthermore, we instrumentalize, whatever the meaning we give to this term.

30 We will return to this point in more detail in Chapter 3.

Let us be clear on one essential point: the fact of recognizing living beings as valuating beings does not necessarily prejudge any moral obligations of humans in respect of these beings. Thus, as O'Neill remarks, the gardener can without difficulty consider aphids as valuating beings, without experiencing the worst moral difficulties in the fact of seeking to exterminate them. I can consider microbes as valuating beings and not as mere biological machines, without yet refusing to wash my hands and, in so doing, get rid of them en masse. In sum, in the words of O'Neill, living beings must be recognized as having an intrinsic value in the second meaning of the term, that is to say in the sense that they value their own life by performing valuations, without it being necessary to assign them intrinsic value in the third sense of the term, that is to say within the meaning of objective value which creates a moral obligation.

That being said, if recognizing valuative activity in all living beings does not imply that we are obliged, morally speaking, towards these beings – in summary: if there is a disconnect between, on the one hand, this recognition, and on the other hand the allocation of moral consideration – do we not fall back into a weakened conception of the relations between care and technology? Care should be taken certainly, but we could very well be content to give this taking care a purely technical meaning, and not a moral one – it would simply mean being careful, meticulous, simply because the living beings that we instrumentalize can only provide their services if certain care is taken with them. If I desire to produce quality cheeses, I must ensure that the microorganisms, precisely because they are valuators and they have specific needs, find in my cellars favorable conditions for them to do what their metabolism has provided that they do, and which interests me. No moral consideration for the bacteria in this case. The man of Song, again, was lacking in judgement by ignoring the basic needs of his plants: he showed himself to be imprudent, incompetent at worst, but not immoral.

In sum: either the extension of valuative activity to all living beings obliges us morally towards each of them, and we are in danger of sinking into both theoretical and practical absurdities; or if this is not the case, but then in what way is this extension actually required for thinking a relationship of care for nature, since this care appears to be limited to the mere fact of being careful, so as to succeed in our enterprises? Either extreme difficulties, or uselessness: we are in the most complete embarrassment.

One possible outcome would be to argue, like O'Neill, that ethical life is all the richer when it assigns moral dignity to a greater number of beings, making their own well-being an object of care. This allocation is in fact not arbitrary, but based on objective considerations, in particular on the state of knowledge in biology or in ethology. In the current state of our knowledge, we are today led to "ascribe" moral consideration to a more and more extended number of beings who were previously deprived of it (Sève 1994). This extension is currently such that our technological action is inevitably implicated in it. To the classical finding, that no living species on Earth is spared by our technological action, it is appropriate to add the corollary and symmetrical finding: no human technological action, whatever it is, can be deployed without telescoping at one time or another the valuative activities of living beings to whom, now, we attribute moral consideration.

That said, Lucien Sève (who takes up Paul Ricoeur's concept of ascription) does not retract Hume and Kant's warnings: from what is – in this case, the fact that living beings are valuators – we will never deduce what should be – our moral obligations towards them.

1.3.2.7. *Innumerable cares*

Ultimately, the difficulty pointed out by Aliénor Bertrand eventually fades away. Let us recall the terms of this difficulty: the very expression of care for nature remains dependent on naturalist ontology; in these conditions, it is a safe bet that it remains very abstract, unlikely to inspire real and relevant practices of care. After Tronto and Fischer's analysis of the implications of the concept of care, let us then say that there is no general concept of care subsuming all species of care, and that there is no place to speak of a care for nature taken univocally. Let us consider rather that all living beings, all valuators, constitute a continuum. This continuum is modulated by the fact that we do not identically ascribe moral consideration to all these living beings – this modulation, says Sève, is a function of the progress of civilization. There is no contradiction between the statement that any living being is a valuator, and the assertion that care for humans is an obligation in a much more compelling manner, morally speaking, than care for such or such a non-human being recognized to be worthy of moral consideration. Descending along this continuum, we find the microorganisms to whom moral consideration is hard to ascribe. Things may certainly change. Three hundred and fifty years ago, Malebranche could explain to Fontenelle that a bitch who is hit and who cries out is only a set of pipes which emits a sound. No suffering in this case, the concept of the feelings of animals is devoid of meaning. In three hundred and fifty years, the wholesale tinkering with microorganisms to which we blithely devote ourselves will maybe also appear as shocking at the moral level as does Malebranche's argument for us today. We are not there, however, and if it is not devoid of meaning to enter these microscopic lives in the field of morality, there is a strong likelihood that today this is simply due to their functional importance for other living beings who, themselves, are recognized worthy of moral consideration. Care is diffracted into innumerable cares, which do not have the same moral density.

In the final analysis, to defend the idea of care for nature, it may not be necessary to face the metaphysical problem raised by Jonas. The latter emphasized that an ethics for technological civilization needs to have responded in advance to the following question, of metaphysical nature: in substance, why is the existence of life morally preferable to its absence? Sève's position may seem to kick this into touch, but in reality it opens up an indefinite perspective of moral progress: we may admit more and more beings into the perimeter of that to which we ascribe moral consideration. However, since human morality, as we have seen, is configured by technology, this position also leads to an indefinite perspective of Technological innovations aiming to agreeably instrumentalize this moral progress – we will return to this very important point at the end of Chapter 3.

The most difficult problem therefore perhaps relates less to metaphysics than to methodology and equipment: what concrete means do we really have to assess the inevitable telescopings between our technological activities and the valuative activities of non-human living beings? A distinction must be made based upon which these telescopings can be situated in space and time (even in an incomplete way), or not. An industry that dumps its very polluting waste waters into a river interfaces with valuative activities that we can more or less identify, if unable to perform an exhaustive breakdown. Let us say that we know, at least approximately, where human technological action causes its ravages. In contrast, in the above example of the environmental cost of data centers, the possibility of spatio-temporally circumscribing the damage is much less certain. Not to mention that, as already said, this obligation of assessment is hampered by the extreme difficulty or even the impossibility of anticipating the entirety of the effects that will result tomorrow from our technological action of today.

Estimating the extent to which our technical action, by disrupting the processes of naturation (*natura naturans*), affects the ability of the constituted living beings (*natura naturata*) to live according to propulsive and not repulsive ways of live: this may prove very difficult, but such is the possible goal of care for nature – to enrich, through our technology, the instruments of our life, without compromising the ability of other living beings (human or non-human, moreover) to enjoy their own instruments of life. The idea of care for nature is precise: to put it like Kant, the expression does not gesture towards a conception of nature as a super-object, as the omni-encompassing object, but rather towards a conception of nature as regulatory Idea. The objects of care are always specific objects, this ecosystem or that river. Grasped as instruments of living beings, these natural objects are objects of care. I can make an effort to conform my technological action to a refusal to alter such a natural object, conceived as an instrument of life for certain well-identified living species. These technologies may however support a defect of care elsewhere; carelessness is moved further away. Care for nature is then the formulation of a regulating principle of vigilance, the obligation to indefinitely renew the question: in this care that I am taking not to degrade this object, am I not at the same time in the process of compromising that one?

Let us note that, of course, these telescopings of valuations mix together non-human beings with, not humanity in general, but well-identified human groups – in other words, they cannot be analyzed and understood in the very abstract terms of a confrontation between the whole of humanity and the living world as a whole. Strictly speaking, there is no human valuation in the generic sense of the term: the valuations of humans are always situated and historically dated, they are always the valuations of specific human groups, which may still compromise the capacity of living beings, and even of other human groups, to conduct their own valuations. Useless to insist further on this point, as it is obvious that the harm of industrialization based on fossil energies can in no way be attributed to humanity as a whole – the Anthropocene, some say, is in reality a Capitalocene, or even a Anglocene given the great role of England in the choice of a certain type of industrialization with disastrous ecological consequences, to the detriment of other possible routes (Bonneuil and Fressoz 2017).

1.3.3. *The difficulty of thinking technology from life*

1.3.3.1. *Technology, a vital activity*

On the level of theoretical representations, the problem of care according to Tronto and Fischer therefore concerns, above all, the conception we have of the human. The problem is of an anthropological order. It also relates to the conception of technology. A concept of caregiving technology, if one draws out all the possible implications of this definition, supposes that we cease to understand technology as the set of means by which a subject cut off from life, living in the kingdom of purely rational ends, relates to external nature, only seeing in it what is determinable according to usefulness. Technology can only perform care in its relationship to the activity by which a living being among other living beings, the human, conducts its own valuations and deploys them in the world (of course always in a situated and dated manner, as we have just seen).

Any human technology, said Canguilhem, "is set within life, that is, within an activity of information and assimilation of material.... Because life is activity of information and assimilation it is the root of all technical activity" (Canguilhem 1978, p. 72). The detour through Canguilhem's concepts has helped clarify the terms of the expression "care *in* technology": ensuring that our technologies are caregiving is on one hand to re-assess them in the light of our vital valuative activities, and it is on the other hand to ensure that they are not predatory and destructive of the valuative activities of other living beings – or even, shall we say further with Dewey and Simondon, that they contribute to the development of the potentials inscribed in nature. A concept of caregiving technology points in sum towards an anthropology that does not cut the human out of the vital, and which does not subscribe in any way to the conception of the human as composed of body and spirit.

In recent literature, Tim Ingold has doubtless gone furthest in this in-depth revision of the conception of the human as technician (Ingold 2000). Going back over the whole of his intellectual journey, Ingold explains that first of all he applied himself to the task – such a difficult one! – of unifying two competing approaches to human action, traditionally designating a division of disciplines on one hand and on the other an apparently insurmountable border: the biological and naturalist approach on one hand, the social and cultural approach on the other. After the publication of his first book in 1986, Ingold engaged with the criticisms which were addressed to him at the time: bridging the gap between the two dimensions, biological-physical and socio-cultural, of human existence implies identifying a third term, mediating between the two extremes, namely the human mind.

However, Ingold remained convinced that this way of posing the problem of the unity of the human being still depends on a dualistic presupposition. Seeking to make a synthesis of the human as an organism taken in its relations with the environment, and the human as a cultural being taken in social relations, amounts in effect to leaving it unquestioned that he has to, essentially, join together two previously separated dimensions through the means of an intermediate term.

More specifically, grasping the human as a compound of biology and personality is according to Ingold to remain glued to an essentialist conception of the organism as well as the person: both are indeed assumed to be defined by immutable characteristics. The development of the organism is understood as the manifestation of biological characteristics which are inherent to its essence: this essentialist conception of development leads to the conception of the organism as a being closed off within its own limits, defined by a series of intrinsic characteristics. As for the person, it is supposed to consist of well identified attributes that a particular organism – the human organism – acquires in the course of its development. Development is development of what is initially in an enveloped state, already given: there is no invention, no creation of novelty.

Contrary to this essentialist conception of human development, which is the common basis for both the biological-naturalist and socio-cultural approaches, Ingold proposes to conceive this development not as the conjunction of separated strata, already structured and well characterized *ab initio* – the biological, social and psychological – but as a process operating in a field of metastability, containing no reality which might be already incorporated and identifiable from the outset as biological, social or psychological. In sum, human development consists in the creation of a novelty; it is a process which generates a being from a reality that contains potentials. Far from being an individual definable as an organism according to inherent characteristics, and an afterthought in relation to a natural environment as well as to its peers within a society, the human being experiences its genesis within a system of metastable reality, a field of relations, as Ingold said (incorporating Simondon's vocabulary), rich in potentials of becoming. The genesis of the human being is not the addition of a stratum of society, culture and personality to an organism assembled in advance; it is a process by which a particular living being, the human, invents a new way of being an organism in relationship with the environment.

This anthropology, alternative to the conception of the human as a composition of a natural part and a non-natural part, is reflected in the conception of technology. Unsurprisingly, Ingold rejects the idea of technology as an operation of the intelligence imposing a rational design upon things and beings. This intellectualist conception of technology, entirely modern, was totally foreign to the ancient Greeks as we have seen. It is also totally foreign to many peoples, as Ingold stresses. Technology is not conceived as the set of means by which an intelligence external to nature seeks to master it: it is a body of skills by which a living being invents for itself a specific way to live, not in opposition to all non-human living beings, but among and with them. The growth of plants and animals is not fundamentally opposed to the artisanal manufacture of things: in both cases, technological activity proceeds from the potentials of nature, locally.

Ingold gives the example of weaving: no idea here of an intelligence forcing a material substance to receive a pre-established form, rather it is a case of the genesis of a form which emerges very flexibly from rhythmic operations of gesture. Form is not imposed; it is generated by the twists and turns of a gesture guiding the threads. Any manufacturing can then be considered according to this model, rather than according to the distinction between intellectual conception and execution. Like organisms, technical artefacts are generated through processes of growth, deployed within fields of relations, or areas of metastability, within which they exert their forces. Manufacturing is not imposing a readymade form onto matter without form: it is actualizing the morphogenetic potentials present in the local situation where the technological gesture inserts itself. Artisans are not intelligent beings who face matter and impose upon it the design they have conceived all by themselves; the artisan is a being taken within a network of multi-level relations, within which their activity is to grow forms.

In these conditions, the term "skill" not only designates the skills of the artisan: it designates more fundamentally a conception of technology when it ceases to be thought of according to the dichotomy of intelligence and execution, to become thought of according to the category of generation. This distinction is a historical and situated construction, it is a phenomenon of Western modernity which must be questioned, in the sense that it does not account for technological activity as an activity of the living being.

Ingold also notes that this dichotomy of intelligence and execution also does not account for technological activities in an industrial context – these are always the implementation of skills, even if, Ingold notes, this experience of work is systematically obscured by the discourse of the industrial era opposing freedom and necessity, design intelligence and mechanical execution.

In summary, Ingold goes very far in the intertwining of these two related tasks: on the one hand, re-evaluating the anthropological discourse defining the human as composed of two predefined strata, the biological and the personal, and, on the other, re-evaluating the concept of technology as the operation of an intelligence exterior to nature. Ingold also associates this double re-evaluation of the human and of technology with the possibility of bringing the latter to care, judgement and dexterity (Ingold 2000, p. 291).

However, in many respects, the contemporary literature on the philosophy of technology accepts, without really questioning them, both the intellectualist conception of technology and a conception of human life which remains finally bifurcated. It is therefore not surprising that this literature finds it hard to integrate a reflection on care.

1.3.3.2. *The persistence of a dualist anthropology in the philosophy of technology*

Indeed, the empirical and pragmatist turn in the philosophy of technology has not really called into question the idea of an exteriority of the human vis-à-vis nature and life. More specifically, a bifurcated conception of human life continues to prevail in the philosophy of technology. Certainly, as we have said, it has become almost commonplace to admit that technologies have their own "agentivity", in that they contribute to configuring the forms of human perception, experience, knowledge and morality – technology shapes the life of humans. However, as it is most often found to be envisaged, this technologically constituted life is basically the intentional life that is deployed in perception, cognition and sociality.

This diagnosis can certainly be nuanced. For example, does so-called emotional design not lead (by destination, so to speak) to dealing with the relationship between humans and their objects at an organic level, consequently refusing any division in principle between the two components of human life? Emotions in fact imply an overall attitude of the organism with respect to its environment. Thus, a supporter of emotional design will in all likelihood orient their approach to design based on considerations of the biological processes. This is in any case what emerges, in the first analysis at least, from an article on affects of which one of the co-authors is none other than Donald A. Norman, the theorist of design famous for having introduced James Gibson's concept of affordance into the discipline (Ortony *et al.* 2005).

The authors of this article argue that the functioning of an organism involves four areas: those of affects, motivation, cognition, and behavior. Behavior refers only to physical action: mobility is only the mechanical resultant of processes that precede it and determine it. These processes appear either in the field of meanings (this is cognition), in the field of tendencies to behave in such or such a way (this is motivation), or in the field of values (this is affects).

At the most basic level of organic life, which the authors call "reactive" in opposition to "routine" and "reflexive", emotional life consists of assigning binary values, positive or negative, to external stimuli. These "proto-affects" serve as fundamental elements in more complex affective and emotional constructions.

Therefore, the co-authors of the article consider the most backward stratum of organic life, not only as a set of given conditions, but also as the site of a specific activity, and moreover of an activity for the allocation of values. That said, they very clearly separate the areas of cognition, motivation and affects on one side, and the domain of mobility and behavior on the other. Affective-emotional valuative life in no way proceeds from the vital activity of an organism faced with its environment; it is the result of reception by the same organism of a set of stimuli likely to be qualified positively or negatively. The organism is not a center of activity as in Canguilhem, but a machine with entrances and exits, inputs and outputs – a machine that deals with information. It is not surprising that technology has no vital significance in this article. On the contrary, it is rather the body which is described as a machine. As it is theorized here, emotional design in no way breaks with the intellectualist conception of technology.

Philosopher of design Jaana Leikas (2009) confirms how little contemporary design is concerned about integrating the biological dimension into the conception of human-technological interactions. According to her, the theory of technological design lacks a really holistic approach – at the same time biological, psychological, and social – to the "forms of life" of individuals. The designers and theorists of technological design are certainly no longer focused only on the question of the usability of technology. Their perspectives have extended, beyond the strictly functional approach, to a more global account of the interactions of human beings and technology, also dealing with their emotional, perceptive, emotional, aesthetic, cognitive and moral aspects. However, Leikas notes, the biological dimension is rarely integrated into these perspectives, though it is undoubtedly constitutive of our forms of life.

In sum, the conception of technology as a vital activity, very present as we will see in the philosophy of technology over the last century (and in the century before that also), seems to have almost completely disappeared off the radars of today's philosophy of technology and technological design. From this situation stems the still-intellectualist framework which characterizes them. It is appropriate here to avoid any misunderstanding and to clarify the situation. By "intellectualist framework", we do not claim that technology continues to be conceived as "applied science". With its scholarly society created in 1976[31], its specialized journals[32], its countless international symposia, its university chairs almost everywhere in the world, it is only recently that the philosophy of technology has managed to impose itself as a new, fully-fledged branch of philosophy, and this, precisely, while claiming its autonomy in relation to the philosophy of science[33]. In addition, by the expression "intellectualist framework", we are not affirming either that even today, technology is still most often conceived according to the duality of ends and means – no need to repeat the unanimous criticisms of this concept. It remains no less however that technology is always defined, in essence, in terms of artifactuality and design, that is to say, by reference to the activity of conscious subjects, gifted with rationality and free will.

For about 20 years in fact, the general orientation of the philosophy of technology in the context of the empirical turn (Kroes and Meijers 2000, Achterhuis 2001, Brey 2010) has taken on two major characteristics. Firstly, the development of an ambitious and pluralist research program designated by the expression "philosophy of technical artefacts", combining analytical philosophy approaches and post-phenomenological approaches (Verbeek 2005). Secondly, as we have already said, the growing interest of the philosophy of technology in issues of design, with a very marked interventionist propensity, as Verbeek emphasizes in particular with his concept of technology accompaniment (Verbeek 2010). In this very rich landscape, technology remains understood as an artifact, made by human hand, and as the result of a design guided by choices of values. Technology gestures towards the activity of an intelligence which represents itself, which deliberates and which decides, but this intelligence is not reassessed as proceeding from the vital.

1.3.3.3. *Thinking technology between intellectualism and organism*

Criticism of the instrumentalist conception of technology may therefore very well accommodate an intellectualist framework of thought. Conversely, criticism of this intellectualist framework does not necessarily imply rehabilitating a conception of technology as a vital activity. Saying that technology is something other than a

31 The *Society for Philosophy and Technology* (SPT).

32 Including the Journal of the SPT, *Technè*.

33 See e.g. (Hottois 2004).

production of the intelligence, this is not to *de facto* admit that it is part of the process of life, that it is a vital thing.

A whole literature inspired by the ideas of Donna Haraway proposes on the contrary to break at the same time with the intellectualist definition of technology, and with its definition as an extension of life. Technology must rather be envisaged from the perspective of the cyborg, that is to say of the proliferation of hybridizations of all kinds between flesh and metal, between the biological and the artefactual – novel and possibly monstrous couplings between the organism and the machine. It is not an outcome of intelligence but an ontological productivity, the generative power of surprising mixtures. The cyborg is our ontology, said Haraway (1991), but this ontology is not primary. The cyborg is not to be understood and assessed in the light of an initial pre-cyborg instance that would be pure, that is to say, non-hybridized with what is not itself. Its terms of reference do not include life, or nature, or anything that would gesture towards an origin, an inaugural power, a source of ultimate values. Relating technology to the organism is to conserve nostalgia for originating totalities, entireties constituted as axiological horizons. It is, says Haraway, fighting the wrong battle, like those whom she calls the progressives (she cites Marcuse and Merchant) who emphasize the need to dominate technology and appeal to an imaginary organic body to strengthen our resistance. Assemblages, composites, connections, encodings: such are the concepts allowing us to describe the cyborg being.

In the same vein, French philosopher Thierry Hoquet distinguishes "organorg" and "cyborg" (Hoquet 2011): the organorg is the unified and active organism that, in humans, extends itself into technology. In the organorg perspective, technology is an extension of life. Cyborg designates a very different relationship of technology and the living being: re-assessed under the concept of cyborg, this appears less as a harmonious whole composed of firmly linked parts, docilely doing what they are intended to do (organs), but as a factory where the biotic components are reconfigured, put into circulation, hybridized with technological components.

The question is therefore the following: can we defend a conception of technology of a vitalist type, oriented by the concept of valuation, without adopting a retrograde organicist position? Another formulation of the same question: can we relate technology to the valuative activity of living organisms, but in conceiving of the organism otherwise than as a harmonious totality of organic parts? Should we not therefore go further than Ingold himself – not just to reassess the concept of the organism in the light of the duality of biological and personalist perspectives, but to reassess it also in the light of the concept of "organ"? We will return to this in the third chapter of the book.

1.3.3.4. *Discussion of the ACT thesis – Anthropologically Constitutive/ Constituent Technology*[34]

There certainly in any case exist thoughts of technology that avoid both the pitfalls of intellectualism and of totalizing organicism, while assuming that technology is part of the activity of the living being. The so-called "Anthropologically Constitutive/Constituent Technology" (ACT) thesis, developed at the University of Technology of Compiègne, merits particular attention from this point of view: it indeed shows the great originality of seeking, precisely, to account for technological fact based upon the activity of the living beings that we do not cease to be. In addition, it also has the merit of seeking, as far as may be possible, to describe the mechanisms of this technological "constitution" of the *anthropos* – which is not the case in most of the competing approaches, which mostly insist on the fact that the human being is shaped by its technology, but without going further into analysis of the details of the mechanisms from which this shaping concretely proceeds. Technology is certainly postulated to be constitutive of human faculties, but we most often lack precise analyses, of a mechanistic character, concerning the "how" of this process. The ACT thesis is conspicuous by its effort in this direction.

Philosophically, this thesis finds one of its inspirations in the so-called theory of enaction proposed by Francisco Varela, closely linked with the theory of autopoiesis developed by himself and Humberto Maturana in the 1970s (Varela 1979). These two concepts intend to account for the behavior of living systems and their interactions with the environment, based on a criticism of the informational paradigm in biology.

At the end of the 1980s, Varela explicitly integrated the enactive and autopoietic approaches to the living being into the framework of a theory of cognition directed against the so-called computational-representational paradigm. While, according to this paradigm (widely shared by the cognitive sciences at the time), cognition is understood as the outcome of the manipulation, by the subject, of mental and symbolic representations, Varela supports the idea that cognition is the result of the sensorimotor activity of the living being. Three concepts are essential here: sensorimotor coupling, operational closure and autopoiesis.

The concept of sensorimotor coupling is opposed to conceptions of living systems modelled on machines with inputs and outputs, like computers. According to this latter model, borrowed from computer engineers, perception is a reception of information from the outside world: this information constitutes an input. The nervous system (among those living beings who have one) then processes the information that it organizes in the form of mental representations, based upon

34 I thank Charles Lenay and Pierre Steiner who graciously re-read this sub-chapter and, through their remarks, have greatly helped me to clarify and to improve it.

which it produces appropriate motor outputs. In this model, action is posterior to sensory perception; the passage of the second to the first is mediated by a representational activity.

In place of this engineering model, Varela substitutes a biological model in which perception and action are inseparable: living is equivalent, for the organism, to building sensorimotor invariants. The organization of the living being determines what, for it, has biological significance: organic activity selects what can be perceived, from the motor to the sensory. In return, perception guides the activity of the organism in its environment, from the sensory to the motor. There is a sensorimotor coupling.

An operationally closed system is a self-referential system: the results of the operations of the system are the operations of the system. By its recursive operations, the system defines its own limits.

An autopoietic system is a particular case of an operationally closed system, such that the system itself ensures the conservation of its own identity through its transformations. Varela draws an equivalence between autopoietic systems and living systems: any living system is autopoietic, and conversely an autopoietic system must be said to be living, even if it does not belong to the class of biological beings such as we know them.

Cognition is therefore not of a representational nature: it is enactive in the sense that it results from an activity of the body which does something, and especially which has something to do – that is, which can move, manipulate or experience resistance. Perception does not consist of treatment, by the brain, of information received by the senses and transformed into representation: perception is inseparable from a motor activity involving the whole of the body. This perspective therefore intends to account for the cognitive faculties without making a hypothesis of internalist character, that is, without presupposing the existence of representations or of mental states: cognition proceeds from the deployment of an activity in the outside world.

The theorists of the ACT thesis (Havelange *et al.* 2002, Lenay 2006, Steiner 2010) add that in living human beings, this activity is constitutively supplemented by technology – a dimension which Varela has completely ignored, technology for him not having any part in the sensorimotor couplings of the organism and its environment. According to the ACT thesis, on the contrary, human cognition in all its forms is always technologically constituted, according to the externalist principle which states that what we produce (our technology) contributes in return to constituting the structures of our experience and knowledge. Varela considered the enactive paradigm as superseding the chicken-and-egg problem. The ACT thesis similarly states that there is no need to maintain the radical separation of the constituent and the constituted, as Kant wished in the *Critique of Pure Reason*: technology is without no doubt constituted by humans, but it nonetheless constitutes the fundamental structures of their experience, their knowledge and their life. It testifies to the fact that an empirical field may have a transcendental scope. In the enactive paradigm, this linking of the empirical and the transcendental takes place in living corporeality, developing its sensorimotor couplings with the outside world.

The concept of the "constitutivity" of technology must be clarified. Varela himself drew attention to this term, "constitutive", asking for the greatest caution. Indeed, to the extent that the input/output model is abandoned, it is impossible to continue to think about the world to which the living being relates as being *the* objective world: it can only mean the world that emerges from the living being itself, by virtue of its specific organization, as the whole of what, for it, has a biological significance. Objectivity (of the world) is necessarily put in parentheses, said Varela. Therefore, we can no longer assume that objects in the environment of the system must be considered as constitutive (Varela 1979). This formulation would indeed lead us to understand that some entities of the outside world have a one-way relationship of constitution, and thus a deterministic one, with living beings, which would contradict the model proposed by Varela (in particular the concept of sensorimotor coupling).

The ACT thesis therefore takes care to specify that the anthropological constitutivity of technology is not precisely a one-way relationship: technology is only constitutive of the forms of human experience and knowledge to the extent that humans "activate" technology – in other words, to the extent that they technologically enact the world in which they live. There is no passivity in this notion of constitutivity. It is not technology as such which is constitutive, but sensorimotor coupling which may be based upon a technological device and not only upon biological structures (proof that an operationally closed system, such as a living being, is not a closed system: it is open to the outside, it incorporates elements that do not belong at the outset to the organic structures characterizing the design of the biological species).

Thus, referring to the work of Bach y Rita on sensory substitution (Bach y Rita 1972), Charles Lenay insists heavily that a sensory perception is devoid of any biological value if it is dissociated from a motive outlet. Bach y Rita's experiment is as follows: a blind person is equipped with a device that converts visual data into tactile data. A camera records images, the image is analyzed and converted into electrical impulses, which actuate a device featuring pins which is positioned on the arm of the blind subject. The pins may occupy one of two positions, lowered or erect. According to the overall pattern of the pins (some lowered, other raised), the subject of the experiment is supposed to learn, gradually, to recognize in a tactile manner the scene viewed by the camera. However, the results of the experience are very disappointing if the subject is purely passive. In contrast, as soon as they become active – by orienting the camera themselves, by inspecting the world with it – the learning is much more rapid and complete. This experience thus allows us to see the development, by the blind subject, of a new sensorimotor coupling. Technology is only constitutive of a new way of being-in-the-world if it is actively "activated" by the living being.

There is no doubt therefore that the biological dimension is central to the way in which the ACT thesis, extending the enactive thesis, deals with the place of technology in human existence. However, the difficult point of the enactive thesis – which we must yet consider whether it is also found, in echo, in the ACT thesis – remains the question of values. Varela's theory of enaction left this already partially sub-determined, although Varela granted it a non-negligible importance.

The concepts of the enactive thesis are normative concepts (in opposition to the purely descriptive model with inputs and outputs of engineers): the living being is a system which builds for itself its own norms of existence. The criterion according to which it is possible to assert that a system is autopoietic is its viability, which is revealed in turn by the adequacy of its behavior. It is this viability which confers value upon a given biological formula. This point is essential: in its own cognitive domain, the living being – every living being – is an absolute. There is no superior point of view that could integrate these different centers of perspective that living systems are: in the living world, no viable organization has any more value than another. The elements mobilized by sensorimotor couplings, and therefore the operational closure of living systems, have no value in themselves: what has value is only the overall capacity of the living being to sustain its identity through disruptions of the environment.

Accordingly, as Lenay stresses, if one admits that a living system defines its own world according to the series of disturbances and structural changes that it may suffer while maintaining its organization, there is no way to introduce any kind of gradation into this. Either the autopoietic organization maintains itself, or it disappears. It is an all-or-nothing situation; that is the principle of operational closure.

However, does the concept of vital value not suppose, precisely, a gradation in a field of possibilities? Is vital value not necessarily differential, that is to say: does it not suppose the capacity of the valuating living being to experience that something – a situation, a phenomenon, an object – has for it more value than something else? Is valuing, for all living beings, not something completely different from following the binary law of all-or-nothing? Is this not rather, precisely, introducing a gradient into a spectrum of possibilities? Such is the meaning of the pathological condition according to Canguilhem: in disease, the living being is not continuously in crisis in the etymological sense of *krisis*, that is to say the point of bifurcation – either life by return to health, or death. A sick living faces a much more complex, much less binary test, in which the idea of vital value which has become shriveled, diminished but not cancelled, takes meaning in reference to the narrowing of the field of possibilities.

What becomes of the valuative activity of the autopoietic living being when technology comes into play? In Varela's perspective – even if Varela himself did not take technology into account – one could respond, Lenay explains, in the following way: in the same way that, based on the possibilities of acting and feeling appropriate to its organization, the organism enacts its own specific world so that this organization may maintain itself; similarly, the introduction of new technologies into the repertoire of means of action of a living system is reflected by the development of new sensorimotor couplings, in such a way that the new living system that results from it will be viable, that is to say that it provides itself with the conditions of its own perpetuation. In this framework, the integration of a technological system modifies the living being's own world but does not add more gradation than previously.

However, it is precisely this point which is the problem: when technology is in play, it is no longer possible to argue that the system formed by the autopoietic living being and its technologies may be assessed only in terms of its viability. There are two reasons for this situation.

Firstly, to the extent that our technological devices produce, as we have said, effects strongly delocalized in space and in time, it becomes impossible to assess the system that we form with them, collectively, without taking into account the system of reality that others than ourselves, possibly very distant in space and/or time, form with these same technologies (which can result in deleterious effects on their own world). What may appear to me to be perfectly viable here and now, from my point of view as a situated living being, may prove to be non-viable elsewhere, for other living beings (human or non-human).

Thus, the French Department of Ecological Transition and Solidarity announced that France has reduced its emissions of greenhouse gases (GHG) by 4.2% in 2018; this result, however, may be explained, at least in part, by the relocation of industrial activities strongly emitting GHGs. In other words, CO_2 pollution probably has not decreased overall, it has become redistributed at a global level. We have delocalized some of our pollution. In sum, when technology enters into play, the concept of operational closure, which only knows the axiological duality of viable and non-viable from the point of view of the system itself, can no longer serve as a term of reference for the evaluation of this system.

Secondly, even limiting ourselves to the system of reality that the living human being forms with its technologies here and now, it does not appear any more possible to consider sustainability as the ultimate criterion of assessment for this system. Indeed, technologies embody in their design, and accordingly in their operation, value choices made by others. These value choices do not result in binary effects – either the system is viable, or it is not – but in the introduction, into the system itself, of a sort of gradient in the liveable: the duality of viable and non-viable is replaced by the continuum of more or less liveable. For example, equipping a blind person with a device of sensory substitution, or rather of perceptual supplementation as Lenay prefers to say, is already a choice in value. This approach may for example suggest that integration of the blind into society can be achieved by primarily (or even exclusively?) technological solutions, which would save us the trouble of broader reflection on the representations and (de)valuations of disability in the said society. Lenay himself provides an indication that goes in this direction. He notes that if we need to develop new perceptual supplementation devices to allow the blind to navigate around a computer screen (for example by recognizing shapes), it is because the new operating systems developed by the micro-computer industry no longer display information in the form of ASCII code, but in graphic form, which is not appropriable by the blind – in other words, it is a question of industrial choices before everything. However, the normative concepts of sensorimotor coupling and operational closure are undeniably insufficient to grasp this type of situation in which the lives of the blind, within the meaning of "living their lives well", are not rendered strictly speaking non-viable, but more difficult to live than in a society where disability would be better recognized and supported, for example, by adaptations of the urban environment.

Undoubtedly, Lenay takes great care to specify that the ACT thesis is not a thesis in the third person: Bach y Rita's blind subject is not a subject "experimented upon", like a rat in a maze, but an "experimenting" subject. Lenay's issue, such as he derives it from systems of perceptual supplementations, is that the value of some technological mediations depends on the emotional values that attach to the very use of these mediations. The way in which the subject emotionally lives its situation is constitutive of the value of the technological mediation.

This being said, Canguilhem delivered, as we have seen, a warning of great relevance: in matters of vital values, the third-person point of view of science can certainly not in any case prevail over the point of view of the subject itself; however, the point of view of the subject can no more take the place of a *terminus ad quem*. I may assign a lot of value to that which does violence to my own life as a living being – we even spend our time on this. That this duality of subjective values and vital values of the being-in-life does not hold when it comes to living non-human systems can easily be understood: among the latter, life does not flourish in "technical plasticity" that constructs an environment according to new values, which are not directly linked with the being-in-life. We should certainly be prudent in stating that this duality does not exist at all in non-human living beings: the thesis according to which there are "animal cultures" may lead us, in these cases too, to find this duality. However, in Varela, who always reasons (so to speak) from the examples of the cell or the bacterium, this duality of vital value and technological value is obviously devoid of meaning.

Lenay is very attentive to the issue of values. He and his co-authors have observed that blind people do not really appropriate devices which are designed for them. They test them with interest, but after initial curiosity, the devices are stored at the bottom of drawers and never used again. How to explain this phenomenon? The response of the proponents of the ACT thesis is very interesting: if the blind ignore these systems, it is because they are unable to help their users assign a value to the things "seen". The use of these devices does not support the activity that is absolutely essential to us: conferring a value on the world that surrounds us. The blind do not necessarily want to "see" by touching, according to the possibilities offered by a perceptual supplementation device: they want to marry, work, have children, travel.

However, it seems that this commendable attention to values that the subject is only competent to define for itself pays a fairly high price: the relegation into the background of vital values properly speaking (in the meaning of the being-in-life), which, as Canguilhem again reminds us, cannot be reduced to the values according to which, as a subject, we want to organize our existence.

The ACT thesis thus surpasses the limitations of Varela's thesis in which vital value is defined only by the viability of an operationally closed system. Despite everything, though, it faces a problem that it does not resolve, concerning the conception of the technologically constituted *anthropos*. One of two things must be true: either the human is in perfect continuity with life as a being-in-life, but then how to account for the constituent character of technology? Or else the human is characterized by the fact that technology is constituent of its sense of being, but then how to account for the articulation, in each one of us, of the vital order and the order constituted by technology? Pierre Steiner summarizes this difficulty with remarkable clarity: if we accept thinking of technology in a non-instrumental, non-anthropological perspective, being ready to see its role as anthropologically constituent (and not only anthropologically constitutive), do we not then risk a tension appearing between the valorization of the living, embodied and sensory dimensions of cognition, and the inorganic or even thanatological character of technology? More specifically: how to reconcile the inorganicity, the virtuality, the intangibility, the acceleration and the creation of simulacra progressively (or at least possibly) introduced by technology with the living materiality, the presence in the world and the bodily, subjective and experiential anchoring of cognition? (Steiner 2010, p. 36).

What characterizes the body is that it is my own *and* that it is in life: in other words, it is the sounding board for the telescoping, always possibly confrontational, of the values according to which I intend to govern my own existence and vital values, those which I cannot deprive myself of completely to the extent that I remain a living being among living beings. This telescoping is always possible, in sum, between the valuating subject that I intend to be as a subject, and the valuating living being that I do not cease to be as a being-in-life. The ACT thesis recognizes this hinge structure of human corporeality, but without considering it from the angle of possible conflict between the technological conditions and the organic conditions of existence, as Simondon put it.

It seems then that the way in which the ACT thesis introduces the question of technology into sensorimotor couplings – which define the invariants that structure the being-in-the-world of living humans – results in a double occultation, from below and from above, in favor of exclusive attention focused at the intermediate level of the subjects' evaluations of its own situation.

We find obscured from above the questioning of the global choices that technology makes effective (like the technological choices which, in terms of the design of computers, move away from the needs of the blind). Varela's thesis also falls under the blow of this finding. Varela recognized this: his biological philosophy dissociates the material structure of systems and their organization, which implies the possibility that there are living (therefore autopoietic) systems which are not beings made of "biological material". An important consequence of this is the perspective, perhaps not so distant, of being able to build living systems at our will. Theoretically, concedes Varela, the possibility exists; it will undeniably demand the greatest care in the technological sense of the term if it is explored. That being said, he admits also, this possibility provides no ethical criterion to justify, or not, this type of initiative[35]. An "engineered" autopoietic system, if it is viable, will constitute an absolute for itself, like any autopoietic system. Nothing, in Varela's perspective, justifies us evaluating it according to exogenous axiological criteria, which is however the major problem with which technological choices confront us: assessing them otherwise than on the basis of their success.

35 Wanting or not wanting to manufacture an autopoietic system is of course an ethical problem. But, if our characterization of living systems is adequate, it is clear that we should be able to manufacture them at will (Varela 1979, p. 44).

We find obscured from below the question of whether, yes or no, such a technology promotes (or not) valuative activities which are deleterious to life within the meaning of the being-in-life, those of human as well as non-human living beings.

1.4. Transition

Technology remained for a very long time what it was for the ancient Greeks, namely an activity taken within the limits of the essential forms, the corporal possibilities and the needs of living human beings. The so-called modern era (from the 17th century) is often held responsible for a complete turnaround in the conception of technology, with this idea that, now, it must be guided by science and its necessary truths. This reversal seems particularly noticeable today in precisely those human activities where the human was until not so long ago only an adjuvant to a process operating without it: agriculture and medicine. Instead of *techné* being limited to creating favorable conditions for *physis* (Nature) to operate in a direction which serves our interests, now *techné* claims to intervene in *physis* by substituting for its own processes an operationality which is the fruit of human thought and knowledge. It is no longer *techné* which is taken into the overall process of *physis*, but on the contrary *physis* which is taken into *techné* and reconfigured by it. We have of course not finished with the piloting of natural processes, but this pilotage is most often integrated into a more global goal: to reconfigure natural processes and more particularly the processes of living beings, according to a logic foreign to them.

This diagnosis is not false, far from it, but we must remove responsibility for this reversal from the thinkers of the Great Century; rather it is in the following century, the century of the Enlightenment, that things really began to switch. A dual vision of the human being, intended to feed political decisions concerning productive activities, then came to be. According to this vision, the human being appears as a rational being which must be strictly separated from the automatisms of the body-machine. This dual vision is reflected in a political project: that of building a society in which part of human reality, the bodily reality, can without cost be given over to automatisms, while the other part, reason, defines the excellence of the human and must thus avoid any form of automation. As Simon Schaffer summarizes (Schaffer 1999), the Enlightenment finally imposed a division between those subjects that could be automated and those which had to be granted the privilege of reason. Technology ceases to be the extension of the living body at work; it ceases to be attached to the unity of a life, inseparably biological and existential, which is deployed in productive activity. It is now folded back onto two clearly disjointed levels: on one hand, that of the automatisms of the body, which is progressively conceived as a machine that could be programmed to mechanically execute prescribed gestures; on the other, that of the intelligence which decides the ends and which rationally conceives the means of achieving them.

What must be called the divorce of technology and life in its unity, in a sense still to be clarified, is without any doubt the major event which has occurred in the immemorial relationships between humans and their technologies. That is what the next chapter intends to examine.

2

Technology and Life: Analysis of a Divorce

2.1. Body, gestures, technology, production

2.1.1. *Work without skill*

2.1.1.1. *A controversial bracelet*

In 2016, Amazon filed two patents for an electronic bracelet designed to track actions performed by the company's warehouse workers[1], for the purpose of detecting possible poor handling for corrective purposes. By emitting vibrations if the hand is incorrectly positioned, the bracelet indicates to the worker how to correct the situation.

Faced with the outcry raised by these patents, Amazon has defended wanting to trace the gestures of handlers to monitor the way in which they carry out their tasks – even to better control and limit their break times, as some have worried – and explained that this device aims above all to relieve workers of a heavy burden. Currently, workers must hold the scanners they use in their hands, for example to manage orders or stocks. It would be a matter of freeing operators from the most time-consuming tasks and assisting them in their actions.

If it were actually manufactured and used, this electronic bracelet would thus be added to the already massive range of wearable technologies for metrological purposes currently in the market. There is, in particular, a very large choice of connected bracelets, also called activity sensors, which record in real time data

1 [Online]. Available at: https://www.nouvelobs.com/societe/20180203.OBS1670/homme-ou-esclave-amazon-depose-le-brevet-d-un-bracelet-electronique-pour-ses-salaries.html [accessed November 14, 2019].

associated with health, such as the frequency of heartbeats during physical exercise, the number of steps taken during the day, or calories burned. These electronic gadgets can allow everyone to trace their vital parameters – we talk about quantified self, that is to say a proactive approach of continuous self-measurement and self-evaluation, now accessible to everyone in order to maintain their "health capital".

However, the Amazon bracelet has two specific features: on one hand, if it were manufactured, it would be imposed on Amazon employees, and not chosen by them; on the other hand, it is not part of the individual quest for "perfect health", but of the long history of the rationalization of work. Between the two, there are undoubtedly connections: it is always a question of measuring in order to be able to control and intervene with a view to optimization, getting the most out of a capital – whether it is "human capital" or "health capital".

That said, quantifiers are primarily interested in their vital parameters, their life as biological life; Amazon is interested in gestures made in the workspace. On the one hand, the organism; on the other, the productive body: these are different targets for applying measurement and control technologies. These two ways of measuring the body even seem to be in opposition, since Amazon's main concern is apparently not the physical and mental health of its staff! In addition, quantifiers may come together and form communities of practice, sharing their data. Self-measurement gives rise to the invention of new socialities – biosocialities, as Paul Rabinow called them (Rabinow 1996). Amazon's connected bracelet is more in line with an aim of traceability of individuals which involves disconnecting their gestures from any form of social context (professional cultures, professional solidarity) to subject them to the metric of maximum performance.

2.1.1.2. *From Industrial Technology to Amazon's Mechanical Turk: deskilling human labor*

Implicit in the controversy surrounding these two patents is therefore an old theme of the critiques of industrialization which is being updated: it is above all about the deskilling of work and the automation of humans. The head of the Italian government at the time, Paolo Gentiloni, reacted strongly to the information disclosed on Amazon's patents, saying that the challenge is work of quality, not work with a bracelet.

In the controversy aroused by Amazon, the connected bracelet thus seems to play the role played in the 20th century by industrial machinery, assailed by diatribes denouncing the attacks made by mechanized, then Taylorized, production against work of quality, that is to say, against work of an artisanal type. To the latter are attached positive values of practical intelligence involved in manufacturing, skills and know-how, shaping the contours of a vibrant and supportive productive community. To the former are attached negative values, dominated by the image of unskilled workers subject to the scientific organization of work, devoid of any form of intelligence and taking no initiative in their productive activity.

It is, however, neither the machine itself which is under indictment here, nor the electronic bracelet in itself today, but rather the project they serve, which is opposed in all points to the idea of work that promotes personal development and fulfilment. The artisan is indeed that one who can objectify their own reality by producing a work. The organization of work represented today by multinationals like Amazon has removed the possibility of objectifying human reality and thought by means of the manufactured object. Simondon wrote in 1958: "The liberating technics of the 18th century is at a human scale because it is of the artisanal type. The technics of the 20th century is beyond the forces of the individual, and constitutes a compact and resistant, but alienated human reality within the industrial world, completely beyond the grasp of the individual just as it was for the previously hierarchized society" (Simondon 2017, p. 119). What can we say then about that of the 21st century, deployed in Amazon warehouses?

Gentiloni's formula, the challenge is work of quality and not work with a bracelet, therefore means this: it is not the use of the electronic bracelet that threatens to degrade the quality of work; it is because the work has already been degraded, that the idea of using this bracelet makes sense. Assuming that Amazon had a personnel policy more respectful of the dignity of its employees, and that the company valued the autonomy of workers rather than trying at all costs to control and plan the entirety of their actions: would the work accomplished under these conditions be however, *de facto*, work of quality? Of course not. The situation is not one in which workers, performing work of quality requiring real know-how, are increasingly controlled, monitored and directed in their activity by electronic systems. The problem is that the activity of work itself seems to have lost all form of intrinsic quality, with or without a bracelet.

The Amazon bracelet therefore takes us far back, into the first half of the 19th century when this vast program of generalized deskilling of work seems to have been enacted, at least if one believes certain texts of the time.

The development of Industrial Technology indeed illustrates, in an emblematic way, the erasure of the figure of the artisan, in favor of a new conception of the body at work as a machine from which it is necessary to obtain the maximum output. As Marx noted, "Technology also discovered the few main fundamental forms of motion, which, despite the diversity of the instruments used, are necessarily taken by every productive action of the human body" (Marx 2019, p. 532). "The principle which it pursued, of resolving each process into its constituent movements, without any regard to their possible execution by the hand of man, created the new modern science of technology" (Marx 2019, p. 532) – a diagnosis which seems completely premonitory of Amazon's connected bracelet.

According to French historian of technology Liliane Hilaire-Perez (2013), this vision well reflects the inflection of Technology in the 1830s in England, taking over from a discipline born on the continent at the end of the 18th century (with, in particular, Johannes Beckmann). Charles Babbage, Andrew Ure, William Rankine and even James Napier thus recommend rationalizing production by breaking down manufacturing into operational segments, allowing either quantification of the skills and the forces necessary for an operation, or to mechanize and combine elementary operations on machines (Hilaire-Perez 2013, p. 8). A space is then created for theorization of technology which aims to abstract, to synthesize work into operations, transferred from humans to machines, programmed and combined at will, arranged, disposed of for best performance (Hilaire-Perez 2013, p. 9).

Ure, the main promoter of this process of rationalization of industrial production, speaks of breaking down operations with a view to their generalized mechanization and the eviction of all dexterity and particular application of the hand, removing manufacturing from the clever worker (Hilaire-Perez 2013, pp. 9–10). Technology as an industrial science is based on calculations applying to bodies; it bears a productivist logic which never ceases to suppress skill, to divide work into operations to oust dexterity and know-how and to reduce the gesture to physical effort (Hilaire-Perez 2013, p. 12). Amazon's bracelet is definitely not a kind of UFO, coming out of nowhere and without history.

The organization of work in large industry, and Technology as a science of this organization, can in these conditions only lead to dissociation between on one side productive activity, and on the other side life, sensitivity and human intelligence, to which the said activity seems to have lost all form of attachment. As Canguilhem pointed out in his review of Friedmann's *Problèmes humains du machinisme industriel* (Canguilhem 1947), the scientific organization of work endeavored to exhaust, as much as it could, any capacity that workers might have to be subject to their own norms of activity. Amazon's electronic bracelet comes to crown the panoply of organizational techniques which have never ceased to deprive individuals at work of any capacity for self-normativity. The evaluation of norms at work is entirely contained within the metric adopted by the engineering offices; today, it passes through connected objects.

Amazon's patented hand movement tracking system thus seems to bring to its peak an enterprise begun in the 19th century, pursued frantically since, which has consisted in emptying what Marx called living labor of all qualitative content and any form of worker self-determination in the performance of technical actions, as well as of any value other than universal and abstract exchange value (equal to the relative labor time materialized in the products). As Marx wrote in 1857 in the

Grundrisse, "Activity, regardless of its individual manifestation, and the product of activity, regardless of its particular make-up, are always exchange value, and exchange value is a generality, in which all individuality and peculiarity are negated and extinguished" (Marx 1973, p. 157).

That hand movements are today being targeted by a major player in the digital economy should not come as a surprise. It is, of course, manual operation which, from the beginning, has been the focus of most efforts to wrest from workers their professional autonomy, by depriving them of what ensured them this autonomy: their "facture", to use Huyghe's term, that is to say their way of doing, their very activity of doing. From the artisanal gesture to those performed by Amazon warehouse workers, a formidable inversion of values seems to have occurred. While the artisan's tool is an extension of their hand, and a machine still had in the 18th century this sense of being an *organon*, it is now the hand that seems to have become the appendage of the technological device. The hand movements of Amazon employees no longer bear an incorporated normativity: they are reduced to a system of coordinates in space and time, capable of being subjected to control and optimization calculations. The head of the Italian government is right, it seems: it is still and always a matter of defending work of quality against the degradation of the worker in the machine.

There is no question here of contesting the legitimacy of the indignation raised by Amazon's patents. It should even be noted that the digital giants in general, and not only Amazon, seem to have opted for a conception of work, as well as of the relationship between the company and its employees, which is nothing other than a hideous magnification of the worst in industrial capitalism. Reduce the worker to a minimum and make of his existence a pure living labor force, according to Marx's formulas: this creed seems to have become that of the GAFA tech giants.

Amazon thus created in 2005 an online participative production platform (Casilli 2019), where thousands of little hands perform more or less complex tasks that algorithms do not yet know how to perform very well. This platform is called the *Amazon Mechanical Turk*, with reference of course to the chess playing automaton built by Wolfgang von Kempelen in the 18th century and to the fact that, just as Kempelen's automaton was a deception (a human was actually concealed in the machine and moved the pieces), so in the case of the *Mechanical Turk*, the tasks sold to clients are not performed by the algorithm itself, but by workers "hidden" around the world, and paid per click made on their computer, for a paltry average hourly wage. The activity of these workers or *turkers*, which is to augment the knowledge database of Amazon's algorithm, is also quite paradoxical since they do what the AI algorithm does not do well (or what it could do for a cost a good deal higher than that of the little "clicking" hands), but that is why they are expected to behave like quasi-automata (click, click, click…).

A reversal has therefore taken place in relation to Kempelen's Mechanical Turk. At the time, the automaton was considered to be devoid of any intelligence; this could only be located in a human endowed with real know-how – hence the conviction at the time that this automaton playing chess, and winning its games against humans, could only be cheating. Today, the idea of giving computers some form of intelligence is clearly no longer outrageous. It is widely admitted that there may be artificial intelligences, but these lack a certain type of know-how: the corresponding activities are therefore passed on to those who have these skills of contextualization and synthesis, but who are not, however, considered to be full contributors to the intelligence of the whole system. They are considered to be only automatisms.

2.1.1.3. *A plural history of Technology and industrialization*

We must therefore recognize that digital multinationals like Amazon today seem to be bringing to its height a program of the deskilling of tasks and the automation of work which, it should be emphasized, has never previously been completely accomplished, at least never so clearly.

Indeed, as Canguilhem pointed out this meticulous attempt to eliminate the workers' own normativity in their work activity was almost constantly met with strong resistance, including in those industries most oriented towards the Taylorization of productive activity. It quickly found itself limited by the fact that individuals at work cannot be assimilated to machines with operating standards decided by engineers. The workers remain living beings, that is to say autonormative beings, and thus working environments must be called "milieux" of work, insists Canguilhem: they are places where the norms of activity are constantly negotiated and renegotiated by those who implement them (as is the case in the relationship of all living beings with their milieu of life).

The sociology of work has, moreover, well documented this reality: work is never the field of pure heteronomy, but a plural arena where workers always have the capacity to set their own norms, up to a certain point, for their own activity. External standards are certainly imposed, but workers can always carry out "renormalizations" as French sociologist and philosopher Yves Schwartz says, that is to say, reappropriation of the conditions of their activity (Schwartz 2012). It is impossible to be a pure and simple executant surrendered to the most complete heteronomy: that would be unliveable.

The works of sociologists of work overlap on this point with those of historians of technology and industrialization, who have clearly shown that in France, from the *Ancien Régime* to the post-revolutionary industrial era, transformations in the organization of professions stand out against a background of deep continuities. Thus, according to Sewell (1980), the French Revolution undeniably abolished the trade guilds (in 1791 to be more precise), but it is clear that this mode of corporatist organization was to a much greater extent transformed than completely disappearing to make room for the deskilled worker in big industry. The vision of an industrial machinery which uniformly and brutally transformed the worker into an appendage of the machines does not stand up to historical analysis: "skilled artisans, not workers in the new factory industries, dominated labor movements during the first decades of industrialization" (Sewell 1980, p. 1). In 1850 still, France was above all a country of small workshops; large industry remained the exception.

If the very idea of an industrial revolution has seen a revival of interest in recent years, following a craze for so-called gradualist approaches (which challenged precisely this idea of revolution in the sense of a brutal change: the transition to industrialization was said to have happened in a rather progressive and incremental way), it is to recognize that the transition to big industry, imposing a generalized deskilling of work and a conception of workers as simple machines, was not at all the irresistible tendency that we generally believe (Hilaire-Perez 2010).

Technology, in the sense of an effort to rationalize productive practices, did not unanimously embrace the productivist logic of Ure. Hilaire-Perez thus shows that the industrial revolution followed multiple paths (especially in England, that is to say in the country which is usually said to have set the tone for victorious industrialism), creating a much richer and more complex landscape than is often said (Hilaire-Perez 2013). The discipline of Technology is indeed traversed by tensions; it is marked by a plurality of orientations which cannot be reduced to the industrialist vision. Between the science of machines and industrial rationalization on the one hand, and the trade guilds on the other, a third way was taken, valuing a rationality anchored in the knowledge of hands and of contact with matter, just as much as in the capacities of organization of work and markets, in the management of subcontracting chains and in the workers' handling of machines (Hilaire-Perez 2013, p. 13). In Charles Dupin, for example, there is a very strong tension between a concern to adapt the productive body to an increase in output and the aim of perfecting workers' intelligence.

The idea of a science of the arts, of a desire to improve manual work governed by skill and by the handling of complex tools, did not disappear at all in the industrial era as evidenced by Louis Sébastien Lenormand's craft reform project. It is undeniable that a profound change affected productive activity, moving it from an economy of the product, based on the development of care and manual intelligence, to the mechanized production of products in mass (Hilaire-Perez 2013, p. 22). However, this technology enhancing the quality of products and their suitability for use, achievable due to professional skills, continued for a long time in the culture and demands of workers. Besides industrial efficiency, emphasizes Hilaire-Perez, there is another technological understanding, based on qualitative, if not aesthetic and hedonic appreciation of the purpose of the objects. It is a facet of technology which has never gone away, even at the heart of mass production (Hilaire-Perez 2013, p. 23). To which we must add that it is not at all true that the machine deskills all gestural savoir-faire. Its mastery often requires skill, dexterity and even virtuosity (Guignard *et al.* 2004, p. 11).

Stories still record, up until the late 20th century, the existence of artisans at the very heart of large modern and Taylorized industry, for whose proper functioning they are essential. Robert Linhart's story, *L'Établi* (The Assembly Line), is instructive on this point (Linhart 1981). Linhart was in the 1960s, at the Renault factories where he was hired, what was called at the time an *établi* – that is to say a Maoist who posed as a worker among others, but whose mission was actually to prepare the proletarian revolution (the mainstream trade union, the CGT, being accused of collusion with the employers). Linhart thus discovers the reality of the industrial world, which amazes him. He says in particular that within the factory itself, near the assembly line, is an elderly man who works in front of a workbench. It is an old wooden workbench, whose function is absolutely essential since without it, the chain stops altogether. His task is indeed to adjust by hand those parts which are slightly defective, and which therefore do not properly attach to the chassis of the vehicle. His workbench is incredible: a thick wooden board, full of bumps and hollows from a history of being used, but the artisan (since this term has to be used here) knows his way around it. However, the factory management one day decides that this old workbench from another age is not suitable and decides to replace it with a brand-new workbench, made of metal, perfectly smooth and shiny. The man loses his bearings and can no longer work. There is panic in the assembly line and the management quickly decides to put the old bench back in place. The man resumes his activity as best he can but something has broken, it is difficult to say what exactly – something of obviously paramount importance, since the man never again found the joy at work which had previously characterized him and, two years later, he died.

Other stories, no doubt less poignant but just as instructive, such as that of Georges Navel (Navel 1949), point to the persistence of craft-type activities in the industrial workshops of the 20th century. Philosophers have a hard time admitting this complex reality, this variegated landscape of industry, which cannot be summed up in the image of a world where humans are purely and simply swallowed up by a mode of existence close to that of machines. A publication of the French Teaching League, dating from the 1950s, illustrates this gap between reality and representations. This bulletin reports a visit made to the Renault Flins factory by teachers. It is a first for each of them and they are surprised to discover a reality which does not correspond at all to the image they had of it in advance: neither the image of the human robot nor that of the unconscious automaton can apply to any of the men they observed in their work at Flins. The worker, they admit, can hardly be likened to a cog without thought who is ignorant of the meaning and purpose of the gestures he performs mechanically: everywhere, the worker can insert the relative importance of his work into the powerful majesty of the whole. The teachers are amazed at the very limited and incomplete nature of automation. The visit showed proof everywhere that the machines required a numerous and competent personnel for their supply, their maintenance, their adjustment or their repair until further notice and for a long time to come.

However, it seems that these representations of an almost completely mechanized worker are on the way to being realized today, precisely in Amazon's warehouses or in other global digital and telecommunications players. The working conditions today of employees of the GAFA tech giants indeed seem dangerously close to what Schwartz considers to be impossible to achieve: an activity consisting of a pure execution of repetitive tasks, accomplished by robotic humans. Where is the ability for negotiation of norms of work of these thousands of small hands recruited by Amazon whose whole activity consists in "clicking" to feed the database of the *Mechanical Turk*? An in-depth investigation would perhaps once again identify possibilities for "renormalization", but it must be admitted that, assuming they exist, we would nevertheless expect them to be very reduced.

Marx's prognosis would therefore only today find its full confirmation: "individuals are subsumed under social production; social production exists outside them as their fate; but social production is not subsumed under individuals" (Marx 1973, p. 158). The old man in Linhart's story, as well as the specialized workers described by Navel, still at least partially subsume social production: this is probably no longer the case with the "clickers" of the *Mechanical Turk*.

2.1.1.4. *The counter-model of the artisan*

Conversely, artisanship continues to be a model of what the activity of work should be (based on know-how, skill, in-depth knowledge of what the activity involves, the ability to master the final goal and all the stages which lead to it, thus the reverse of the division of tasks associated with the industrial worker's labor, etc.). The recent success of works, sometimes tinged with a somewhat old-fashioned romanticism, glorifying a direct relationship to the material and the technological object, bears witness to this still very high valuation of artisans and their activity – consider for example Matthew Crawford and his *Shop Class as Soulcraft: An Inquiry into the Value of Work* (Crawford 2009).

According to Simondon, the artisan is the tool bearer, the one who, through their body, coordinates productive operations and therefore the one who assumes the role of the technological individual. The rise of industrial machinery is said to have changed the situation. In large industry, the bearer of tools has been marginalized, their role being only to operate, supply and regulate the machine; the machine has become the real tool bearer, the technological individual (Simondon 2017, pp. 21, 78–80).

In exact opposition to the singular, idiosyncratic character of the artisanal product coming from the hands of the artisan, the industrial product would be devoted to the most advanced standardization possible. It would involve erasing within it, in its appearance (that is to say aesthetically), all traces of the operations involved in its production. As Huyghe points out (Huyghe 2015), in the case of artificial objects in general, that is to say objects made by human hands, "facture" is decisive: if the manufacturing activity is regulated at every instant and in each of its elements, then the technological dimension of this activity is reduced to running a pre-established program, a *pro*-gram in the sense of a completely planned script for the different sequences of the productive activity. The object is literally *pro*-duced, it is only the materialization of a preformed concept – of a concept which is, moreover, not only a concept of the object, but also a conception of what the society and order that characterize it must be. Industrial machinery is therefore said to have lost that which is experienced above all by direct contact with matter, through craft activity and tactility: a "facture" which goes beyond complete planning and flawless correspondence to a pre-given ideality.

It should however be noted that the idealized vision of a craft industry before industrialization, dominated above all by the search for creativity and originality, as well as by the desire to escape any programming of gestures, is simply wrong. The arts & crafts were in no way thought of in the 18th century according to the romantic vision we tend to project onto them today. As historian William Sewell points out, art (that is to say all of the craft trades) was not at the time "a matter of originality, inspiration, and genius, but of rules, order, and discipline" (Sewell 1980, p. 22). Art still fully subscribes to the Aristotelian definition: it is a method of properly performing a thing according to certain rules. This definition applies of course to the mechanical arts but also to the liberal arts, and even to the fine arts which are also subject to compliance with strict rules.

That said, this submission to rules should not be confused with the heteronomy to which the worker in large industry is subject. The apprentice or journeyman who enters a guild undeniably places himself under the authority of a master. Yet Sewell specifies that if the relationship of dependence of the apprentice on the master is clear, this is less the case for the journeyman who is older than the apprentice, more experienced too, and less inclined to accept this authority – conflicts between masters and journeymen are therefore very common.

The guild is not structured by a desire to give priority to individual autonomy, far from it. Nevertheless, it does not seem that the control exerted over tradespeople involved a rigorous control of gesture. Sewell explained that jurors in each guild were responsible for verifying the quality of the work. "The specific standards to be applied by the jurés in these inspection tours varied from trade to trade, of course, but they regularly included restriction as to the types and quality of raw materials, the kinds of tools to be used, and the design of the objects to be produced" (Sewell 1980, p. 29). There is apparently no question here of controlling the gestures and the progress of technical operations, of imposing standards external to the technical practices themselves, as advocated by Ure. In the corporatist organization of the *Ancien Régime,* technical "facture", to use Huyghe's expression, does not seem to fall within the scope of what must be controlled from the outside, re-assessed in a metric allowing for calculations of optimization, and if necessary rectified.

This irreducibility of technological "facture" to rational planning has moreover been widely recognized throughout the ages. The Greeks certainly already approached their technologies on a theoretical level, with the aim of rationally anticipating the effects of the action. To dig his tunnel, Eupalinos used complicated triangulation procedures. *Architecton* means the person who directs the work by providing knowledge. There is undoubtedly a theory of technology among the Greeks. It is found for example in pseudo-Aristotle's *Mechanika problemata* in the form of a theory of simple machines (lever, pulley, winch, screw, wedge) and their combinations (Schiefsky 2007). That said, the further technology moves away from theory and towards practice, the more *empeiria*, the trial and error of experience, becomes dominant; so much so that this gap between a science, inspired by a logical ideal, and an *empeiria*, reduced to the trial and error of observation, could not be bridged by Greek thought (Vernant 1985, p. 314). Kant still admits, at the very end of the 18th century, that true art begins where science ends. The transformation consisting in entirely submitting the *technê poiêtiké* to computational rationality therefore seems to have occurred with industrialization.

2.1.2. *Control and discipline of technology*

2.1.2.1. *Diderot and Alembert's Encyclopedia project*

That being said, the invective against the Amazon connected bracelet, although touching on a very important dimension of work – its "quality" as well as the ability of workers to negotiate the norms of their activity – may be missing the essential point. What does this outcry against Amazon actually entail? Control, monitoring, and the fact that management has the means of tracing an employee's slightest gesture, to schedule breaks and to know exactly where each individual is at a specific time.

This problem of controlling those who exercise technological activities is not new. It originates from the idea that technology is potentially destabilizing and dangerous, and that it must be scrupulously managed. However, this idea is specific neither to our time, nor to the societies that chose industrialization. On the contrary, it is both very old and widespread. As we have seen, according to the ancient Greeks, manufacturing activity must be imperatively contained within very strict limits, precisely because it may always disturb the distribution of the sensible upon which the order of the city rests. As Huyghe says, technology involves the "contestation of what is", hence the need to contain it politically (Huyghe 2015). To regulate or to deregulate technology is to regulate or deregulate society: let us therefore be cautious; the Greeks were not entirely wrong.

While in the 18th century, a taste for artifice, for the design of new materials that did not aim to imitate nature but to surpass it, came to light (Nègre 2016), the suspicion continued to weigh on technical design and production: artifice is what does not exist and, consequently, to bring it into existence by human design and production is to risk a confrontation with the unknown and to subvert the laws of the established order, including the natural order (Vérin 1993). There is always a need to circumscribe this dangerousness, for example by conforming productive activity to well-established problem-solving procedures, and by linking the arts to the sciences that provide knowledge and principles (Vérin 1993).

Of course, political containment does not necessarily mean obstruction, or abolition, that is to say, purely and simply submitting technological activity to the "unleashing of the political" which aims to frame and govern everything – in the way that the Nazis tirelessly sought to destroy the Bauhaus experiment. On the contrary, it is a matter, as Huyghe specifies, of bringing into existence a field of practice which is not under the influence of this illimitation of politics.

In some ways, the *Encyclopedia* of Diderot and Alembert perfectly illustrates this effort to build technological know-how into a field of practice organized on new bases. The Encyclopedists indeed intended to release technology from the secrecy of the trade guilds (let them know that to conceal a useful secret is to be guilty of larceny towards society, writes Diderot in the article *Art*) and to promote public technological knowledge, a kind of *open technology* in the sense in which today we speak of *open science*[2].

2 Bertrand Gille also points out that the *Encyclopedia* did not happen in isolation. There was at the time an already long-standing interest in the mechanical arts, and many treatises had been published since the 16th century. As a major contributor to the *Encyclopedia*, Diderot exploited a large quantity of existing writings, of accumulated descriptions. In addition, in the

In the prospectus of his methodical encyclopedia, Panckoucke, in 1782, was grateful to Diderot for having wanted to bring about the knowledge and the combination of the Mechanical Arts, which were so difficult to extract from the workshops and to retrieve from the mysterious practices of workers of all kinds[3]. To do this, explains Diderot in the preliminary discourse of the *Encyclopedia*, he addressed himself to the most skillful in Paris and the Kingdom: in other words, far from seeking to subject technological practices to a rationality foreign to them, in order to control them, Diderot's approach was on the contrary to learn from the practitioners. The main difficulty he encountered was the inability of many of them to describe their activity, due to the complexity and sharpness of the tasks.

That being said, the program of the *Encyclopedia* is not, however, completely foreign, far from it, to the aim of disciplining technological know-how, as emphasized in particular by Foucault in his course at the Collège de France in 1976. "The Encyclopédie is usually seen only in terms of its political or ideological opposition to the monarchy and at least one form of Catholicism. Its interest in technology is not in fact a reflection of some philosophical materialism; it is actually an attempt to homogenize technological knowledges, and it is at once political and economic" (Foucault 2003, pp. 180–181). According to Foucault, four mechanisms presided over this disciplining of technological knowledge and know-how: a selection separating knowledge which is legitimate from that which is not and is disqualified; standardization of this dispersed knowledge; a hierarchical classification of this knowledge; and finally, a pyramidal centralization which allows control.

2.1.2.2. *The article Art of the Encyclopedia: entering technological know-how into a new rationality*

The project of liberating the arts, by freeing them from the tutelage of trade guilds, thus appears, in reality, as a subjugation of artisans to a new type of power, whose outlines are indicated by the article *Art*.

Diderot begins by saying in this article that all art is inseparable from a system of instruments, or of rules (whereby it remains ultimately very close to the Aristotelian definition of *technê poiêtikê*, as a disposition to produce accompanied by rules). Art is both a practice and a theory: without practice theory is blind, without theory practice is limited. Thus, Diderot explicitly recognizes the impotence of theories which are not based on any experience: "In every art there are many particulars concerning its material, its instruments, and its application which can only be learned through practice". "How many bad machines are suggested every day by

18th century, there was a whole stream of technological progress among the practitioners themselves. See (Gille 1952).

3 Cited by (Gille 1952).

men who imagine that levers, wheels, pulleys, and cables perform in a machine as they do on paper! Because they have taken part in practical work, they have never learned the difference between the effects of the machine itself and of its section." On the other hand, an art which does not signal towards any theory does not progress; it gropes and stagnates. If it is not possible "to have a thorough knowledge of the speculative aspects of an art without being versed in its practice", reciprocally: "It is difficult if not impossible to go far in the practice of an art without speculation."

Admittedly, this theory of the arts cannot be reduced to any mathematical formalization: "Everyone will readily agree that there are few artists who can dispense with the elements of mathematics. Yet here we have a paradox, although its truth is not immediately obvious: in many situations knowledge of these elements would actually hamper an artist if, in practice, the precepts of mathematics were not corrected by an extensive knowledge of physical circumstances; such as location, position, irregular figures, materials and their qualities, elasticity, rigidity, friction, consistency, duration, as well as the effects of air, water, cold, heat, dryness, and so forth."

That being said, as irreducible to mathematization as it is, the theorization of practice in the form of physical knowledge and explicit rules remain essential for progress. In the article *Art*, Diderot militates in favor of a "grammar" of the arts, that is to say of a rigorous formalization of artisanal practices. As Sewell points out, commenting specifically on the *Art* article, "However untutored the artisans may be, and however habitual and unsystematic the rules by which they work, if these rules are efficacious they must necessarily conform to general natural laws discoverable by science. The philosopher's task is to render the rules of the arts comprehensible to all by converting the craftsmen's conventional gestures, half-verbalized rules of thumb, and traditional lore into scientifically demonstrable natural laws – and in the process discard those practices that have no scientific foundation" (Sewell 1980, p. 69). This task first requires the establishment of adequate classifications by choosing clear and unambiguous terms to designate things and gestures. It would therefore be desirable, writes Diderot, "if a good logician, well versed in the arts, undertook to describe the elements of a 'grammar of the arts'".

Descartes still recognized the irreducibility of the arts to clear and distinct ideas, an excess of the former over the latter, and Kant still, in 1790, also recognized a constitutive excess of art over theory. But now Diderot – and through him the whole project of the *Encyclopedia* – intends to convert the know-how of artisans into a formalism which, of course, proves to be irreducible to mathematical mechanics in the strict sense, but which nonetheless sets out an organization of knowledge and technical practices in such a way that these, from now on, must be entrusted to the care of a new category of experts (a certain type of logicians, to quote Diderot). It is not up to artisans to create the theory of their art, but to the people of the Academy: "We would wish that from the halls of the academies there would emerge a man who would go into the workshops, record everything noteworthy about the arts, and set it forth in a work that would induce the artists to read."

In short, to bring technology out of the secrecy of the guilds, it was necessary to start by bringing them out of the secrecy of the bodies of the craftsmen. The political dimension of the *Encyclopedia* lies in a program of formalization of technology leading to the expropriation of the artisans' know-how, and more precisely to "ex-corporate" these, in the words of French historian Jean-Baptiste Fressoz. Gestures are not supposed to have any opacity; only an appropriate vocabulary is required to lay them out in reproducible sequences: "It is the lack of precise definitions and the great number, not the diversity, of movements in various operations that makes it difficult to speak clearly about the arts." The artisanal gesture is thus deployed outside, formalizable, and in the end subject to a new form of heteronomy: "A particular workman will spend his lifetime performing one single operation; hence each operation is carried out quickly and well, and moreover the best-made product is also the cheapest." We may therefore understand, as French historian François Jarrige explains, that workers at the start of the industrial era opposed technological trajectories that threatened to accentuate discipline and erode the control they held over their know-how and the organization of work (Jarrige 2014, p. 38).

2.1.2.3. *Towards a new concept of work*

It is, moreover, possible to spot the precursors of this disciplining of the arts & crafts from the beginning of the 18th century, with the first formulations of the concept of "mechanical work" by two members of the Royal Academy of Sciences, Guillaume Amontons and Antoine Parent, as well as by Daniel Bernouilli a little later. This concept is usually attached to the names of Charles-Louis Navier, Gaspard Coriolis, and Jean-Victor Poncelet, in the 19th century. Nevertheless, we must start from the 18th century to look for its premises, according to the historian Yannick Fonteneau (Fonteneau 2009).

The ambition of these precursors was at the time to be able to compare different types of production, and to be able to compare the efficiency of human and machine work respectively in these productive activities (in terms of cost or of maximization of effect), by identifying a common standard for measurement. For the purely qualitative notion of work, Amontons and Parent therefore substitute a measurable notion, "mechanical work", providing criteria for evaluating human work compared to that of machines in various productive activities. The notion of work ceases to have the purely qualitative meaning it once had, by virtue of its attachment to the activity of an individual living body; it now acquires quantitative significance. Work becomes measurable.

It is certainly not a question of considering Amontons, Parent or Bernouilli as having paved the way for the disciplining of artisanal gestures. However, the idea of "decorporealizing", so to speak, the concept of work with the aim of optimizing it, announces a general orientation that is being implemented.

2.1.2.4. *The specifics of Encyclopedic Technology*

However, if a disciplinary aim at least partly guides the project of the *Encyclopedia*, there can be no question of reading and judging it in the light of Industrial Technology and finding there, already, the outline of a concept of production based on the notions of the industrial machine, the capitalist entrepreneur and a market open to free competition. As French historian Anne-Françoise Garçon points out, nobody in the 18th century could imagine that the machine would become the prescriber of the gestures of the industrial workshop, and with it, the industrial engineer, who brought the regime of technology to its peak in the scientific management and organization of work (Garçon 2014, p. 87). Diderot cannot of course be accused of having voluntarily passed over in silence the future misdeeds of industrial machinery, or even of having planned the scientific management and organization of work! To argue that would be meaningless, and we can easily subscribe to Georges Friedmann's warning: please, do not ask the people of 1750 for the sensitivity and experience that we possess today (Friedmann 1953, p. 58).

The Encyclopedists, Friedmann points out, still had an artisanal conception of the machine; it was for them only a more sophisticated tool, the extension of the tool held by the hand of the craftsman. Analyzing the practices of artisans, extracting their schemes, their principles of action, creating their theory and making them adaptable to all trades: this is the approach that underpins the project of the *Encyclopedia* – and not the administration of trades and their knowledge for the purpose of improving economic productivity. It would be totally wrong to unreservedly associate the project of the *Encyclopedia* with the industrialist conception of Technology defended by Ure. The technological thought of the Enlightenment did not arise only from the enlightened vision of modern engineers, laying the foundations of industrial science and the rationalization of work. It is part of a culture of artifice (Hilaire-Perez 2013, p. 34) as well as of attention to the particular. Far from being the intermediary of a purely analytical vision of productive work, breaking down gestures into elementary sequences with a view to optimization, the technology of the Enlightenment (which remains very much alive in the following century) is part of a "synthetic and artificial thought", as Hilaire-Perez writes. Manufacturing gestures are not entirely subject to the norms of scholarly rationality: they are transcribed in a language of action resulting from reflection on practices.

The article Bas[4] (Bonneterie, & autres marchands, comme peaussiers, &c.)[5] in the *Encyclopedia*, also written by Diderot, illustrates very well this specific form of technological rationality, incompatible with future industrial science in the manner of Ure. The activity of manufacturing stockings on the loom is divided into seven main operations, whose goal is the formation of meshes. Each operation is broken down into separate movements, and each movement into combined actions. The description is therefore very analytical. The second operation is particularly instructive. Here is how Diderot describes the first movement of this operation:

> It is done with the foot ... & with both hands. The worker takes the handle with both hands, so that his thumbs are applied against the parts called thumb-pieces. See Pl. VII. fig. 1. His hands are at AA, & his thumbs at BB. He then does three actions at one time; he presses with his foot the moving piece 15, fig. 1. Plan. II. whose bridge he has picked or moved from right to left ; he pulls the handlebar down perpendicularly AA, fig. 1. Pl. VII. & he presses his thumbs strongly against the thumb-pieces BB, fig. 1. Pl. VII.

This precise description seems to bring the gestural repertoire of workers into a logic of optimization and control, consisting in clearly specifying how the hands and feet should act. However, this is not the case: here, Diderot does not say what are the right gestures to execute, he is content to describe as precisely as possible, with an illustration as support, what workers at the loom actually do. He is an observer (or possibly a reader of memoirs written by others, as Bertrand Gille points out), and not a prescriber.

In addition, a little further on, he summarizes the effects of the first movement executed for the accomplishment of the second operation: the three combined actions of this movement therefore tend to produce two opposite effects; one to lower the lead sinkers, the other to raise the wave plates. What then is the second movement of this second operation? To gently manage these two opposite effects, to combine them finely. The vocabulary used is no longer that of wholly rational mechanics (*metron*), but that of the dexterity of the experienced worker who can act with care and finesse (*metrion*).

It is in the following century that a change in the trajectory of Technology seems to have brought new perspectives to control productive activities. As French

---

4 Stockings.

5 [Online]. Available at: https://fr.wikisource.org/wiki/L%E2%80%99Encyclop%C3%A9die/1re_%C3%A9dition/BAS [accessed on November 14, 2019].

historian Michelle Perrot, quoted by historian Jacques Gleyse, points out, to require this reticent, unruly workforce, accustomed to controlling their process of work, to bend to regular schedules and rhythms, respect for order and hierarchy, economy of gestures and words, the fixity of the body, to enact a real industrial training through discipline: this is the major function of the factory (Gleyse 2000, pp. 113–114). It is indeed a question of submitting, of making docile, of controlling, but more radically than the simple implementation of a surveillance system would do: rather, by the reduction of any possibility of appropriation and control, by the worker, of their own work process. From the 19th century, this implacable law of the capitalist economy of which Huyghe speaks is said to have consequently imposed a new mode of supervision of technical know-how, such that workers found themselves deprived of any capacity to control their own activity and to define their own norms.

From Ure's Industrial Technology to Amazon warehouses, it is always the same problem: how to discipline technical know-how in order to first stifle its destabilizing power, and then put it fully at the service of the productive system? Everything becomes a matter of surveillance and control. The history of industrialization reveals the extent to which workers have in reality been resistant and reluctant to allow themselves to be swallowed up as mere performers of repetitive tasks. Amazon today is taking the Taylorist ideal of control over workers to a never-before-equaled level. The electronic bracelet would indeed be the latest in a long line of attacks on the "quality of work" and on this worrying professional autonomy of workers.

2.1.3. *A rupture in the conception of technology: the divorce of technology and life*

2.1.3.1. *When work ceases to be the activity of the living being, and technology to be a vital activity*

However, this is not the essential point, we would suggest. Indeed, this focus on "quality of work" assumes that work remains a major issue for the productive system. Workers seek to defend their professional autonomy; capitalists on the contrary seek to make them more and more heteronomous in order to better control them. Everything would thus play out around work as the main source of value, which is, moreover, perfectly in line with the labor theory of value, which poses equivalence between the latter and work: the exchange value of a product is equal to the average (i.e. social) working time it took to produce it, says Marx.

However, Marx very spectacularly saw in the *Grundrisse* the inadequacy of this model, thus seriously undermining the labor theory of value. Here is what Marx wrote in this text which dates from the end of the 1850s: "just as production founded on capital creates universal industriousness on one side – i.e. surplus labor, value-creating labor – so does it create on the other side a system of general exploitation of the natural and human qualities, a system of general utility, utilizing science itself just as much as all the physical and mental qualities" (Marx 1973, p. 409).

Marx therefore recognizes that capitalist exploitation is not limited at all to the appropriation of the surplus labor provided by the industrial worker: it relates much more generally to the whole of natural and human resources, which includes (if we understand correctly) the intellectual, emotional and even spiritual skills of the human resource. The source of value is ultimately much less in the surplus labor provided by a deskilled worker than in all this mass of individual and collective skills, as well as in the natural resources which cannot be limited to raw materials and energy (let's think today of the tremendous boom in the all-round exploitation of the metabolic activity of living beings, which is increasingly massively captured by the productive system, notably through genetic tinkering). Science has become a dominant productive force, integrated into capital in the form of technological applications.

Let's be clear: the Amazon electronic bracelet certainly says a lot about the dubious practices of this company in terms of labor law and respect for workers, but it is an absolutely insignificant gadget in comparison to the extraordinary capacity that these digital platforms have to capture on a very large scale, for their benefit, the collective intelligence generated by social interactions, as well as the activity of living creatures as a whole. Marx summed it up in the *Grundrisse* in a disarmingly simple sentence: capital has given itself the means of "the universal appropriation of nature as well as of the social bond" (Marx 1973, p. 409).

Something much more considerable has therefore happened than putting the holders of technical know-how under control, by exhaustion of their professional autonomy and the forced deskilling of work. What has happened is more fundamentally a general escape from control of technology, which has gradually become independent of the productive activity of human individuals. Marx explains that "in the production process of capital… labor is a totality – a combination of labors – whose individual component parts are alien to one another, so that the overall process as a totality is not the work of the individual worker, and is furthermore the work of the different workers together only to the extent that they are [forcibly] combined, and do not [voluntarily] enter into combination with one another". "The combination of this labor appears just as subservient to and led by an alien will and an alien intelligence – having its animating unity elsewhere – as its material unity appears subordinate to the objective unity of the machinery, of fixed capital, which, as animated monster, objectifies the scientific idea" (Marx 1973, p. 470). This foreign intelligence and will are diffuse, dispersed and sedimented in the functioning of machines. These are the result of a capture of science by capital. Science is indeed reified in the form of technological applications with a productive aim. It has become an anonymous process, foreign to the producers, but literally captured by the machine system: "The science which compels the inanimate limbs of the machinery, by their construction, to act purposefully, as an automaton, does not exist in the worker's consciousness, but rather acts upon him through the machine as an alien power, as the power of the machine itself" (Marx 1973, p. 693).

As a result, "the production process has ceased to be a labor process in the sense of a process dominated by labor as its governing unity. Labor appears, rather, merely as a conscious organ, scattered among the individual living workers at numerous points of the mechanical system; subsumed under the total process of the machinery itself, as itself only a link of the system, whose unity exists not in the living workers, but rather in the living (active) machinery, which confronts his individual, insignificant doings as a mighty organism" (Marx 1973, p. 693).

Defending the "quality of work" of the Amazon handlers, if this really has a meaning, is certainly an excellent demand; however, this obscures the fundamental fact that technology has long since ceased to have the main meaning, precisely, of situated practical knowledge and qualitative know-how that should be preserved at all costs against the aim of automation of work. For Amazon to give up making its bracelet, and even for the company to grant greater professional autonomy to its employees, will greatly improve their working conditions and that is of course already significant. But that will in absolutely no way change the fact that technology has manifestly become so foreign to any form of concrete living and human activity, that Amazon and others can today convert all vital and psychosocial processes into an absolutely phenomenal market capitalization. Marx's formula was decidedly prophetic: the epoch is effectively that of "the universal appropriation of nature as well as of the social bond".

While productive activity had always been an activity of the living being, Marx in the end saw very clearly that industrial capitalism had the essence of enacting a radical detachment between the living being and productive activity. The origin of capitalist accumulation lies precisely in the contradiction "between labor force as capacity, as labor power, and living labor as subjectivity submitted to the logic and discipline of capital[6]". As Marx says in an unpublished chapter of *Capital*: "As an effort and an expenditure of vital force, work is the personal activity of the worker, but, as it creates value, when it is engaged in the process of its objectification, the work of the worker, as soon as it enters the production process, is itself a mode of existence of capital value, an integral part of it[7]." It is no exaggeration to say that

6 Marazzi C., preface to (Fumagalli 2015).

7 [Online]. Available at: https://www.marxists.org/francais/marx/works/1867/Capital-inedit/kmcapI-6-2C.htm [accessed November 10, 2019].

this contradiction in the relationships between life and work is the very principle of the accumulation of capital.

However, Marx takes remarkable note of the fact that this duality of meanings of life and work is itself dependent on a more fundamental split which has arisen in the relationship between technology and life. We thus read in the same unpublished chapter of *Capital*: "It is not living labor which is realized in material labor as in its objective organ, but material labor which is preserved and grows, by absorbing living labor, so that it becomes value creating value, capital in motion." In other words, technology is no longer an extension of the life of the living human being and of its own movement: having become industrial machinery, it is what captures the living force of the worker to convert it into something dead.

To return to Amazon's electronic bracelet, the decisive point is in the fact that the activity of work has been so methodically redacted of any dexterity and of any particular application of the hand that a complete control of gesture ends up appearing both possible and even logical. In order for the gesture to be subject to such close metrological monitoring, it first had to be detached from the faculty of *metrion* which, from the Greeks to the industrial era, never ceased to be an essential characteristic of the technological gesture. Deprived of this dimension of *metrion*, that is the dependence of technology on a faculty which is not unrelated to sensory experience and life, the technological gesture can be effectively handed over to *metron*.

When exactly does this radical transformation of the relationship between technology and life take place? How did we go from a situation, not so long ago, where technology was still conceived as deeply rooted in the needs and activity of the living human being (still in Descartes), to a situation where technology seems to have broken its bond with the living being, to be carried away in a dynamic of generalized appropriation of the processes of life, whose manifestations are multiplying and today taking on dizzying proportions?

It seems perfectly wrong to locate in the 19th century only this great transformation in the relations between technology and life, in the sense of the life of the living being and its sensory faculties, and to attribute sole responsibility for it to big capitalist industry. If technology, including the machine, still preserves in the 18th century its classical sense of an artificial organ extending the gesture of manufacture, the fact remains that already in the middle of the Enlightenment, a decisive shift began in the relationships between the living body, technology, productive activity and the type of rationality in which this activity can be captured. The *Encyclopedia* doubtless does not lead in a straight line to the capitalist entrepreneur and the market of free competition: nonetheless, the fact remains that the era was already moving towards an unprecedented project of the rationalization of productive practices.

2.1.3.2. *New representations of the body at work: towards the rationalization of productive practices in the 18th century*

Representations and practices around the body were the first to change in depth (Gleyse 2000). Up until the 17th century, the body was still conceived as a piece of nature which must be shaped and straightened (Vigarello 1978) in order to transform it according to accepted standards of elegance and civility. For this it was necessary to apply artificial forces to it, to subject it to corrective practices such as compression (by corsets and swaddling).

In the 18th century, a whole new interventionist logic took the body for its object: it is less a question of instituting a worldly body whose norms are those that civility and politeness required (Vigarello 1978, p. 99) than to shape an organization according to criteria of functionality and rationality. The body is no longer the sign where affiliations are marked with care, but the machine where the proper functioning of organic devices can be decoded (Vigarello 1978, p. 93). According to Gleyse, the first outlines of this new representation of the body begin to take shape from the 16th century. An ordering of body movements takes place in gymnastics. Jérôme de Monteux de Mérybel, and after him Hyeronimus Mercurialis, set a new frame of reference, a reading grid for movement which introduces a first regularization of gestures (Gleyse 2000, p. 89).

Gleyse combines this aim of rationalizing the body and its movements with a growing concern for physical health, but also with the development of manufacturing. Indeed, exercise strengthens the hardness and strength of the members, they are more tolerant of effort and work, and ... do their job better (Gleyse 2000, p. 90). The new bodily prescriptions seem to be born from the imperatives of the structure of labor. Referring to the Weberian interpretation of the spirit of capitalism, Gleyse thus suggests that it is perhaps, more than Protestant ethics, the desire to put corporality in order for the purpose of better efficiency in work that becomes embedded in all areas of discourse and in all of social space (Gleyse 2000, p. 145).

Of course, factories still only involved relatively few individuals in the 16th and 17th centuries. Nevertheless, their development heralds what Gleyse calls a progressive "manufacturalization" of the world and of humanity. The body is more and more led to submit itself as an object of instrumentalization and not as a living subject (Gleyse 2000, p. 173).

In this new economic context, the idea is gradually being imposed that the body must be disciplined in its movements in order to be made more productive. That which, Gleyse argues, totally differentiates the body of the craftsman of the Middle Ages from the manufacturing worker of the 17th century is the obligation to which it will be put to produce standardized products, with precision, rigor, and method (Gleyse 2000, p. 114). Economic concerns explain the growing desire to rationalize work, as early as the 17th century. Thus, in order to pay the diggers a fair price for their labor, the military engineer Vauban is led to identify and evaluate the tasks of workers on construction sites. Wishing to set prices, he finds himself on the construction sites of fortifications to measure times, volumes and distances, to assess forces, to establish averages (Montel 1992, p. 136).

These observations on construction sites continued throughout the 18th century, extending beyond the framework of military engineering to also involve civil engineering. Time-measuring instruments are more and more used by engineers, such as for example Bélidor, who in 1750 timed the driving of piles (three hours, five minutes, and thirty-four seconds for a "pile with sixteen feet of plug"). In 1738, French architect and engineer Jean-Rodolphe Perronet, first director of the future École Royale des Ponts et Chaussées founded in 1747, visited the manufacture of L'Aigle, in Normandy, and gave an extremely meticulous description of all the elementary operations required for the manufacture of a pin. In his activity as a bridge and pavement engineer, Perronet demonstrates a very modern vision of work organization, based in particular on the application of science to construction, to the extent that some, like the Taylorist Charles de Fréminville at the beginning of the 20th century, could see in him a precursor of Taylor. Perronet, enthuses de Fréminville, really performed the scientific organization of work based on a meticulous study of elementary operations and of human work (de Fréminville 1927). Perronet describes the workshop as he would a machine, with its assembly of organs and elementary devices; he breaks it down, isolating each operation, each gesture (Girard 2015, p. 8).

Control over gesture is also motivated by the desire to develop standardization. Colbert already imposed rules on factories in the 17th century; this was to force them to standardize their production (in particular, of fabrics). The process intensified at the beginning of the 18th century and as historian Ken Alder (Alder 1998) points out, it was not the capitalist industrialists of the 19th century but indeed the military engineers of the *Ancien Régime* who first promoted the need to rationalize artisanal gestures through technical descriptions (plans, technical drawings, gauges, models). It was particularly imperative at the time to impose upon gunsmiths precise, objective manufacturing standards (so that the army could receive uniform equipment).

Alder specifies that this normalization was not accomplished smoothly, that conflicts broke out and that difficult negotiations were necessary. Quoting Alder, Hilaire-Perez also recalls that gunsmiths opposed the idea of calibration of pieces and mechanization, highlighting their skills of adjustment, finishing, and adaptation of objects to uses.

In an 18th century progressively dominated by concern for rationalization of manufacturing and construction tasks, it was undoubtedly not yet a question of treating the body according to a metric which aims to break down movement to its simplest dimensions, to split motor skills into movements made almost abstract, circumscribed and fragmented by analysis (Vigarello 1978, p. 138). The 18th century had not yet plunged into the atomization of gestures and the desire to mechanize them to the maximum, and undoubtedly gestures were not yet being very directly monitored (Vigarello 1978, p. 141). The rationalized geometrization of their path is not yet underway.

While stressing that among some authors of the *Encyclopedia*, an approach to work based on more demanding quantification than in the past is sketched out, and that in this perspective, human effort tends to be assimilated at best to a resource to be spent (Picon 1992, p. 132), French historian Antoine Picon thus specifies that this vision is relatively marginal and that as a general rule, Diderot's collaborators are still far from this conception which will not really impose itself until the 19th century. According to Vigarello, the *Encyclopedia* of Diderot and Alembert still celebrated industrious and skillful hands, captured on engraved plates (Vigarello 1978, p. 140), and still paid attention to very particular and often individual methods.

However, to limit ourselves to the article in the *Encyclopedia* devoted to pins (which takes up the essay published by Perronet in 1740), it is clear that the description and the illustrations accompanying it are not at all driven by a desire to magnify the incorporated skills and know-how of craftsmen, nor of workshop solidarity. Useless to search the engravings of the *Encyclopedia* for testimonies as to the décor of the workshops, the dressing of the workers, the rituals of the guilds and apprenticeship, the secrets of manufacturing; there is only an abstract representation of operations and tools (Girard 2015, p. 8). Vigarello may be displaying excessive generosity in attributing to the Encyclopedists the desire to celebrate the prestige of the skillful hand. Although the authors of the *Encyclopedia* are not clearly aware of this, explains Picon, what emerges through the criticism of the trades, their cumbersomeness, but also, it must be said, the barriers which stood in the way of the desire of the State and big entrepreneurs for rationalization, is the possibility of pure work, which can be shaped at will, of work quantifiable as an expenditure of energy and simultaneously as a basis for economic value (Picon 1992, p. 134).

According to historian Jean Ehrard (1984), the *Encyclopedia* is not really interested in workers; they remain ultimately anonymous, abstract, very stylized servants of the machine, sometimes reduced to the essentials. Sewell also (Sewell 1980) states that Diderot "portrays the very artisans that are to be admired as human automatons, incessantly repeating a prescribed set of mechanical operations" (Sewell 1980, pp. 68–69; see also Schaffer 1999).

By examining the plates of the *Encyclopedia* which represent workshops, it is not ultimately impossible to think, already, of Lavoisier, of Coulomb, of all these scholars of the end of the 18th century who assimilate man to a machine and bodily activity to measurable phenomena (Girard 2015, p. 8), even if, specifies Picon (1992, p. 136), this conception is still little represented among the authors of the *Encyclopedia*, where engineers remain quite rare. At the end of the 18th century, Charles Augustin Coulomb indeed reports in his Theory of Simple Machines (Part 2, chapter III) on the results of several experiments intended to determine the quantity of action which men may provide by their daily work, according to the different ways in which they use their forces (Coulomb 1821, p. 255). The human body here is clearly conceived as a machine, and even as a particularly flexible and interesting machine to use. The human body, writes Coulomb, made up of different flexible parts, moved by an intelligent principle, bends to an infinity of forms and positions: considered from this point of view, it is almost always the most convenient machine that can be used in compound movements, which require nuances and continuous variations in the degrees of pressure, speed and direction (Coulomb 1821, p. 255). Human effort is detached from a simple turn of the hand and its efficiency is subject to precise calculation, more precisely to a ratio between the effect produced and fatigue, which is to be optimized.

Coulomb's text actually teaches us something essential: on the one hand, the body at work has not yet been re-assessed as a pure machine emptied of all reference to intelligence, quite the contrary. Coulomb thus explains the preference that should be given to the use of human force compared to that of animals, in particular by the fact that men's intelligence makes them save their strength, moderate their work, in response to the resistance they must overcome. In other words, it is because men can use their intelligence that their strength can be used optimally.

On the other hand, however, Coulomb says very explicitly that this intelligence is not an intelligence immanent in the body, an intelligence enveloped in the incorporated know-how of the artisan. It is an intelligence separate from the body, a "principle" external to the machine and which governs it. It is no longer a question of a technological intelligence comprising skill and dexterity in the gestures of manufacturing, but of a simple capacity to manage itself, to save effort, to adapt physical expenditure to the circumstances (depending on the nature of the resistance to be overcome). This body at work is more that of laborers ("*gens de bras*"), those who lift heavy loads, than that of artisans. As Vigarello certainly underlines, such a study [that of Coulomb] was attached to a very global context of portage ... and did not really question an atomization and possible mechanization of its gestures (Vigarello 1978, p. 141). That said, the desire to extirpate any form of intelligence from working bodies, including those involved in manufacturing tasks (and not only those of laborers), is not an ideal that remained completely foreign to the 18th century. If this ideal is clearly stated in later texts, in the 19th century, it is already beginning to be formulated in the middle of the Enlightenment.

An analysis by Scottish philosopher Adam Ferguson gives an unambiguous signal, already in 1767, of a conception of productive activity in which intelligence has no part. Ferguson indeed points out, not without concern: "Many mechanical arts, indeed, require no capacity; they succeed best under a total suppression of sentiment and reason; and ignorance is the mother of industry as well as of superstition. Reflection and fancy are subject to err; but a habit of moving the hand, or the foot, is independent of either. Manufactures, accordingly, prosper most, where the mind is least consulted, and where the workshop may, without any great effort of imagination, be considered as an engine, the parts of which are men" (Ferguson 1767, p. 305).

The article Bas (Bonneterie, & autres marchands, comme peaussiers, &c.), already quoted, ultimately goes in the same direction. The craft of making stockings is indeed an extremely sophisticated machine, to the operation of which it seems that the worker contributes only very marginally – or rather, one should say, to the operation of which he does not *have* to contribute: the workforce, Diderot writes, is very little; the machine does almost everything by itself: its mechanism is more and more perfect and delicate. The whole sequence of operations of the loom is indeed a single and unique reasoning, of which the product is the conclusion; so there is between its parts such a great dependence, that to take away only one, or to alter the shape of those that are considered the least important, is to harm the whole mechanism. In other words, the intelligence required to manufacture the final work is no longer in the form of an aptitude to execute the task with accuracy, being attentive to present circumstances; it comes from a calculation and a regulated path of elementary sequences. *Metron* and no longer *metrion*.

Diderot quotes Perrault, who is plunged into admiration by the stocking machine: when we see the knitting of stockings, we admire the flexibility and dexterity of the worker's hands, even though he only makes one stitch at a time; what then is it when we see a machine which forms hundreds of meshes at a time, that is to say, which makes in a moment all the various movements that the hands do only in several hours? How many small springs pull the silk towards them, then let it go to take it back, and make it pass from one stitch to the other in an inexplicable way? And all this with the worker who moves the machine understanding nothing about it, knowing nothing about it, not even thinking about it. The worker is no longer required, neither for his manual dexterity since the mechanism replaces it nor for being endowed with intelligence since it has no influence on the operation of the machine. The worker does nothing but move the machine: his involvement in productive activity is reduced to almost nothing.

Ultimately, as noted by Alder (1998), though the *Encyclopedia* undoubtedly intends through its work to free the mechanical arts from the condescension of the literate and the nobles, it does so indeed, despite everything, according to an ideal of streamlined work processes under the direction of enlightened managers. The 18th century, continues Jarrige, referring to David Landes, saw both the beginnings of industrialization … and a new way of conceiving of technological achievements. The old tacit knowledges, manual, olfactory, visual, transmitted by gesture and speech, have given way to more and more formalized knowledges inscribed in new institutional logics, borne by new actors – the engineer, the entrepreneur – and subject to new ends (Jarrige 2014, p. 38).

It should be emphasized once again that this expropriation or "ex-corporation" did not happen overnight and smoothly. Resistance was common in the 18th century. Jarrige evokes Vaucanson who planned to automate the looms and came up against the hostility of weavers and practitioners who denounced the abstract character of science, unable to compete with patient learning and daily practice of a trade. In the 19th century, the figure of the deskilled proletarian is indeed a reality, but does not sum up all productive activity, far from it. Production, as we have seen, remains largely artisanal in nature, while the workforce employed in industry continues to firmly defend its "gestural knowledge" and its "sensory cultures". As Jarrige points out, in the iron industry, workers relied on their senses and their incorporated experience to control production. Where the engineer sought to impose fixed measures determined by scientific instruments, workers often resorted to their senses (Jarrige 2014, p. 55). In the 20th century, testimonies still emphasize this "gestural knowledge" and these "sensory cultures" which are far from having disappeared from productive activities. Primo Levi's beautiful story, *The Monkey's Wrench*, is testament to this (Levi 1995).

The fact remains that a profound transformation began in the Age of Enlightenment, whose repercussions were considerable: a new heteronomy was introduced into the field of human practices. Anne-Françoise Garçon has analyzed this heteronomy (Garçon 2014), by identifying its premises already in the 16th century in the multiplication of technical prints aimed at giving a "*lecture raisonnée*" of the arts & crafts, by their visibility in writings describing technical gestures and their sequences (what is called "reduction en art", *redigere in artem*). Technical gestures were hitherto part of an operating regime that Garçon identifies as that of "practice". They were learned orally in a "gestural milieu", to use the expression of French ethnologist Blandine Brill, defined by local standards of technical activity.

"Reduction en art" inaugurates a new mode of existence of technical gestures by moving from the operating regime of "practice" to that of "technique". The gestural practices of artisans, Garçon writes, are therefore embedded in an intellectual environment that escapes the control of the practitioner and gives these practices an efficiency of a new kind – thought of, grasped, conceived in a strict relation to technology beyond all religious, cosmological and even socio-professional considerations.

At the end of the 17th century and especially in the 18th century, this new operating regime for artisanal gestural practices is articulated with an order of preoccupations of ever-growing importance, relating to improving technical productivity through the rational organization of practices. What the 18th century invented, continues Garçon, is the inscription of technical rationality into an economic science still in the process of constitution, starting from this technical culture that engineers and scientists had developed over a century… The regime of technology, the trades of engineer and scientist, and economics as a science of production, have deep interrelations. As Sewell points out, undoubtedly somewhat excessively and abruptly, the capitalist vision of a progressive application of scientific technology to private enterprise is a logical development of Diderot's thought (Sewell 1980) – in the same way, moreover, he adds, as socialism.

2.1.3.3. *Is the human machine an automatable worker?*

How then do these developments in the field of technical practices translate philosophically? Can we in particular maintain that this new conception of the body, oriented towards the rational control of productive practices and the growing concern for yield, finds its philosophical justification in the Cartesian theory of the animal-machine or in the materialist theory of La Mettrie – the famous theorist of the Machine Man (La Mettrie 1996)? The answer is no. No more than the Cartesian thesis of the animal-machine, the conception of the Machine Man cannot be invoked to provide a philosophical foundation for the idea of applying to the productive body an "ethics of performance", according to Simondon's expression (Simondon 2020).

By rejecting the Stoic representation of the material universe, according to which the place of each thing is fixed for eternity in a perfectly finalized hierarchical order, Descartes no doubt wanted to justify faith in the creative efficiency of technology[8] and defended the value of indefinite technical progress. That said, as Canguilhem

8 "In the doctrine of Descartes, as in that of the atomists, a matter without real qualities, a universe without teleological hierarchy are the metaphysical reasons for faith in the creative efficiency of technology" (Canguilhem 2011).

explains remarkably, Descartes's position on the subject of technology is more complex and nuanced than what philosophical tradition has retained.

Any good technology is undoubtedly, for Descartes, a technology informed by science. Only science can liberate technological practices from what Canguilhem considers rough empiricism and blind routines, with the aim of opening them up to continuous progress. However, Descartes knows very well that the passage from scientific knowledge to technological practice poses difficulties that theoretical intelligence alone is unable to solve. Technology does not allow itself to be completely grasped by science and this gap is not a pure negation, a deficit that might be completely filled: the excess of the technological over the theoretical has a positivity; it signals towards the existence of a practical intelligence capable of solving problems by itself and which has its own value.

There is more: by recognizing the existence of an approach to solving practical problems in which science is not sovereign, Descartes reintroduces into his conception of technological progress limiting principles which are internal and even constitutive of technological activity itself. Science can enlighten us, point out directions for us; it is very precious, but the indefinite nature of its progress does not completely apply to technology. This latter has its own limitations of which Canguilhem, commenting on Descartes, gives examples. Thus, the formulation of a complete theory of the steelyard balance would not allow us to construct one that was perfect; there will always be imperfections, a gap between realization and theory due to internal limitations of practice, so much so that Descartes, notes Canguilhem, suggests to proceed by empirical trial and error. The construction of flying machines is theoretically possible, since birds exist and they are similar machines; however, there would be needed springs so subtle ... that they cannot be made by men.

In theory, technology should therefore be framed in the perspective of indefinite progress that science opens up to it, but in fact it is not so, and technological activity knows intrinsic limitations to its exercise. These limitations may come from matter, which may ultimately prove to be non-compliant and even recalcitrant; from the manufacturer, whose bodily possibilities are not commensurate with the manufacturing subtlety required by theory; and also from the ends. For Descartes, technology is above all at the service of life; it is life itself which defines what the needs and the directions to take may be, and not science.

Descartes has certainly turned his back on the Stoic world view; nevertheless, it is not certain that he has completely freed technology from the very tight network of limitations, intrinsic to technological manufacturing activity, within which the ancient Greeks had framed it. The idea that Descartes cut technology off deeply from life cannot be defended in any case. Descartes, notes Foucault, undoubtedly started to write the great book of the Machine Man; its register, anatomo-metaphysical, was however that of explaining the functioning of the body. It was for the next century to write its technical and political pages, concerned not with explanation but with control and correction of the body's operations, with a view to "exercising upon it a subtle coercition, [to] obtaining holds upon it at the level of the mechanism itself – movements, gestures" (Foucault 1995, p. 137). No longer the analyzable body, but the manipulable body.

Neither does La Mettrie's concept of organization, which underlies his thesis of the Machine Man, imply that the human being must let its bodily activities be organized according to external norms, with reference to a rational metric and a goal of maximum yield. It is even quite the opposite: the law of the Machine Man is not that of yield according to La Mettrie, but that of pleasure and enjoyment. Moreover, it is not a law imposed from the outside: it is the law of the body itself, that of its biophysical organization. The Machine Man thesis does not at all mean that the human being has the vocation to behave like an automaton: it means that the human being is subjected to a law as imperative as that of the universal mechanism for automata. Yet, for humans, this law is that of pleasure. Pleasure is the essence of the human; it governs the human body Machine as inexorably as the laws of classical mechanics govern the movements of clocks.

The materialism of La Mettrie, as with Cartesianism, does not lead directly to the formulae that are found in Coulomb for example, and which announce the industrialism of the following century. La Mettrie caused a scandal with the publication of *The Machine Man*; as a matter of fact, he was undoubtedly one of the last representatives of the century, along with Rousseau, to firmly defend a conception of the human that was refractory to these aims of rationalization.

2.1.3.4. *Technology and needs*

Rousseau was indeed one of the few at that time to glimpse prophetically the dangers and potential ravages of machinery and the division of labor (Deneys-Tunney 2010, p. 27). There is with him, as with the Greeks, a very strong awareness of the danger of technologies given their potential to transform the human condition.

That said, Rousseau appreciates artisans. Because he is forced to limit his desires and stay *in his place*, the artisan represents in Rousseau a form of neo-Stoic wisdom (Deneys-Tunney 2010, p. 37). Stay in place: precisely because his activity is intrinsically framed by the *eidos* of the thing manufactured, by material constraints, by the needs of the user and by the bodily and intellectual possibilities of the manufacturer. The power of technology must be contained. Avoiding the double pitfall pointed out by Huyghe, of an answer in terms of unlimited political coercion on one hand, and in terms of rational theorization of practices in the service of economic return on the other, Rousseau mainly invokes needs, such as the need for the human being to remain the subject of its technologies – such is the teaching of *Emile*.

Technology can always become a field of heteronomy; in this respect, it must be kept within narrow limits so that it does not compromise the subject's autonomy. Humans cannot help but produce; however their production must not be such as to transform the human condition. The natural man, like all living beings, have vital needs, and these needs fix once and for all the framework within which there can be a certain technological progress. An industry which progresses, without leaving this framework and without modifying it, does not present any danger: a simple and lonely life, very limited needs, instruments invented to provide for it, all this does not fundamentally upset the balance of human nature. On the other hand, because they seek to obtain ever newer amenities, things which were originally a simple search for pleasure but which ended up degenerating into real needs, humans themselves fell under a terrible yoke, the first they imposed upon themselves without thinking about it.

Needs signal towards life in its biological dimension. Taking them into account is above all not to forget that we are living beings and that biological life has its own requirements. As Bergson says in the same vein, on the subject of diet (Bergson 1977, pp. 300–301) – an area in which, in rich industrial societies, artificial needs are legion – science and medicine alone are really competent in matter. Our biological constitution, which we have inherited from a long evolutionary process and which we may not modify as we please, imposes its requirements. Life has needs that we cannot bend at our will.

If Bergson takes up the definition of technology as a projection of the organic, which was very widespread at the time ("If our organs are natural instruments, our instruments must then be artificial organs. The workman's tool is the continuation of his arm, the tool-equipment of humanity is therefore a continuation of its body" Bergson 1977, p. 309), the fact remains that for him, technology does not only extend being-in-life. It is not, to parody Clausewitz, a simple continuation of the movement of the life of the living beings we are, and that we do not cease to be, by non-organic means: technology is an activity creating new values, resulting in the constant appearance of new needs. Now these needs can telescope the needs required for the perpetuation of the being-in-life, which continues to assert its own requirements under the developments of civilization. Meat consumption may be highly valued in industrial societies (less and less today, no doubt), but it poisons us.

When he militates, with very Rousseauian accents, in favor of a frugal existence contenting itself with satisfying real needs, Bergson does not maintain a nostalgia with regard to a bygone golden age when human life was indeed only an extension of nature. He recalls rather that the escape of the human species from the framework fixed by nature for it, did not after all reduce the requirements of biological life to nothing, and that these must be considered in the assessment of technological developments.

Need therefore has an eminent status. This is explained by its constitutive dual nature: it is a fact related to the biological organization of the living being, and it is at the same time the manifestation of a choice of values – such as for example when we choose to eat meat. Need signals both towards a mechanism and an end. If certain needs (most of them) can be considered as artificial and illusory, it is precisely because they dissociate mechanism and finality. They testify to a finality which ends up running idle, unless articulated with the requirements of biophysical life and its own mechanisms.

2.1.3.5. *The body-instrument*

Rousseau's genius, however, is to bring the analysis beyond the reference to needs. He also considers that it is necessary, for humans, to have their eyes fixed on what they are handling and not to be too theoretical and abstract, turning away from their gestures. The arts have an impact on the body, on perception. Industry must be bodily experienced, felt; wresting technology from any possibility of experiencing, within this body which had been seen both as organism and as instrument since Aristotle, the alliance of a nature and a finality, is to surrender with feet and fists bound to the worst form of heteronomy.

By defining slaves as instruments of instruments, or as moving instruments as opposed to non-moving instruments, Aristotle had certainly defended a political organization of slavery which has fortunately become indefensible. That said, he therefore introduced an absolutely decisive conception of technological activity which deserves examination. The activity of the slave is above all the activity of a living body, taken up not as an organized body with parts and accessible to the inspection of knowledge, but as an instrument whose handling escapes what can be described and understood in rational terms. The instrument is inseparable from its handling, and all this is enclosed in the folds of the living body, inaccessible to *theorein*, that is to say, to contemplation for the purpose of knowledge. The *organon*, that is to say the non-moving instrument, is a dependent instrument which must be "animated" by the living being. However, this animation, according to Aristotle, cannot be recaptured as a sequence of operations clearly identified and capable of being reproduced outside the living being.

To extract technology from the living being, the body of the technician therefore had to cease to be considered as an instrument, in order to be assimilated to a machine performing regulated sequences of operations: such is the perspective that we saw taking shape in the 18th century. It would then be necessary to wait for Marcel Mauss and his famous article on the Techniques of the body (Mauss 1973), until the technical body can, once again, be philosophically considered not as a machine but as an instrument – that is to say, so that technological activity no longer appears to be entirely relocatable and transferable to machines. Only perhaps an anthropologist like Mauss, gifted with a wide and deep knowledge of non-Western cultures, could rediscover, after Rousseau, this non-mechanical but instrumental meaning of the technical body. The body is "man's first and most natural instrument", says Mauss: this definition is not so far from the Greek concept, still alive in Rousseau, certainly rid of any reference to slavery.

However, it is important to avoid misunderstanding: by qualifying the body as an instrument and not a machine, Mauss does not mean to recover a body held to be inaccessible to disciplines and to the imperative of performance. It is even quite the opposite: "The techniques of the body can be classified according to their efficiency," says Mauss, "i.e. according to the results of training. Training, like the assembly of a machine, is the search for, the acquisition of an efficiency. Here it is a human efficiency. These techniques are thus human norms of human training. These procedures that we apply to animals, men voluntarily apply to themselves and to their children. The latter are probably the first beings to have been trained in this way, before all the animals, which first had to be tamed" (Mauss 1973). Mauss remembers his childhood, when his teacher reprimanded him for his improper approach: "Idiot! Why do you walk around the whole time with your hands flapping wide open?" There is also this young girl whom we know to have been brought up in the convent, because she always walks with closed fists. And what about table manners: "you can be certain that if a child at table keeps his elbows in when he is not eating, he is English. A young Frenchman has no idea how to sit up straight; his elbows stick out sideways; he puts them on the table, and so on" (Mauss 1973).

To define the body as an instrument is therefore to see in it the very opposite of a reality inaccessible to normalizing powers, an unsocialized territory and a refuge for the individual. It is rather to admit that the human being is a matter that is indefinitely malleable according to external norms and goals, that its body is not a refuge for its irreducibility and idiosyncrasy, but that it is, on the contrary, the fulcrum of social training.

However, Mauss also insists on another aspect, one just as essential, which makes the body something other than the anchor of social norms in the individual. He remembers, having been wounded in the war and hospitalized, being surprised by the particular gait, unfamiliar to his eyes, of the nurses. His explanation is as follows: young women had just discovered American cinema and tried to walk like the actresses they wanted to imitate due to fashion. They had changed their habitus, that is to say their incorporated disposition.

In other words, the body is that by means of which the social sinks its teeth into individual existences by imposing normalized behaviors on them, but it is also that by means of which individuals always have the possibility of breaking free from established norms to act otherwise. Nothing is strictly natural in us, but nothing is either strictly social, that is to say entirely determined by the social. The body has a pharmacological mode of existence, in Derrida's sense: it is poison and remedy, remedy because poison. Mauss specifies: "The body is man's first and most natural instrument. Or more accurately, not to speak of instruments, man's first and most natural technical object, and at the same time technical means, is his body" (Mauss 1973). Thus, the instrument splits: in that it is malleable, the point of application of training for normalization, it is an object; insofar as it allows for appropriation, doing differently, that is to say invention, it is a means. As Canguilhem says, in a very Maussian vein: "Man, even physical man, is not limited to his organism. Having extended his organs by means of tools, man sees in his body only the means to all possible means of action… organic vitality flourishes in man in the form of technical plasticity" (Canguilhem 1978, p. 118).

The technical body is not a machine but a means, an instrument, it displays plasticity and deploys a range of possibilities. Note the inversion of values: according to the thesis of the neutrality of technology, to consider technology according to the dichotomy of means and ends is to submit it entirely to the representations and social evaluations of the moment (society decides the usage to be made of technology). To reject this thesis, freeing technology from the representations and social evaluations of the moment, restoring its transformative potentials – its "non-social" part as Simondon says – is to see in it something other than a means – rather, an object of the world characterized by a certain autonomy. However, in the case of the techniques of the body, the situation is symmetrical and opposite: to make the body an object is to see there the point of application of training; to consider it as a means is to see in it something that the individual may use to develop a range of possibilities that do not fall entirely within the social norms of the moment. Mauss's great merit is to have maintained that there is not one without the other, the means without the object. The body may open the way for individual autonomy by the very fact that it has previously been constituted as a field of heteronomy.

It is therefore understandable why the techniques of the body are so important to Mauss: it is that they provide privileged access to the fundamental meaning of the technological act, to the articulation of the biological, the psychological and the social. Whereas technologies involving an externalized tool, or a machine, can give the impression that the intelligence which plans is sovereign, techniques of the body reveal this essential fact of the vital anchoring of technological behaviors. They touch on the articulation of technology and biological life. They more precisely touch on the following fact: if the exteriorization of life in technological artifacts can open up creative possibilities, it is on the condition that the technical body has already experienced a kind of proto-exteriorization, consisting of becoming an instrument for itself. In short, the technical body is the exteriorization of the interior, that is to say not yet mediated by a tool or a machine.

It is precisely this conception of the body not as an organism projecting its organs into externalized tools, as Ernst Kapp conceives it as we will see later (Kapp 2018), but as an instrument which I must learn to use in order to be able learn to use a tool, which was stifled at the same time as the industrialization of Europe was taking off. The rupture introduced by industrial machinery thus consisted in technology being taken, on the one hand beyond the sphere of needs, and on the other hand beyond the body which experiences, endures and regulates human industrial activity. The body itself is found forced back onto the level of pure mechanisms emptied of any reference to the very idea of finality: the gestures of the worker are grasped as a sequence of mechanisms. The degradation of the body itself, capable of enduring the work situation, in a sequence of operations which may be formalized in mechanistic terms, is the exact counterpart of the expansion of needs beyond the framework where life as being-in-life could exercise control. These two processes, characteristic of the industrial era, ultimately become one: they make effective the divorce of technology and life, in favor of a purely intellectualist conception of technology.

Now it is precisely this which Kant perfectly grasped, on the threshold of the first industrial revolution, leaving us some decisive pages on the question of technology.

2.2. The intellectualist conception of technology: the Kantian turning point

2.2.1. *Difficulty in thinking of the artisan's activity*

2.2.1.1. *Art in general: paragraph 43 of the Critique of Judgement*

The place occupied by technology in Kantian thought is difficult to define. A quick glance at the Kantian corpus may suggest that things are not so complicated: on the one hand, the question of technology is certainly not at the heart of Kant's reflections. The passages he devotes to it are rare. On the other hand, when he approaches it, he does so without apparent originality and in very conventional terms at first. The passages in question are essentially taken from the *Critique of Judgement* (Kant 2007), where Kant evokes the idea of a technology of nature (mobilized in judgements of taste on natural forms) or of a nature that is technological by its own power (mobilized in teleological judgement on the condition of possibility of organized beings), or because of paragraph 43 of this same work where Kant endeavors to circumscribe with clarity and precision the domain of the arts, that is to say of techniques, by distinguishing it at the same time from the operations of nature, scientific knowledge and the domain of handicraft.

It is indeed in this famous paragraph 43, entitled "Art in general", that Kant provides if not a definition, at least an overall characterization of the domain of the arts. Three distinctions allow this area to be circumscribed, in contrast to activities or operations that do not fall under art.

First, art is distinct from nature, in the sense that doing (*facere*) is distinct from acting (*agere*). Art is therefore primarily characterized as a domain of manufacturing, as opposed to nature which produces effects (*effectus*) but not works (*opus*). As a domain of artificial manufacturing, the arts imply the reference to a will which is determined to produce something according to a final intention (*nexus finalis*). If we find a piece of hewn wood in a swamp, we will not say that it is a production of nature, but a production of art. This domain only concerns human beings, to the exclusion of other living things: only human beings are capable of artificial manufacturing.

Second, art is distinguished from science as power is distinguished from knowledge, as technic is distinguished from theory. Kant explains that a thing that one can do simply because one has the knowledge of it and its construction, is not art: art implies an excess of doing over knowledge, an irreducibility from the first to the second. In other words, art begins where theory turns out to be powerless. Though Camper had the most complete theory of shoes, he would have been unable to make one: there is therefore no intersection, no overlap between the theoretical and the technological. Technology has an autonomy compared to theory – an idea which was not original at the time, the problem of the designer and producer of fireworks, including the engineer of the 18th century, being in fact the identification of the threshold at which simple deduction becomes inoperative, the threshold at which science becomes useless for the various practitioners and the particular situations they have to face (Vérin 1993). However, Kant does not really specify the very nature of this autonomy of technology: in what does it consist, in its positivity?

Thirdly, art is distinguished from handicraft in the sense that an activity which is in itself its own end is distinguished from an activity whose end is outside of itself, for example in the search for gain. Note that this distinction does not exactly replicate the Aristotelian distinction of *poiesis* and *praxis*: in Kantian terminology, both arts and handicrafts fall under *poiein*. The difference is that the arts have no purpose apart from the activity of producing and the thing produced themselves, while handicraft activities aim for heterogeneous ends both in production and in the product (earning money, etc.). Kant's distinction between the arts and the handicrafts therefore covers the distinction between the free and the constrained.

Unlike the first two distinctions, however, the third seems less clear, more problematic in Kant's eyes. First of all, in the very heterogeneous field of handicrafts, that is to say mechanical arts, can we not distinguish more precisely between those who participate in art and those who fall within the category of handicraft alone? While blacksmithing is a profession, isn't watchmaking an art? The *Encyclopedia*, already, draws a distinction between artists on the one hand, that is to say workers who excel in those mechanical arts which imply intelligence, and artisans on the other hand, that is to say workers who profess those mechanical arts which require the least intelligence. The watchmaker is an artist, the blacksmith an artisan. However, the watchmaker exercises a trade which allows him to earn his living, like the blacksmith: the distinction between what is its own purpose and what aims for something else is blurred.

To define the extent to which art in involved in the domain of handicraft, a new criterion must be introduced, suggests Kant: namely, talent. A profession participates in art in proportion to the talent it requires – a criterion which is, let's face it, somewhat imprecise. Is there really more talent in watchmaking than in the forging of steel in Damascus or Toledo? How to establish the maximum and minimum in this area?

Admittedly, the distinction between art and handicraft remains valid, but human activities are not pure, which makes them difficult to identify. In addition, the distinction between liberal arts and mechanical arts seems confused; certain liberal arts may indeed, explains Kant, be related to sciences, while others rather are closer to handicrafts.

Finally, the freedom enjoyed by the arts, as opposed to occupations dominated by constraint, is a freedom that must necessarily flow into a mechanism. Kant thus castigates pedagogues who claim to expel any laborious dimension from the liberal arts, by making them into pure games: all art has a necessary relationship to work, to constraint. So here again is a distinction that seemed clear and is now becoming blurred: the artistic dimension of human activities is the exercise of a freedom, but this freedom is not pure and there is in all art something forced.

2.2.1.2. *Technology and mechanics*

The broad domain of the arts therefore has three main characteristics: the making of artificial objects according to the representation of an end; an excess compared to the theoretical; the mixture of freedom and constraint (which makes it clear why the arts are neither on the side of the natural sciences alone – necessity without freedom – nor on the side of morality – pure freedom, without constraint).

According to some commentators, these few remarks seem to indicate that Kant has profoundly overhauled the concept of technology, so that this concept is now closely associated with the concept of life. In other words, far from validating any idea of a divorce between life and technology, Kant is rather credited to have redefined each of the two concepts to make them correspond to each other.

As French philosopher Philippe Huneman notably analyzed (Huneman 2008), Kant's position on technology changed over time. In a first, pre-critical phase, Kant links technology to the production of a contingent order in nature, which involves the intervention of a divine artisan and therefore the implementation of a divine technology. Opposed to this type of order is the necessary order of the laws of nature. "Technology" is distinguished here from "system": technology signals towards the activity which intentionally arranges materials for an end; system follows from the universal laws of motion, without referring to an intention. As Huneman points out, the examples that Kant gives to illustrate the contingent order in nature produced by divine technology essentially relate to the realm of living beings (Huneman 2008, p. 96). The concept of divine technology thus allows Kant to attribute a specific status to living beings and, at this stage, technology qualifies life in its own right (Huneman 2008, p. 176): for the parts of animals and plants to result in such subtle arrangements, and testify to an order that is both admirable and contingent, it was necessary for a wise original being to officiate.

A radical shift in meaning occurs later when, in the *Critique of Judgement*, Kant no longer opposes technology and system, but two types of systems: the system of universal laws of nature on the one hand, the system of empirical laws of nature on the other side. This latter system signals towards a finality of nature in its form and not in its products: in other words, it does not indicate a suitability of the products of nature in relation to particular ends (that which previously motivated the concept of divine technology, by analogy with human art which aims at specific ends), but a suitability of the empirical order of nature in relation to our faculty of understanding.

Huneman then explains that this reworking of the concepts of the order and purpose of nature requires a reworking of the concept of technology. A distinction is now introduced, between technology on one side, and mechanics on the other. Technology signals towards the work of the artist, which consists in the creation of a harmonious object whose initial design only imposes an overall line, without predetermining the detail of operations to achieve it. Conversely, Mechanics consists of the execution of a set of operations regulated in all their details, with a view to achieving an end which is clearly identified at the outset. For this, it makes use of instruments which may be used for several purposes (Huneman 2008, pp. 267–268).

The stumbling block of the analogy between technology and living organism is therefore no longer the question of intention, which is purely and simply discarded. Thus dissociated from mechanics, technology now designates an activity whose productivity is not limited to the execution of a primary intention, since only the general plan is given but not the detail of the constructive operations. The concept of technology of nature is no longer involved with the analogy with human art that was previously implied: the existence of an intentionality determining each stage of fabrication, like a watchmaker who not only has an overall idea of their purpose, but who knows very precisely what successive operations are to be carried out to achieve it. The distinction between technology and mechanics thus neutralizes the reference to the technological theology of artisanal understanding (Huneman 2008, p. 309).

However, even if the relation between intention and result finds itself loosened in the concept of technology, while a strict execution rigorously connects them in mechanics, the fact remains that all human technology proceeds from intention.

In addition, the inapplicability of the technological analogy to the living organism is also shifted onto another front; it is also explained by the fact that any technology (and not only mechanics) consists in arranging "surfaces", by pressures and shocks, but without creating the interior of the materials. Living organizing processes, on the contrary, go to the most intimate parts of matter, through the use of formative forces (Huneman 2008, p. 308), as explained in paragraph 65 of the *Critique of Judgement*.

In summary, the analogy between technology and organism can only work on condition of having been neutralized beforehand, says Huneman, as to these two aspects which are, firstly, the inevitable reference to intention in human art, and second, the inability of human art to produce its own materials, descending to the very depths of their production process.

Taking up Huneman's reading, Italian philosopher Emanuele Clarizio (Clarizio 2019) also argues that the *Critique of Judgement* gives itself to a decisive change in the concepts of life and technology: the analogy of life to a "technic of nature" allows Kant, according to him, to import into the conception of the living being the idea of a new causality, irreducible either to mechanical causation or to final causality. The two causalities known to us, *nexus effectivus* and *nexus finalis*, indeed share the same premise: the result of the action contains nothing more than what was already there at the start, in the initial impulse. This is evident for the *nexus effectivus*. Indeed, the laws of the universal mechanism govern in a perfectly deterministic way the process which leads from the impulse to the result, so that at the limit, there is no need to actually carry out this process to know its outcome: I can deduce what will happen, given perfect knowledge of the initial conditions. This is also true with regard to *nexus finalis*: according to the requirements of this type of causality, the result is the faithful material translation of an initial representation. It contains nothing more than what was included in the representation. In both cases, the causal process brings nothing new to the world.

However, technological activity does introduce novelty, precisely because it does not consist of mechanically executing a pre-established plan, fixed at the outset in every detail. Technology therefore produces originality; it does not arise from mechanical causality alone, nor from final causality alone, even if the technological action implies both a regulated sequence of mechanical processes and the representation of an end.

This remarkable affirmation – technology produces originality – however, only makes sense on condition of having previously distinguished technology from mechanics. Mechanical action is entirely deterministic, not productive but simply executive; technological action, on the contrary, has a certain spontaneity, producing a result that is impossible to anticipate in every detail. This, moreover, is also the characteristic of living beings: they possess a productive spontaneity.

In summary, both technology and living beings testify to a causality which is neither that of mechanics nor that of ends – a causation unknown to us, says Kant, but which it expresses by the concept of "organization". As Clarizio rightly puts it, the domain of life and that of technology now communicate thanks to the notion of organization: living beings and works of art are the unpredictable results of organizing powers that mingle with the material and which do not impose upon it a prefabricated project ready to be applied. The creation of genius, like biological morphogenesis, does not pre-exist the creative act and it is therefore a question of forms of organization which are immanent in matter (Clarizio 2019, p. 219). Note that it is by bringing technology close to art, as Huneman already noted, that it is possible to attribute to the first the creative spontaneity usually devolved to the second.

In short, Kant is said to have introduced an unprecedented possibility of thinking about technology in close connection with life. The distinction between technology and mechanics, as well as the irreducibility of technical causality either to *nexus effectivus* or to *nexus finalis*, therefore seem to oppose Kant to this whole movement which, as we have seen, begins to take shape in the 18th century and which aims to dissociate the productive act into two clearly separated dimensions: the rational dimension, located upstream in the activities of programming (*nexus finalis*), and the machine dimension, assumed by a body-machine which carries out the activities of pure execution of programs designed by others (*nexus effectivus*).

However, two points should be underlined which lead us to qualify this diagnosis of a Kant standing against the current of his time.

Firstly, Kant does not really theorize this new concept of technology – in particular, he does not really explain what precisely is this excess of technological action over theoretical knowledge, which he illustrates with Camper's shoes. It is clear that, comparatively, Rousseau has a more positive and elaborate conception.

Secondly, by distinguishing between technology and mechanics, Kant precisely confirms the fact that a certain type of technological action is now purely mechanical. The concept of mechanics indeed here covers, not only the domain of phenomena governed by the universal laws of nature, but also part of the human productive domain. It crystallizes the concept of work that we have seen gradually taking form in the Age of Enlightenment: the producer who manipulates instruments is about to be assimilated to a simple executant, devoid of intelligence. The distinction between technology and mechanics thus confirms the divorce of productive activity from life, much more than it opposes it.

Kant has the undeniable merit of pointing out the irreducibility of any technological action to its degraded form, which is mechanical in the sense of execution without intelligence. However, he clearly could not have pushed any further his reflection on the deep meaning of the relationship between technology and life. More precisely, if we reformulate the theses of Huneman and Clarizio, it seems that Kant sought to redefine the concept of technology other than as *organon* – that is to say otherwise than with reference to the body-instrument, that is, the body which the living being possesses as a proto-instrument so as to be able to handle other, artificial instruments.

The conceptualization of life as "organization" would precisely make possible, according to Clarizio, an alternative conceptualization of technology: it would pass from the body-instrument to the body-organism, the latter deploying in the productive domain a creative spontaneity analogous to that of living organisms.

However, as we will see in the next chapter, the body-organism is susceptible to two antinomic interpretations, which are themselves a function of two conceptions of the concept of "organ" which radically differ. Because he definitively gave way to the conception of the body as an instrument, Kant is led to conceive of the organ not, therefore, as an instrument native to the living being, but as part of a whole – and it is indeed in these terms that he defines the body-organism: the organized being that organizes itself according to the reciprocal causality of the whole and the parts. Causation is doubtlessly immanent in the parts themselves, and not external as is the *nexus finalis*. Nevertheless, in this conception of the technical body as an organism and no longer as a body-instrument, something decisive has apparently been lost: the possibility of understanding the activity of the shoe manufacturer, and of the artisan in general, that is to say, one who has, precisely, learned to use his body as an instrument, “the first and most natural of all instruments”.

2.2.2. *Technology excluded from the field of aesthetics*

2.2.2.1. *Technology and the two meanings of life*

In the introduction to the *Critique of Judgement*, Kant draws a distinction between the domains of the technically-practical and the morally-practical: only the second belongs to the practical part of philosophy, the first has a relationship of dependence upon the concepts of nature and therefore finds its place in the vast field of theoretical philosophy. Indeed, since the mechanical arts apply rules which arise from the natural sciences, technology is a province of science.

However, as we have just seen, paragraph 43 is clear: technology goes beyond the theoretical; from complete possession of a science, one cannot deduce the capacity to produce technically. The technically-practical is not an extension of the theoretical. In addition, the cleavage between the technically-practical and the morally-practical can be, if not disputed (Kant does not return at any time to this distinction), at least slightly nuanced in the idea that a technician implements maxims of prudence: their activity is not yet moral, but it is already concerned with the question of what to do or not to do.

If, consequently, technology does not arise from the faculty of understanding (the theoretical), nor from the faculty of desire (the morally-practical), must it be said that it falls under the jurisdiction of the faculty of feeling pleasure and displeasure? Pleasure and displeasure in the judgement of taste, says Kant, are what makes the subject “feel his life” (Kant 2007, p. 36). They lead more precisely to the feeling of a pure, immediate life, whereas sensation has only a mediated relation to life (material sensation gives access to the feeling of living by way of an object).

Technology does indeed have an essential link to life, in a sense, however, which must be clarified. In *Anthropology from a Pragmatic Point of View* (Kant 2006), Kant explains that for the human being, to feel alive is nothing other than feeling continually forced to leave the present state, that is to say to withdraw from a life defined as absolute contentment in immediacy. Living for the human being is to mediate its relation to its own life as a living being, and this is what defines work, whose relation to suffering is constitutive. “Pain is the incentive of activity, and in this, above all, we feel our life; without pain lifelessness would set in” (Kant 2006, p. 126).

In other words, according to this anthropology with Rousseauist accents, technology and work, or even the government of things through technical predisposition for manipulating things, as Kant says (Kant 2006, p. 226), only serve life and provide it with vivid feeling through pain, the suffering resulting from them. Technology therefore indeed has a constitutive link to life. However, it is not appropriate to conclude that technology involves the faculty of pleasure and pain, inasmuch as in the two cases – technology on the one hand, feelings of pleasure and pain on the other – the concept of life obviously does not have the same meaning.

The wide range of experiences covered by the feeling of pleasure and pain indeed introduces the feeling that everyone has of their own life as an individual. The three modalities of Kantian aesthetic judgement – the agreeable, the beautiful and the good (both what is useful, i.e. good in view of something else, and what is considered good in itself) – refer to the individuality of the subject. The agreeable distinguishes me in my subjective preferences; these even involve only myself. The beautiful may certainly be universalized, but no rule can be applied to decide *a priori* whether such a form or such a work is or is not beautiful: only feeling decides, which again means that the judgement of taste is based in an irreducible individuality (an individuality which, however, is already disturbed by what is greater than it and, as shown by Arendt, who promotes the plural community in public space, contrary to the agreeable in which the individual is satisfied with its isolated singularity). As for the morally good, it is undoubtedly determined by the Reason which dictates the moral law, but in the feeling of pleasure related to the fact of having acted by duty, the subject experiences itself both as a rational being and as a sensitive individual, that is, as a person.

On the other hand, technology as Kant conceives it in *Anthropology* relates not to the individual but to the species-level dimension of life, to a life which is attached to the human being as a living species, as a biological formula, and not as an individual. In this sense, pain and suffering have anthropologically opposite meanings: pain is experienced by the individual in its process of individuation, while suffering has above all in the text of the *Anthropology* a meaning for the future of the species, to avoid, as Kant says, the extinction of life as life of the human species.

In short, technology appeared to possibly pertain to the vast domain of aesthetics, which in Kant introduces the feeling that the subject has of its own life. However, this is not the case, and technology does not ultimately belong to any of the three sub-domains of aesthetic judgement: as an experience of mediations, its link to life is quite other than this pure, immediate feeling, which brings the subject back to itself in the individualizing experience of its own life. Neither does technology know the distribution of the agreeable and the unagreeable around which material sensation is organized. Finally, it does not allow itself to be caught up in the pleasure that comes from acting morally, out of duty.

This exclusion of technology from the vast domain of aesthetics, in all its components, leads us to understand in a different light the famous Kantian distinction between the beautiful and the useful. There is no reason to deny the reality of this opposition in Kant: the useful is essentially linked to an interest, while the judgement of taste (on the beautiful) is a disinterested judgement. That is to say that, unlike the two other types of aesthetic judgements relating respectively to the agreeable and the good, the judgement of taste has no interest in the existence of the object. However, in paragraph 5 of the *Critique of Judgement*, relating to the comparison between the agreeable, the beautiful and the good, Kant introduces a decisive remark: neither the agreeable nor the good concern human beings alone, to the exclusion of all other beings. "Agreeableness is a significant factor even with animals devoid of reason"; "the good is good for every rational being in general" (which may extend its perimeter beyond the human domain). Beauty on the other hand "has purport and significance only for human beings", says Kant, "i.e. for beings at once animal and rational (but not merely for them as rational beings – as spirits for example – but only for them as both animal and rational)" (Kant 2007, p. 41). The feeling of pleasure and pain thus unites, in each human being, life as the life of a living being and reason.

Consequently, situating technology in relation to the aesthetic domain as a whole – not only the beautiful but also the agreeable and the good – makes it possible to shift the terms of the problem and to situate Kant's real theoretical decision concerning technology, the one most fraught with consequences for the future, not in the opposition of the beautiful and the useful, but in the theoretical decision, which is considerable, consisting in emptying technology of any reference to the faculty of pleasure and pain; that is, to the feeling that the subject has of its own life insofar as it is both, inseparably, a living being and a rational being.

Ultimately, as revealed by bringing together the *Critique of Judgement* with *Anthropology from a Pragmatic Point of View*, this exclusion is carried out by performing a split in technology. This latter is restricted, on one hand, to final representations – that is, the existence of a rational being – and on the other hand to the life of the human animal, which is supported by technology, the work of the hand in its arduousness, to ensure its perpetuation. Clarizio says it excellently: the type of causality which, in Kant, circumscribes the technological territory, evades both mechanism and finalism, while being somewhat mechanical and somewhat finalized. Somewhat mechanical and somewhat finalized: one could not say better that technological causality, irreducible to either the *nexus effectivus* or the *nexus finalis*, can still only be understood as a kind of strange mixture, of hybridization between these two causalities known to us: the faculty of representations on the one hand, life in its dimension of accomplishment of pure vital processes on the other – that is, in its functioning as a machine (because if in Kant the organism cannot be assimilated to the machine in terms of its production, it nevertheless remains a machine in its functioning).

The opposition of the beautiful and the useful is not minor, of course, but it stands out against the background of a much more fundamental problem: that of the relationship between technology and life, and the way in which technology is only related to life in one sense (biological life, that of the species) so as to be better dissociated from life in another sense (that of the feeling that the subject has of its own life).

2.2.2.2. *Difficulty locating technology in the critical edifice*

The domain of technology is, in short, according to Kant, a question of pretense, of deception – from which Kant finds the Greek etymology of *mechanè*, cunning. Technology derives its value only from its dependence on the theoretical, by applying the concepts of nature; but technology is also in excess over theory, as the example of Camper shows. Technology is linked to prudence; it implies a form of practical wisdom, but Kant is forced to firmly establish, against frequent confusion, the difference in nature between the technically-practical and the morally-practical. Finally, technology has a link with life, but this life is not that of a subject who experiences its own life and whose aesthetic faculty (the feeling of pleasure and pain) circumscribes its own individuality.

Undoubtedly, Kant introduces, in *Anthropology*, a case where technology is linked to the faculty of judgement: the inventor may have genius; their activity may be such that nature (and not science) gives its rules to art. Kant specifies, however, that the inventor can introduce genius into what is not exactly a matter of genius: the mechanical arts (Kant 2006, pp. 119–120). So, there is again an impression of pretense, almost irony.

Furthermore, the inventor, even a genius, must conform their products to ideas; they must follow what Kant calls mechanical rules. Invention is always haunted by the theoretical, not necessarily in the sense of the sciences, but at least in the etymological sense of the contemplative (*theorein*): the inventor, even if possessed of genius, cannot free themselves from models and rules, or they would fall into extravagance. Here we encounter the Greek vigilance again, consisting in circumscribing precisely, by respecting models and rules, the possibility the artisan has of varying without limit the forms they inscribe in the material.

We therefore understand better the embarrassment in which we are seeking the meaning and scope of technology in Kant: it is certainly not absent, but it seems to find its place with difficulty in critical philosophy. It is again in the *Anthropology from a Pragmatic Point of View* that the fundamental reason for this difficulty finds its explanation – and we must be grateful to Kant for having very precisely indicated it.

2.2.3. *Technology, a synthetic activity without representation and without rule*

2.2.3.1. *An embarrassing organ player*

Kant opens the *Anthropology* with a paragraph devoted to examining the definition of the human being as a person, that is to say as a self-conscious subject unlike animals. From this anthropological or even anthropocentric commonplace, Kant goes on to explain in substance that it is not good for human beings to indulge in too much introspection, lest they come to observe themselves internally as if what was happening within themselves owed nothing to their activity, as if their representations were not the product of their activity. Indeed, there is a very large field of obscure representations in us, of which we are not aware, but these representations do not belong either to theoretical philosophy or to practical philosophy – which means that it is not really a question of representations. What is it exactly? Kant gives the example of the organist, who is able to play "a fantasy on the organ with ten fingers and both feet and also speaks with someone standing next to him" (Kant 2006, p. 24): he thus plays without being aware of it, as if mechanically (which is reminiscent of Diderot's paradox of the actor). The example of the organist is valid for any human being who possesses an art, and who can exercise it without thinking about it, as if they were setting in motion a mechanism without being at all present in what is being done by any representation whatsoever.

To clarify things, Kant introduces a distinction between two ways of exercising a skill, of acting technologically in the safety of a habit: skill and ease. Skill designates the capacity to produce without having to reason about causes and effects; it suffices to remember the cases observed, the regularities empirically observed. Ease refers to the ability to produce an effect without any representation of the relationships of cause and effect being involved. The difference seems minimal, but it is capital. In the first case, it seems that Kant wants to speak about this form of learning by accumulated experience – I know that if the hardening of the steel does not respect such or such a condition, the manufactured object will be fragile; it will break. The production of a steel blade is very dependent on a whole knowledge of causes and effects, which could moreover find confirmation in the developments of chemistry. Simply put, this body of knowledge is available in the experience acquired, and directly mobilizable; it does not need to be redeployed by the artisan at every moment. The shoemaker Kant speaks of probably does not apply theoretical knowledge; his activity consists rather of a practical judgement, attentive to the particular facts of the situation. He is very cautious and focused, totally present in what he does, even if of course his experience allows him to execute his various gestures with great sureness. In addition, he never loses sight of the purpose of his activity: the manufacture of a pair of shoes which must meet the needs of a user. We therefore find here the virtues of the Greek artisan whose manufacturing activity is an art of contextual judgement and takes place within narrow limits. However, this practical judgement, involving the "handling" of a body that has become a means to achieve a practical goal (a body-instrument, in short) is not really understandable within the framework of Kantian thought. Kant also leaves completely obscure the very mysterious process by which this body achieves, precisely, its status as an instrument, by acquiring dexterity and sureness of gesture, that is to say, a form of embodied, immanent intelligence, for which there is no name.

In the second case, which involves the organist who performs very easily while doing something else which in reality is very difficult (playing the organ), the activity does not seem to mobilize any knowledge of causes and effects, even sedimented in experience. The organ player doubtless learned to play the organ and experienced difficulty in doing so. However, by playing without paying attention, he does something other than what the shoemaker who makes shoes does. In the latter case, the productive activity is indeed dependent on representations, but these do not need to be recalled in detail in their sequences; in the case of the organ player, the productive activity (of an effect) does not seem to depend on any representation. The musician does not seem at all focused on his music; he is not present in what he does; his activity does not seem to mobilize the faculty of representations. He runs through his score like an automaton, just like those musical automatons of which the era was fond (think of the extraordinary timpani player of Kintzing and Roentgen, which was built in the 1780s and which may be admired at the Museum of Arts and Crafts in Paris). The sureness of his gestures is of course, like that of the shoemaker, the result of long apprenticeship, but the performance of the score has this destabilizing effect that at no time does it seem to demand what any technical activity requires: attention and care.

The most obvious characteristic of the organ player's activity is the impossibility of identifying any form of external intelligence. What might it then indicate? A form of technological intelligence immanent in the body, foreign to the faculty of representations? An automaton's intelligence in the subjective sense of the possessive? The difficulty is precisely that Kant cannot grant any positive status to these technological activities where the body alone seems to guide and execute the operations, without any intervention from the faculty of representations.

In this sense, ease is not the prerogative of what Kant calls in *Anthropology* "common sense" (*sensus communis*), as opposed to the savant: to have common sense is to know the rules only in their application, in context – the savant being the one who knows the rules before their application. This is what Kant also calls "logical tact". The organ player, or the possessor of incorporated knowledge allowing easy production of a difficult effect without even paying attention to it, and above all without applying rules, is distinguished both from the savant and from the one who possesses this "logical tact": these latter have an essential relation to representations, the former does not. From the point of view of the critical enterprise, which deals precisely with representations and their relation both to a possible object and/or to the subject, this is unthinkable.

Ultimately, Kant closes in the *Anthropology* a large gap opened up in the *Critique of Judgement*, at paragraph 43. If art is signaled by its excess over any enclosure of productive activity by the faculty of concepts, it is because this productive activity, technological activity, cannot be fully understood as a synthesis of sensible data according to rules. More precisely, it must be concluded that technological activity refers to a synthesis that consists of adding to the sensible data a regularity which is not of a representational order – which is completely confusing since technology has been relegated to the category of the useful, which means that it does imply the representation of an end.

In short, Kant is very close to identifying in paragraph 43 the existence of a non-intellectual but sensorimotor type of synthesis activity, which cannot be assimilated to a synthesis which adds rules to the sensible data. Now, must we not admit that this synthesis not made according to rules implies that prior to the manipulation of an externalized tool (like a musical instrument), a pre-intellectual synthesis, of precisely a sensorimotor nature, has already taken place in the depths of the body – a very enigmatic synthesis consisting of ensuring that the operations of the body are made homogeneous with the sequences of technological operations, so that the former may later assimilate the latter in the execution of safe and well-controlled gestures?

If the organ player can play without thinking about it, literally as if he were the organ and there was no distance between them, it is because his body is, so to speak, built like an organ. For the organist to be able to play without paying attention, the operations of which his body is the seat must be homogeneous with the operations required for a melody to spring from the organ. Consequently, its sensorimotor arrangements must take place in operating sequences which are homogeneous to the operating sequences implemented in the technological activity.

In this embarrassment which leads to the idea of a productive activity relating at the same time to the *nexus effectivus* (the machine) and to the *nexus finalis* (that is to say, the representation of an end: playing and interpreting a melody), but nevertheless not assimilable to a synthesis according to rules, in short, a conception of the living body takes shape, not as an organized and self-organizing being, that is to say as an organism as posed masterfully by the third *Critique*, but, to use Mauss's formula, a conception of the living and acting body as "man's first and most natural instrument", as an irreducible proto-instrument with the status of a means to an end. The body-instrument, endowed with a non-representational and immanent intelligence, is thus distinguished from the body-machine, directed in its learning and its operations by an external intelligence. Clarizio could rightly speak of a causality that is somewhat mechanical and somewhat finalized. It is clear, however, that with the organ player, we are dealing with a type of productive causality which seems *completely* mechanized, without ceasing to be "somewhat finalized" – impossible, indeed, to assimilate the musician to a real automaton. Therein lies the difficulty.

2.2.3.2. *The technical body reverting to the machine-body*

Kant therefore identifies a very broad field of human activity, pertaining to the arts & crafts, which belongs neither to the theoretical field (here, there is no knowledge providing rules for action), nor to the practical field prescribing what humans should do, with which the *Anthropology* is concerned. Again, this vast field of technology does not seem to consist in applying rules, but on the contrary of producing an effect without any representation of rules for its realization.

Kant then says something crucial: this vast domain of human activity without representation, carried out with ease, does not concern pragmatic anthropology but physiological anthropology (Kant 2006, p. 25). This concerns the human as a living species, which is not effectively addressed in *Anthropology from a Pragmatic Point of View*. The gap is in a way closed by the rejection, as outside the limits of pragmatic anthropology, of this body which proves capable of incorporating a regularity close to automatic mechanisms, but without this incorporation consisting of a synthesis by addition of an external rule – without there being, in short, a manufacturer of the automaton.

In short, Kant stops where it would have been necessary, no more no less, to do away with the Aristotelian definition of technology: a disposition to act in accordance with rules. Kant remains fundamentally faithful to this definition and, very consistently, he denies the activity of the organist the qualification of technical. It is indeed the split between technology and living corporeality which is established here.

This considerable theoretical decision, which consists in refusing the status of technician to this living and acting body (which is therefore capable of becoming an instrument without reducing itself to the status of a means whose use would be conditioned by the representation of an end), then opens the way to the possibility of a reinvestment of this body, but this time from outside, by the addition of rules which are imposed on sensorimotor assemblies by subjecting them to the most complete heteronomy.

From Descartes to Kant, an inversion thus seems to have occurred: in Descartes, the organism is assimilated to a machine, but the body remains the instrument of a "technical plasticity", in the beautiful words of Canguilhem, that reason cannot grasp, and which is not intended to enter into planning and calculations. In Kant, on the contrary, the organism is conceived as unassimilable to the machine; on the other hand, the body is haunted by its own mechanization. It proves to be potentially open and accessible to programming, risking a tendency towards the automaton.

The Kantian analysis therefore signals towards what had to be presupposed for artisans to be expelled from their own know-how, and for this to be made fully accessible to formalization and transfer to machine processes. This presupposition resides in the idea that a body whose operations are not enveloped by representations cannot in any case, in Kant's terms, *facere* an *opus*, but only *agere* an *effectus*; it is indistinguishable from a machine. What distinguishes on the one hand an *opus* produced by a technological *facere*, and on the other hand an *effectus* of nature, is representation – a conviction that Marx moreover took up on his own in his famous remark that however sophisticated the constructions made by animals (for example beehives), there will always be an impassable trench between them and the humblest, crudest product of human technology; only the human worker imagines what he does before doing it. (What distinguishes the worst architect from the best of bees is that the architect raises his structure in imagination before he erects it in reality [Marx 2019, p. 198].)

So, it is very logical that Kant opens *Anthropology* by circumscribing the anthropological field by reflexive consciousness, that is, by the field of representations. This latter must be very clearly delimited by isolating it from the obscure regions which border it, where processes operate whose link to representations is weakened.

All this analysis of technology is moreover consistent with the characterization of the living being in paragraph 65 of the *Critique of Judgement*. As we know, Kant defines the living being as the organized and self-organizing being, that is, an organism whose genesis cannot be made intelligible by any technological model known to us. Kant explains that the living being is not only endowed with a "motive power", but also with a "formative power" (*Bildende Kraft*) which does not operate like the artisan watchmaker assembling the parts of the watch according to a construction plan. Kant specifies that in this way, the organism is not only a machine – "not a mere machine". Here, every word counts: the organism is not a mere machine, that is to say that it is also a machine, from a certain point of view in any case. This point of view is that of the completed, constructed organism, whose operations can effectively be assimilated to the functioning of a motor machine. This paragraph was rightly deemed essential for the development of a philosophy of the organism; what it establishes for the technical body has been less noticed, although it turns out to be just as essential. Considered from the point of view, not of genesis, but of functioning, the body is comparable to a motor machine which operates by enacting automatisms. What *Anthropology* establishes with the example of the organist is just as fundamental. This motor body-machine, performing automatisms, can take on new automatic assemblies; it is, so to speak, a programmable machine, a machine whose operating sequences can be reconfigured. Kant does not go so far as asserting this, of course; nevertheless, the way is now open in this direction.

2.2.3.3. *Beyond organon: redefining the links between technology and life*

In short, we indeed see a divorce here, a rupture of the deep link between technology on one hand, and life in its unity on the other – that is, life in so far as it indicates, inseparably, both being-in-life and experiencing the feeling of one's own life. Far from continuing to conceive of technology as it had been since the ancient Greeks – that is, as *organon* – Kant proceeds to a split in the very concept of technology, related to the split enacted within the concept of life – between on one hand, the biological life, and on the other hand, the life of the subject as a rational and emotional being. Technology gestures towards the representation of ends, but it also ends up sinking into the depths of the body and becoming housed in the automatisms of the living machine, and by making itself so to speak indistinguishable from these same automatisms. Somewhat mechanical and somewhat finalized, indeed, but since we only know two types of causality, the mechanical and the final, this "somewhat" becomes an "at the same time" which dissociates the registers.

The Kantian conception of technology thus echoes, in critical philosophy, the socio-political project identified by Schaffer and Sewell: one part of human activity seems to be able to be automated, the other not. The organ player is doubtless not completely an automaton; however, it is possible to build automata that play the organ. The activity of which this body-machine is capable can really, by construction, be transposed into real machines.

Admittedly, Kant also mentions in the *Anthropology* an old Aristotelian theme, that of the hand which is no instrument in action and can become all instruments in power. This theme puts technology in connection with exteriority, which is consistent with the idea that the human is the living being which could not remain a simple living being, in absolute contentment with its life. The human is the living being which must constantly force itself to leave its present state, that is, in short, to exteriorize itself with respect to this life in the state of nature, characterized by nudity and immediacy.

As Kant also says in the third proposition of the *Idea for a Universal History with a Cosmopolitan Purpose*, "Nature has willed that the human being should produce everything that goes beyond the mechanical arrangement of his animal existence… The invention of the means of his nourishment, his clothing, his external safety and defense (for which nature gave him neither the horns of the steer, nor the claws of the lion, nor the teeth of the dog, but merely his hands)… should be entirely his own work" (Kant 2009, p. 12). Kant thus indicates a path for anthropology which has subsequently been traveled, considering technology as that which supplements human life by externalizing its main functions (muscular, motor, intellectual functions) into artifacts.

It would nevertheless be wrong to maintain that Kant already shares the premises based on which Ernst Kapp introduced, a little less than a century later, the concept of technology as "organ projection". At first glance, Kant is more related to Rousseau, whose anthropology is also based on the idea that the human condition is ultimately very precarious. To provide for its needs, the human being does not have instinctive animal features; it is engaged in a process of indefinite perfectibility and as such, technology finds its meaning and its place: it is what compensates for the deficiency of nature in humans. The theme is also very old since it is already expressed in Plato in the famous myth of *Protagoras*.

Thus contained within the limits of the human condition, finding its legitimacy in its role of provider of the necessities of life, that is to say, the essential needs of the living human being, technology becomes suspect and dangerous precisely as soon as it exceeds this anthropological framework. The Kappian thesis of organ projection, as we will see, does not owe much to this anthropology of needs; it rests on other premises and opens onto a different anthropology than that of Rousseau.

That said, Kant's connection to Rousseau's anthropology of technology must be qualified, in the sense that the question of need is not a guiding one in *Anthropology from a Pragmatic Point of View*. Kant rather indicates a new perspective (in a very negative way, however) from which the question of the intertwining of technology and life can be posed, no longer on the basis of the concepts of need and deficit in the human condition, nor in connection with the future thesis of organ projection.

This intertwining, the possibility of which is closed off as soon as it is sketched, is linked by Kant to a whole other theme: that of the ease of doing what is difficult, that is, the theme of an action which, in a way incomprehensible, is effective but without representations and without rules. Neither Rousseau nor Kapp: Kant thus seems to open up a third path in the anthropology of technology in which it is above all a question, in and through technical activity, of articulation within each of us of the biological and of what exceeds it by circumscribing what is supposed to be "proper" to the human (reason, symbolism, etc.).

This orientation is indicated by Kant, but by omission and as if in the negative only. On the one hand, it excludes technology as such from having any relation to the faculty of pleasure and pain – and therefore to the feeling that the subject has of its own life. On the other hand, consequently, it also rules out that this vast repertoire of activities which are in a way mechanical, likely to be confused with the *agere* of nature, may place us at the junction of the biological and the anthropological, at the birthplace of this feeling that the subject has of its own life. These activities are rejected in physiological anthropology, in the study of the human being as a living being, but they do not give access to this enigma of a deployment of the anthropological, in the sense of this primordial feeling, at the very heart of the biological.

It was not until Mauss that technology (as a technique of the body, and not as a means of satisfying needs or as an organ projection) was finally, positively considered as an introduction to the articulation, in each of us, of the biological and the social, the life of the being-in-life and the life of the subject.

Another remark concerns the machine. This, says Kant, bears a danger: by making in the place of man, it defeats the dialectic of ease and difficulty. The organist easily does what remains fundamentally difficult to execute; this is still his merit. As for the machine, it does not know difficulty, so it does not know this anthropological experience of the conversion of the difficult into the easy, it even makes it impossible. The industrial machine, which (Kant does not say it quite yet like that) deprives humans of their role as bearer of tools and coordinator of technical tasks, cuts them off at the same time from this fundamental experience of difficulty converted, without rules and without underlying representation (here is the enigma), into ease.

Now, made foreign to this dialectical experience of the difficult and the easy, technology becomes indistinguishable from pure and simple automatism. For a human being to produce effects without any conscious representation, without any rule involved, is to be relegated to the animal, as Kant forcefully recalls from the first paragraph of the *Anthropology*. Technology belongs properly to humans when it is accompanied by representations, when it consists in the application of rules (even if these rules, as we have seen, may be sedimented in accumulated experience, in successive trials and errors). When it is disconnected from the sphere of regulated actions, technology falls back on the biological, to animality in the human – on the living machine therefore, that is to say on this machine that the human also is, as a living being.

We said of the organist that he illustrated a concept of technology housed in the automatisms of the living machine. It is now necessary to qualify: because he easily does what is difficult, the organist (or the technician who has what is called a knack, which allows him to make, to produce, without thinking about it) is deserving insofar as, precisely, it is not strictly speaking the living machine which acts within him; it is an art which has in a way put at its service both the living machine and its assemblies. The human always dominates nature; it does not dissolve in its operations. On the other hand, if the human being no longer knows this dialectic of the difficult and the easy in its technological activity – like, for example, the

Taylorian worker who must perform only one gesture – it can no longer be distinguished from the living machine: it is only that machine. By destroying the conditions of possibility for this anthropological experience of the easy and the difficult in industrial production, the machine destroys the person within the human being and shrinks it back to its animal being-in life, understood as a living machine – a machine of a particular type, certainly, but a machine all the same.

From there arise two lines of questioning whose extent and importance we have perhaps not yet-fully realized, more than 200 years later.

Firstly, beyond the fallacious opposition (long since rejected as such) of the beautiful and the useful, can we ask the question of how to (re)associate technology with the feeling of pleasure and pain, that is, with an aesthetic which, as Kant wanted in his definition of aesthetics, would deal in technology not with the relation to the object, but only with the relation to the subject? By the very fact that he excludes technology from the field of aesthetics, Kant invites us, as if by omission, to ask this question of technology in its relationship to the subject.

The question was asked, of course, mainly within a philosophical *topos* dominated by the concepts of prosthesis and exteriorization. Man makes his tools and is made by them in return, there is an anthropological constitutiveness of technology – this is a well-known and well-marked theme, the idea of a circularity co-constituting the subject and its objects, as we saw in examining the ACT thesis.

However, Kant invites us to ask this question in a different light, not in connection with the theme of exteriorization but with that of pleasure (and pain). On the sidelines of the great path of biological philosophy of technology whose common thread is precisely the concept of exteriorization of life into technology, and whose big names are Ernst Kapp, Alfred Espinas, Henri Bergson, Edouard Le Roy, André Leroi-Gourhan as well as Georges Canguilhem, as we will see in the next chapter, another path has been traced; one more confidential in a sense, with William Morris, John Dewey, Lewis Mumford, Laszlo Moholy-Nagy but also Gilbert Simondon, who have all posed the question of pleasure from a perspective primarily concerned with the question of the links between technology and life, and marginally with the theme of the organ projection or exteriorization of the biological in artifacts.

Kant may not have devoted important developments to the question of technology, but he was nonetheless the one who, undoubtedly in a subterranean way, underlined the philosophically elusive nature of technology, inasmuch as, in any case, we refuse to see in it an area of experience introducing everyone to the feeling of their own life, through the articulation of biological life and what exceeds it – an area of experience, in short, where the subject can constitute itself as a subject, both theoretical and practical. As we know, Kant reserved this privilege for the experience of fine art.

Secondly – the second line of questioning opened by the Kantian treatment of the question of technology – should we take note of the danger pointed out by Kant, and take for granted that indeed, the machine has ousted the human being from any possibility of knowing this dialectic of the easy and the difficult – or can we at least ask the question of how to restore this possibility today? How to return to the human being the full possession of this territory, from which the industrial revolution is said to have ousted it – a territory comprising all these activities which must be qualified as technologies, but avoiding the domain of final representations (*nexus finalis*) without being dissoluble into strictly natural operations, that is into the operation of the living machine (*nexus effectivus*)?

These two lines of questioning are, moreover, connected and one has to wonder if one should not go through them together – a hypothesis which ultimately amounts to questioning the distinction established by Kant between the two meanings of life: the feeling that the subject has of its own life, and life in the biological sense of the term.

The question of technology in Kant might at first seem undetermined or even without originality. The Kantian characterization of technology in paragraph 43 of the *Critique of Judgement* indeed revisits many motifs from the ancient, Greek conception of *technè*. Firstly, the Aristotelian distinction between *physis* and *technè*, and between natural and artificial beings. Secondly, the idea that technology is certainly not refractory to theory, but that no theory can anticipate all the elements to be taken into account, in a particular context, in order to ensure technical success: in short, no theory can circumscribe the domain of *empeiria*, experience, which requires practical wisdom, an ability to grasp the relevant elements of the situation and to be cunning with nature, rather than theoretical knowledge. Thirdly, let us finally recall Kant's perplexity in the face of this domain of human activities which seem to dispense completely with representations, and come dangerously close to the automatisms of the living machine, if not for their connection to an eminently anthropological experience, that of ease in accomplishing the difficult.

For the Greeks already, an activity unable to account for what it involves, and its own operations, is not even a technology; it is excluded from the domain of *technè*, and whoever is given to it is *atechnos*. This is also basically what Kant seems to be saying in *Anthropology from a Pragmatic Point of View*: these activities come from a physiological anthropology and not from a pragmatic anthropology. Technology in Kant is technically-practical; what is excluded from the practical field cannot, logically, be qualified as technological. However – and therein lies Kant's genius, by which he opens up an entirely new line of questioning – these activities cannot, however, be completely excluded from the technological domain, and be reduced to the level of purely natural operations (*agere* as opposed to *facere* in paragraph 43). Because they consist in doing easily what is difficult, they undeniably belong to the anthropological sphere, they are not comparable to what animals do by instinct – hence the notion of merit which attaches to it.

In short, the embarrassment is considerable. Here then is a type of activity which is not theoretical, which on the contrary seems to have no relation whatsoever to the theoretical; a type of activity which is not of a moral nature, which is even almost indistinguishable from something precisely opposed to moral action: automatic conduct, without freedom. Finally, it is also a type of activity which does not belong to the arts as Kant characterized them in paragraph 43, insofar as, firstly, there is no production of an artifice (the organist does not produce an *opus* but rather an *effectus*); secondly, there is no application of rules (the shoemaker does what Camper cannot do, but he does it by following rules: all art must mobilize what Kant calls "mechanical rules"; however, this is apparently not the case for the organist who plays without thinking about it); thirdly and finally, work and its own arduousness, from which art is inseparable, seem here to be erased, cancelled in the easy accomplishment of what is however difficult.

Neither science, nor art, nor moral action: where are we then? In pure biology? No, Kant finally seems to say, at least not entirely: that is the mystery. Though Kant does not remove the embarrassment, his great merit is to have indicated it, to have specified the reasons for it; otherwise to have circumscribed, or at least indicated, the existence of a nebulous, obscure, incomprehensible domain, which belongs according to him to the specifically human to the exclusion of animals, but which however does not seem to belong to any of the human faculties which, in critical philosophy, delimit the contours of anthropological difference.

Of course, rejecting this domain in favor of a biological anthropology, as Kant does, does not solve anything and only displaces the difficulty: it would be necessary to understand in what way the human biological formula differs from those of other living beings, a sort of second order difference. How, within the biological domain, in the immanence of the operations of nature, can a properly anthropological domain, not shared by other animals, already be outlined? Kant stops at this point; this question is not his but, as we will see, that of Leroi-Gourhan. From there only, in the space of thought opened by this question, could a conception of technology as inseparably both a biological fact and a specifically anthropological reality be proposed.

As such, as we have already indicated, an essential milestone is the famous text by Mauss on the techniques of the body, opening up a conception of technology at the hinge of the biological and the social, but devoid of any reference to the theme, then very much in vogue, of organ projection and exteriorization. I have the impression of being a hen that gave birth to a duck, Mauss is said to have commented about his pupil Leroi-Gourhan (at least that is what Leroi-Gourhan confided towards the end of his life). Over this duck – that is, Leroi-Gourhan – no doubt hangs the shadow of Bergson as we will see; but behind this hen – Mauss – looms the great figure of Kant and his difficult legacy.

2.2.3.4. *Technology beyond the artefact*

The embarrassment at which Kant arrives can be explained, as we said, by a conception of the arts which is still classical, in many respects Greek. The first characterization of paragraph 43 of the *Critique of Judgement* is therefore decisive: art is the production of a work, of an artificial reality, as opposed to nature which produces only effects. This production of artificial reality, for the Greeks, goes beyond the idea of artifice, of an artificial object.

Plato indeed distinguishes two kinds of technology, which we have already encountered above, and which we may call respectively manufacturing and piloting. Manufacturing results in a product, a work in the proper sense, a tangible thing which is added to the furnishings of the world: it is the field of handicraft but also of fine art. Piloting does not produce this type of objective reality, nevertheless it produces its own reality: the pilot of the ship, for example, produces the safe arrival of passengers and/or cargo to port.

In other words, piloting is not a fabrication in the true sense, but it is not a simple production of effects in the sense that nature produces effects: it leads to a tangible result in its own way; not a pure *effectus,* but already an *opus*. In Kant's terms, piloting still involves *facere*, making, even if it is not the making of an object. In any case it does not involve *agere*, in which Kant remains altogether faithful to the Greeks. *Technè* and *physis* are not for the Greeks two domains which can be united under the same genre, that of production – hence the scandal in describing as technologies these operations within us which seem to be nothing more than a surge in *physis*, of an active nature, of an *agere* and not a *facere* (the organist who plays with such ease that his activity seems to be controlled by nature alone, not by a nature which provides rules for art and which may be reflected under a concept of finality, but by a blind nature, whose *agere* within us cannot be judged according to rules).

Kant thinks of technology in an intellectualist framework: technology is necessarily dependent on representations, even if these do not achieve clarity and distinction and retain a degree of obscurity and confusion – like the shoemaker whose activity goes beyond what can be theorized by Camper, but nevertheless does not derogate from this dependence of any technology on representations. On the one hand, nothing contradicts the fact that technological activity is motivated by a goal, a purpose – *nexus finalis*: the artisan has a representation of an end. On the other hand, in the Kantian shoemaker, even if they go beyond the theorizable and the formalizable, representations are nonetheless assumed. They are simply gathered up, combined in the action (remember that in the *Encyclopedia*, the shoemaker is the typical example of the craftsman whose mechanical art contains a weak measure of intelligence), sedimented in these gestures of which Kant does not speak and for good reason: these gestures disturb the representational conception of technology, but they do not find an appropriate language to justify it.

Paragraph 43 is therefore ambiguous: on the one hand, it confirms the excess of the technological over the theoretical, but on the other hand, in Kant, this excess does not positively sketch out a domain of human action which would not involve representations. Paragraph 43 is ambiguous because it indicates an excess, but it immediately closes it off by widening the sphere of the representational. The shoemaker still acts by representations, firstly of the ends that he aims for, and then of the operations that he fulfils, even if these largely escape him in the sense that they do not meet, to use Cartesian terms, the criterion of clarity and distinction.

The case of the organist at the start of the *Anthropology*, on the other hand, poses, as already said, a much more difficult problem: that of an activity which does not involve any representation and which Kant does not call technological, but which in any case cannot be compared with the activity of the bee (it is not an *agere*). The organist in the *Anthropology* and the shoemaker in paragraph 43 seem to fall under the same scenario, an excess over any possible theorization. However, on closer inspection, this is not the case. Unlike the shoemaker, who has nothing to do with an automaton, the organist plays like an automaton although he is not an automaton, precisely because he has experienced this conversion from difficult to easy, while of course, the automaton does not know the concepts of easy and difficult.

To recognize these operations – those of the organist and not those of the shoemaker – as being fully technological, we must therefore begin by relaxing the well-defined perimeter of the domain of the arts, that is to say of the artificial; firstly, by redefining what Kant calls *opus*, a work. More specifically, it means tackling the fundamental problem of intellectualism upon which Kant remains dependent.

Kant has clearly seen the difficulty in relation to the subject, the organist who cannot be considered as a technician, the organist whom it is convenient (at the moment when he plays without thinking about it) to reduce to the level where bees deploy their effects – in terms of the *agere* of nature. Unfortunately, as Kant knows well, the organist is not a bee playing the organ. He recognizes this (not necessarily in these terms of course) but stops at this point.

The first paragraph of the *Anthropology* sets the distinction of the human from the animal. The animal is pure biological life; the human has reflexive consciousness. The organist plays as if only nature were operating in him, like the bee, and yet he can legitimately be the subject of admiration, of recognition of merit, because he easily does what is difficult. Shall we say that a bee has merit? Of course not. So, the organist must be doing something other than allowing nature to *agere* within him. However, no representation is involved. So, it is not a *facere*, a technology. Neither *agere*, nor *facere*: what is it?

Kant answers without answering – this is part of biological anthropology – which means that he leaves the question to us. His genius is precisely to have given an answer that is not one, that is, one that requires us to consider the question itself from a completely different angle.

What Kant could not see, however, is the relationship between, on the one hand, his problem of human activity that is automatic but not to be confused with the *agere* of nature, and on the other hand, his definition of *opus* as an artefact – as something manufactured as a means for a definite end, or as a finalized activity like piloting. We are in the 18th century, technological activity effectively boils down to these two activities which both lead, as we have seen, to the production of a certain type of *opus*. However, as Hannah Arendt forcefully emphasizes, the industrial era, with electricity produced by thermal (coal) and then nuclear power plants, opens up something new in human history (Arendt 1958): from now on, acting technologically no longer consists of *facere* (*opus*) but of *agere*, triggering processes in nature. The need to abandon the intellectualist framework therefore faces the technological subject – this organist that Kant could not understand – as well as the technological object – which Kant could not of course anticipate.

Recent work has insisted on the fact that our devices behave more and more like natural realities: with nanotechnologies, biotechnologies and synthetic biology for example, our activity consists less in manufacturing objects which will take their place in the human world (*facere*) than in triggering processes indistinguishable from the processes of nature (*agere*). By emphasizing the incomprehensible nature of what the organist does, and by kicking the question into touch to remain consistent with his frame of thought, Kant opens the way to a redefinition of the technological subject, but also, by necessity, of the technological object. It is indeed the relationship that counts: if the subject operates by interposed representations (even if they are forgotten, collected, sedimented in gestures), then the object (*opus*) is also dependent on a representation. This representation, Kant tells us, can only be of two kinds: *nexus effectivus* (the technological thing, the shoe for example, is always dominated by theoretical ambition) and *nexus finalis* – technology is something useful. Carrying further the problem left to us by Kant is therefore both to recognize the activity of the organist as an authentically technological activity, and to loosen the relationship between the technological object and these two causes known to us, mechanics and finality, which Kant discusses.

This is not a cosmetic, marginal revision: if critical philosophy excludes from its field these activities which are very close to the automatisms of instinct, then the widening of the technological domain beyond what Kant defines as the arts calls for an exit from critical philosophy, with a view precisely to making full use of this enigmatic irruption of anthropology at the very heart of the biological, in the most obscure of the organism's operations.

According to Kant, the human faculties – faculty of understanding, faculty of desiring, faculty of pleasure and pain – start from this feeling that the subject has of its own life, a feeling that is not of the order of a sensation (this feeling does not really sense anything), nor of the order of a knowledge that the subject has of itself, nor either a beginning, an embryonic version of morality. These three themes indeed, sensation, self-knowledge, and morality, involve representations of objects – the object which gives rise to a pleasant or unpleasant sensation, the ego as the object of my knowledge, the good as the object of moral action. However, the feeling that the subject has of its own life does not pass through any representation of objects, including itself as an object. This feeling places us at the heart of anthropological experience (echoing the first feeling of man which was that of his existence in Rousseau). If self-awareness, as Kant recalls in the first paragraph of the *Anthropology*, distinguishes humans from animals, this self-awareness is not primarily a reflection of oneself upon oneself, an inspection of one's own representations (remember that Kant explains on the contrary how, and why, engaging in an inspection of this type can be dangerous). It is originally this feeling without object, this pure feeling of its own life.

The problem left to us by Kant is therefore to articulate, on the one hand, this feeling as the inaugural deployment of the anthropological field, which the three faculties of the soul will come to organize, and, on the other hand, the domain of biological processes. These activities which have no name, no status, which are not thinkable as human activities in Kantian anthropology, and which however are only encountered in humans (trivially, a non-human animal is not a skilled organist) – these activities at the limits of the automatisms of the living machine therefore introduce the challenge of articulating these two meanings of life. This is the challenge, considerable it will be agreed, of a philosophy of technology: going beyond the framework of a classical, still Greek, characterization of the arts (paragraph 43 of the third *Critique*), to inscribe the anthropological field within the thickness of biological life.

We must therefore start by reconsidering the reassuring, apparently well-mastered, well-defined area of the arts and crafts – understanding technology other than in reference to *opus* as defined by Kant in paragraph 43 of the third *Critique*, and before him a long tradition of thought that goes back to the ancient Greeks. However, in Kant, the characterization of the arts, though it seems to be carried out under conditions of security, in the reassuring shade of this old tradition, is already bothered by a kind of anxiety.

Technology is indeed, we said, of the order of pretense; it conceals its true nature. Its theoretical nature is always tainted with *empeiria*; it involves foresight, a form of practical wisdom, but in this it only simulates acting out of duty, moral action. It implies the faculty of pleasure and pain (does the watchmaker not take pleasure in manufacturing a watch with subtle and impeccable mechanics?), but what is properly technological in the arts relates exclusively to work, and is therefore unpleasant and arduous: pleasure can be mixed with technological activities, but it is not part of their essence. Technology therefore has an ambiguous relation to the theoretical domain, to the moral domain and to the feeling that the subject has of its own life, in which it seems to participate only in a truncated, mutilated way – painfulness, never pleasure.

In summary, expelling technology from aesthetics goes hand in hand with a definition of technology as *opus*, as well as with an anthropology which cuts off the human from the vital. Under these conditions, the only possible evaluations of technology relate, on one hand, to its conformity with the rules of the art, and on the other hand to its capacity to satisfy the expectations and needs of the users. From these premises, technology can only have a very narrow relationship with the moral requirement of care: it is simply a matter of being careful in the way of doing things, of being careful with the materials transformed and also concern for customers – which, it will be said, is already a lot and even somewhat demanding from a moral point of view. In no case, however, can there be any question, in technological action oriented towards the outside world and manufacturing, of "taking care" and "caring" for the overall processes of nature and living beings in general.

Or rather, more precisely, this intellectualist understanding of technology can lead, at best, only to a vertical conception of care, in the sense of sheep herding of which Haudricourt speaks – an attitude which, on the one hand, may prove to be fundamentally devoid of any moral value, and which, on the other hand, does not measure up to the challenges we face.

2.3. Transition

Kant consequently introduces us to the most general terms of the problem of care, in the negative so to speak. To extend the problem of care to our technological action upon the natural world and all living beings, it is necessary to return to the terms of a dualistic anthropology which began by cutting humans off from their own being-in-life – more precisely, which concentrated upon one dimension of human reality (reason) that is supposed to define the human. According to this dualistic perspective, valuations proceed from an extra-natural rational activity (the reign of ends). Knowable external nature, including that of the body, is emptied of any reference to anything that affects values: it is only a useful mechanism and resources.

To admit on the contrary that valuations proceed, not only from reason, but from the human being as a living being, is to *de facto* admit that human valuations are formed within nature itself. Another way of saying this is that technology, which is today more than ever the channel for human valuations, is not the work of humans as rational beings, but as living beings.

It must be noted, moreover, that after Kant, philosophical reflection on technology has taken great pains to relate it to life as being-in-life, the life of the organism. From this common premise, however, theoretical positions diverge very strongly and not all of them, far from it, really introduce a thought of care into technology.

Thus, German philosopher Oswald Spengler affirms very explicitly the vital anchoring of all technicality (Spengler 1958): technology is immemorial and widespread throughout the animal world, it is defined as the set of "tactics" that the latter uses to defend itself against nature. A plant is completely immersed in nature, it cannot be detached from it; an animal is mobile, it has greater or lesser independence from nature and must defend its integrity. Technology is the expression of the struggle for life; it is above all a weapon, and from this point of view the situation of humans is no exception: the human being is a beast of prey, it too must fight, dominating nature.

As we can see, this frankly anti-rationalist conception of technology, which justifies predation and domination of nature, will probably have little to teach us about care. It turns away from it without ambiguity, which leads to this observation: linking technology to the activity of the organism, on the trivial observation that human beings, since they are still living beings among living beings, belong to nature, is in no way a sufficient condition to direct technology towards care. More specific conditions are needed, and this is what the next chapter intends to examine.

3

The Conditions of Care in Technology

3.1. Vitalist approaches to technology

3.1.1. *The concept of technological evolution: contributions and limits with regard to care*

3.1.1.1. *Origin and development of the concept of technological evolution*

It is paradoxical at first sight that the 19th century was, at the same time, that in which the divorce of technology and life manifested itself most clearly, the premises of which we have examined in the previous chapter; but also that in which theoretical proposals for thinking about technology from the basis of life in its biophysical dimension were particularly numerous and stimulating. These attempts to link technology to life have moreover followed two major paths: on the one hand, evolutionary approaches, stipulating that technology follows lines of evolution along the same lines as living species; on the other, approaches in terms of the "organ projection", according to which the human being is a very special living being which, in order to perform the functions necessary for its own life (using muscular force, moving around, keeping traces of its past, calculating, etc.), has acquired artificial organs external to its organism. The two perspectives are not, however, mutually exclusive and could be combined.

What does it mean exactly, following one approach or the other, to link technology to life in the sense of being-in-life? Is this enterprise absolutely unambiguous with regard to the requirements of care?

The publication of *On the Origin of Species*, in 1859, was undoubtedly a seminal event in the vitalist orientation of conceptions of technology in the second half of the 19th century. The idea of applying the Darwinian theory of evolution to artifacts and their development dates back to the 1860s, immediately after the publication of

Darwin's work. It is found for the first time in a novel by Samuel Butler, *Erewhon* (Butler 1970), published in 1872 but whose first drafts date back to the early 1860s with the publication of an article, in 1863, entitled *Darwin among the Machines*.

Meanwhile, Marx invokes, in Book I of *Capital*, a possible analogy between, on the one hand, the specialization of tools in the service of productive operations which are themselves more and more specialized and fragmented, and on the other hand, the development of the organs of living beings. Indeed, writes Marx, "the manufacturing period simplifies, improves, and multiplies the implements of labor, by adapting them to the exclusively special functions of each detail laborer" (Marx 2019, p. 375). However, in a footnote, Marx refers to Darwin: "Darwin in his epoch-making work on the origin of species, remarks, with reference to the natural organs of plants and animals: 'So long as one and the same organ has different kinds of work to perform, a ground for its changeability may possibly be found in this, that natural selection preserves or suppresses each small variation of form less carefully than if that organ were destined for one special purpose alone. Thus, knives that are adapted to cut all sorts of things, may, on the whole, be of one shape; but an implement destined to be used exclusively in one way must have a different shape for every different use'" (Marx 2019, footnote 1 p. 375). In other words, from the point of view of the process of specialization and differentiation, human technology can be compared to natural technology.

Commentators have, however, underlined the limits, even the erroneous character, of this analogy, from the very perspective of Marxian political economy (see for instance Guerpillon 2015). The application of the Darwinian model of natural selection to analysis of the development of the factory comes up against the problem of teleology: an organ is not oriented towards a function, unlike the technological tool[1]. This difficulty is also found in the mainstream of evolutionary economics. In the 1970s and beyond, the evolutionary conception of technology met with growing success, under the impetus of economists (Nelson and Winter 1982) and economic historians (Mokyr 1996), but also thanks to historians of technology like George Basalla (1988). As shown by what is certainly the most documented collective work on the issue (Ziman 2000), the aim of these authors was to give greater theoretical consistency to the use of Darwinian concepts of variation and selection in the field of technology. This involved passing from purely descriptive analogies between technological evolution and biological evolution, to a realistic model of technological evolution, explaining the mechanisms according to which technologies actually succeed each other in time (Solée *et al.* 2013). The new theorists of technological evolution tackle in particular the thorny problem

1 Guerpillon cites the work by Naccache, B. (1980). *Marx lecteur de Darwin*, Vrin, Paris, in which this critique of the analogy between natural selection and the development of manufacturing is developed.

underlined by commentators on Marx: can we really propose a realistic model of technological evolution, when it is obviously not possible to do without the concept of intentional design in technology – a concept which obviously has no place in a realistic model of biological evolution, variation there being blind? How to integrate intentional factors into a model of technological evolution?

In addition, the application of the Darwinian theory of natural evolution to technology also appears very early in archaeology, in particular among archaeologists John Evans (who proposes to apply the theory of natural selection to numismatics, Evans 1864) and Pitt Rivers (who undertakes to think about the best way to classify ethnographic documents in museums scientifically, according to the connections they show between forms, Pitt-Rivers 1906).

Evolutionary approaches also guide efforts to propose theories accounting for the development of lithic industries over the long term of prehistory, as evidenced by the work of Charles Frémont (1913), Leroi-Gourhan (1943, 1945) or, more recently, the overhaul of the evolutionary theory of Paleolithic tools by Eric Boëda (2013).

The evolutionary conception of technology is, finally, found in machine theorists like Franz Reuleaux (1876), Jacques Lafitte (1980), Simondon (2017) or Yves Deforge (1985).

However, what emerges from these different specialized approaches, in archaeology, prehistory or in general theories of machines, is that the analogy between technological evolution and biological evolution takes on a mainly methodological significance: it must allow us to build a coherent framework for classifying technologies, beyond their diversity of forms, sources and uses, by applying a rational system to them. Leroi-Gourhan is particularly clear on this point: he presents his work on technology as a logical enterprise, and not as a historian's book concerned with finding the real succession of technological inventions and the ways in which tools have been disseminated. Thus, he confides about his work, "one should not read in these pages the history of technology", and except when "there appears a case of origin, of innovation" which "will be welcomed with the enthusiasm justified by its rarity", the documents will be ordered "not historically but logically" (Leroi-Gourhan 1943, p. 18).

The idea that there is technological evolution, by analogy with the evolution of living forms, is thus based on considerations of logic and even, in Leroi-Gourhan, of mathematics: "Paleontology ... notes that by very numerous routes, the plan of construction of animal species changes to achieve more effective interactions with the outside environment. These modifications always lend themselves to a mathematical analysis of shapes ... We have extended these observations to Ethnology ... each tool, each weapon, each object in general, from the basket to the house, responds to a plan of architectural balance whose main lines are under the influence of the laws of geometry or of rational mechanics" (Leroi-Gourhan 1945, pp. 337–338). To clearly mark the meaning of the concept of technological evolution as classificatory above all else, Leroi-Gourhan adds: "The fossils [and thus the technical objects] that we arrange to constitute the genealogical line are not necessarily direct ancestors of each other, but constitute a logical assembly of increasingly older forms. This approach does not lead to the establishment of a historical picture, but to a reconstruction whose very high likelihood is practically equivalent to the real genealogy, made inaccessible by the scarcity of documents" (Leroi-Gourhan 1943, p. 21).

In other words, the problem of teleology is embarrassing only if one undertakes to propose a realistic evolutionary model, explanatory of the history of technology as it actually took place. On the other hand, this problem recedes if the concept of evolution is no longer explanatory, but classificatory – if it is a non-descriptive but logical concept. This has indeed been the case.

3.1.1.2. *Ambiguity of evolutionary approaches with regard to care*

This brief[2] overview of evolutionary theories of technology calls for two remarks.

The first concerns the value, with regard to the question of care, of these evolutionary approaches, understood as projects of rational systematization of the domain of technology. These approaches have the common feature of going beyond the current instrumentalist and intellectualist perspective, which deals with technologies according to the categories of end and means. Organizing technologies into evolutionary lines is indeed, *de facto*, to detach them from this perspective by refusing to reduce them to the status of means subject to final representations. Simondon could not be clearer on this point (Simondon 2017). He thus opposes two ways of conceiving of technology. The first consists in seeing only the aim of a goal, and in this case, according to the example he gives, we will place in the same utilitarian category of "motors" various technological objects whose operating principles are very different (a thermal motor, an electric motor, a spring motor or a weight motor). The second consists on the contrary of freeing oneself from uses and goals, to envisage technical objects according to their operating schemes – in which case, the objects will be classified according to purely technological criteria and, to use the aforementioned example, the different engines will be sorted into different categories. Such is in any case for Simondon the scope of the analogy between machines and living beings. Like the second, the first are inscribed in evolutionary lines whose overall unity is that of a general scheme of functioning, and not that of a use determined by the aim of a specific goal.

2 For more details, see (Guchet 2005).

As a result, the evolutionary approach gives technologies a form of autonomy with respect to final representations; it attributes to them a specific rationality which cannot be reduced to these representations, which makes it possible to consider them not as simple means with a view to prefixed ends, but as beings intended to produce effects in the world, resulting from their internal scheme of functioning. The way is therefore open for a conception of technology which disregards the usual intellectualist and finalist framework, and possibly introduces an attitude of vigilance with regard to what our technologies do to the world.

The second remark, however, qualifies the first.

Indeed, it should be noted that the evolutionary point of view is primarily intended to make it possible to render judgement on the history of technology. Historically, it is very common to see technological innovations succeed, and prevail over others, for psychosocial reasons independent of any consideration of technological rationality. Effects of prestige or fashion, industrial and commercial interests: so many factors explaining technological choices which distance them from the pure rationality of evolutionary lines.

Admittedly, this dichotomy between "pure technology" and "psychosocial overdetermination", to use Simondon's terms, appears today to be very questionable, if not untenable. This work of purification, as Latour called it, is no longer considered relevant in analyses of technology. That said, an even more essential insufficiency seems to relate to the evolutionary perspective, with regard to the problem of care. A technological object which wonderfully exemplifies the potential for evolution of a technical line – in Simondon's words, which "concretizes" the operating scheme defining the lineage – is not necessarily a caring and careful object with regard to the world. According to an often-cited example from Simondon, the diesel engine is more concrete than the petrol engine – that is to say, it has a higher degree of interdependence of the parts and of operational synergies between them than petrol engines. Maybe. The fact remains, however, that diesel vehicles pollute as much as, if not more than, petrol vehicles. Technological "concretization", as an orientation with normative value for the engineer, in no way implies a caring conception of technology.

The evolutionary perspective here appears self-referential. The judgement of rationality which is based upon it is not encumbered with extra-technological considerations: the only thing that counts in the judgement is the value of the internal coherence of the lineage. We are poles apart from the requirements of what could be called the hetero-referentiality of care.

We might think that an evolutionary approach to technology would distance us from the dominant intellectualism, but in reality, it is nothing of the sort. It seems that the opposite is true: in introducing a systematic and rational organization of the very heterogeneous domain of technology, the concept of technological evolution seems rather to distance us from life.

3.1.2. *Technology as an "organ projection": contributions and limits with regard to care*

3.1.2.1. *In what sense is technology an organ projection?*

What about the other vitalist approach to technique – the so-called "organ projection" theory? Is this theory, proposed by German philosopher Ernst Kapp in a work published in 1875 (Kapp 2018), a better introduction to the idea of care in technology? This theory was a great success and percolated into a good number of analyses of technology, as we will see. However, it should be emphasized that its reception generally happened at the cost of an amputation.

The theory of organ projection is indeed not one but two theses; Kapp himself insisted on this essential point which has generally not been sufficiently recognized. More specifically, this theory has two aspects: a genetic aspect of the unconscious transposition outward of something bodily, in the manufacture of tools and machines (this is projection in the true sense); but also an aspect of reflexivity, on the other hand, which makes technology a model of intelligibility when applied to the organic. Projection, reflection: one does not go without the other, which suggests that Kapp has indeed attempted the double overcoming of a form of vitalist immanentism (projection alone) and pure rationalism (modeling of the living being by technological analogies alone).

We undoubtedly owe it to Canguilhem for having given his stamp of approval to the tradition of "biological philosophy of technology" stemming from Kapp, in which he saw a true philosophical elaboration of the technological fact, creating an alternative to the dusty intellectualism which at that time, that is to say in the 1940s in France, characterized most philosophical contributions on the subject (Canguilhem 2008). However, Canguilhem himself had clearly seen this duality of movement in Kapp, this back and forth of projection – from inside to outside – and of reflection – from outside to inside.

In "Machine and organism", Canguilhem indeed formulates criticisms against the intellectualist conception of technology – more precisely, against the Cartesian conception of the animal-machine – by pointing out precisely that this conception of the living being rests entirely on forgetting the first movement of the thesis of Kappian organic projection, retaining only the second. Descartes thus claims to find in the existing machines of his time good explanatory models of vital phenomena (reflection), but in doing so he completely obscures the fact that it is precisely living beings who build machines (projection). Vital activity consequently precedes that which is supposed to model it; the living being comes first before the intelligence which attempts to grasp it.

Canguilhem adds a vital precision: the intellectualist conception of the animal-machine contributes to justifying the economic exploitation of living beings. Indeed, if animals are comparable to machines, then their instrumentalization in human economic activities raises no more moral difficulties than the use of any machine. By restoring the complete circuit of projection and reflection, Kappian theory may therefore, if we understand it correctly, support a moral and political criticism of the forms taken by this instrumentalization, with regard to the values linked to vital activity which is now conceived, no longer as a mechanical functioning, but as a valuative activity. Kapp's thesis would thus distance itself from human predation on environments and the beings that inhabit them, and would introduce the requirement of care into the relationship of humans to other living beings.

It should be added that the theory of organic projection is something else, Kapp is very clear on this point, than a pure and simple transposition of the apparent forms of organs into technology. This would suit the most rudimentary tools, which extend the hand by reproducing its shape (for example, the hammer reproduces the shape of the arm and clenched fist). On the other hand, as soon as we consider more complex machines, in particular industrial machines, the idea that organic projection is reduced to a kind of morphological mimicry is no longer tenable. Kapp writes, "The products of organic projection were initially, as we have seen, of crude sensuous simplicity, serving only to facilitate difficult manual work. Gradually, we watched them assume, in multifarious configurations and sophisticated designs, the form of scientific tools and apparatuses to serve the ends of spiritual activities. At this point, the formal after-image begins to give way to the representation of inner relations and proportions" (Kapp 2018, p. 79). Kapp explains that the hand makes, using the first tools, other tools which end up losing the very approximate conformity which they originally maintained with organs, to the point of hardly showing any morphological resemblance to them. Despite everything, this is still a projection of the organs. Kapp says, "If anything, the projection is all the more powerful the more essentially it reveals to us the organism's proportions and relations – which themselves appear purer and more spiritually transparent the less our attention is distracted by too great a fidelity of the plastic form to the natural prototype" (Kapp 2018, pp. 58–59).

Organic projection therefore does not designate a pure and simple transfer of apparent forms from the inside of the body to the outside; it means the exteriorization of relationships, functional principles, operating chains. It is not structures that are exteriorized, but operations. Thus, "as the steam engine so clearly illustrates, progress toward higher mechanics consists less in the unconscious reproduction of organic forms than it does in the projection of their functional image – that is, the image of the living and of spirit operating as organism" (Kapp 2018, p. 102). To build a machine capable of producing a convergent functional activity, says Kapp, the machine builder was forced to do much more than project the determined form of an organ into exteriority: the builder had to seek within themselves the model of the unitary living articulated totality of the bodily organism. It is not this or that determined form of an organ which is projected into the machine, but the overall functional organization of the organism.

The Kappian theory therefore has the advantage of linking technology, not to a simple morphological and mechanistic mimicry, but to the activity of humans as living beings – which strengthens a Canguilhemian reading.

3.1.2.2. *Limits of the theory of organic projection with regard to care*

That said, it still seems that Canguilhem has extensively extrapolated from the letter of Kapp's text, drawing organic projection towards a conception of technology as a valuative activity, with the moral and political implications already mentioned.

In reality, Kapp does not express at any time in the *Grundlinien* that his theory might have the potential for criticizing the predation that humans exercise over living beings and nature in general. In particular, nowhere does it deal with possible conflicts between the valuations of humans as technicians and those of humans in that they do not cease to be living beings, as Canguilhem does so well.

Kapp does not understand organic projection as a movement oriented by values according to which humans organize their environment, but which therefore can always risk doing violence to the valuative activities of human and non-human living beings. At no time, in short, does Kapp conceive that technology may plunge human existence into a "fundamental absurdity", in the words of Simondon, due to the incompatibility that is always likely to arise between the technological and organic conditions of its existence.

In Kapp's eyes, the industrial era clearly does not seem to introduce any new data into the appreciation of the values of technology with regard to the needs of living beings (human as well as non-human): the first knapped flint is an organic projection, just like the machine tooling of the industries of the 19th century, as well as large socio-technological networks on the scale of a country or a continent (railways, telegraph). One would hardly find in Kapp arguments serving a critical analysis of industrialization and its lack of care. The Kappian philosophy of technology ultimately comes to support a conservative politics, illustrating the totalizing and nostalgic organicist posture so well criticized by Haraway (Haraway 1991).

The Kappian conception of technology has therefore closed off rather than gotten to grips with (much less solved) the problem left open by Kant: namely that the solution to the great problem of articulation, in humans, of biological life and the feeling that everyone has of their own life as an individual, cannot be found in the unfolding of the organic processes themselves (whether "projective" or not), but in an activity creating new values – a valuative activity – which for Kant the model is provided by art, not by technology.

In all likelihood, therefore, Kapp's theory does not immediately impose itself as a good candidate for an evaluation of technology with regard to care, and the situation seems ultimately discouraging: the possibility of moving towards a caring conception of technology seems resolutely barred, both on the side of evolutionary approaches and on the side of the Kappian theory of organic projection. Should we give up?

Before that, the reasons for this situation of deadlock remain to be examined in detail.

3.1.3. *The utopia of Erewhon: analysis of an aporia. First condition of care in technology*

3.1.3.1. *Two opposing positions concerning technology in Erewhon*

It turns out that before Marx, before also the first writings of archeologists proposing to apply the evolutionary model to technology, before even the publication of Kapp's book, Butler had already given a remarkable account of the philosophical ins and outs of the two great vitalist approaches to technology (Butler 1970).

Erewhon is not a work of philosophy properly speaking, but a utopian novel (Erewhon is an anagram of "nowhere"). Erewhon designates a country cut off from the rest of the world, governed by an absolute law which may be surprising: machines which were invented less than 271 years ago have all been banned there. This technophobic decision, curious at first (and the narrator is surprised), is explained both by the conception that the Erewhonians have of human beings and by their evolutionary conception of machines: the human is a being of nature, it is a form of life among others and it is deeply influenced and modifiable by all kinds of external factors. Machines have always been such factors. Humans have shaped machines, but machines have also shaped humans. The human has thus become dependent on machines in its constitution as a living being. The problem is that machines, the solution to the great problem of the adaptation of living human beings to their environment, are on the way to becoming autonomous and acquiring a life of their own. A great danger lies in wait for humans: the reversal of the relationship between ends and means. While the machine is today the means by which humans have adapted to their environment, tomorrow it is humans who risk becoming a means at the service of the machine. To prevent this supreme risk, the Erewhonians have therefore decided to block, by collective decision, the somewhat natural evolution of machines towards their autonomy. Less advanced machines, simple tools, guarantee the maintenance of technology within the perimeter of possible human control.

Butler's analysis so far could not be more classic. However, if *Erewhon* is known mainly for this evolutionary conception of technology, Butler also introduces a completely different idea, also passed to posterity, of the relationships between technology and human life: this idea is precisely that of organic projection. The word is not there, but the thing is undeniably there. The human is in fact described by Butler as a prosthetic being, engaged in the adventure of what he calls dephysicalization: by his prostheses, the human is about to break with what natural evolution has made. Some Erewhonians see it as a very interesting and desirable future; however, the majority of the inhabitants of Erewhon do not see it like that. They consider this perspective rather from the angle of the danger of enslavement, when the human is somehow reduced to being only the prosthesis of his prostheses, a simple cog in a field of prosthetics considerably widened and freed from any link with living corporeality.

In the great controversy that pitted the Erewhonians against technology, enthusiasts were in the minority. The evolution towards ever more sophisticated machines, exteriorizing human bodily functions ever more fully, has been stopped altogether.

It should be noted, moreover, that this diagnosis of a decorporatizing and debiologizing prostheticism, if one may put like this, finds its exact counterpart in the symmetrical and inverse diagnosis made a few decades later by philosophers like Adorno, Horkheimer or even Arendt. Let us quote Adorno and Horkheimer: "The human species, they predict, including its machines, chemicals, and organizational powers for why should they not be seen as a part of it as teeth are a part of the bear, since they serve the same purpose and merely function better? – is the last word in adaptation" (Adorno and Horkheimer 2002, p. 184).

Arendt is not to be outdone, and worries about the "danger signs that man may be willing and, indeed, is on the point of developing into that animal species from which, since Darwin, he imagines he has come. If, in concluding, we return once more to the discovery of the Archimedean point and apply it, as Kafka warned us not to do, to man himself and to what he is doing on this earth, it at once becomes manifest that all his activities, watched from a sufficiently removed vantage point in the universe, would appear not as activities of any kind but as processes, so that, as a scientist recently put it, modern motorization would appear like a process of biological mutation in which human bodies gradually begin to be covered by shells of steel" (Arendt 1958, p. 322). "If we look down from this point upon what is going on earth and upon the various activities of men, that is, if we apply the Archimedean point to ourselves, then these activities will indeed appear to ourselves as no more than 'overt behavior', which we can study with the same methods we use to study the behavior of rats. Seen from a sufficient distance, the cars in which we travel and which we know we built ourselves will look as though they were, as Heisenberg once put it, 'as inescapable a part of ourselves as the snail's shell is to its occupant'" (Arendt 2006, pp. 273–274).

In other words, generalized prosthetics make humans appear no longer as the being of meaning and political action, but as one species among others, which has simply succeeded better than others in its adaptive strategies. At the antipodes of the prospect of dephysicalization which worried the Erewhonians, Adorno, Horkheimer and Arendt point rather to the danger of a reduction of the human to its dimension of "simple" living, in which all activities would be only manifestations of the vital metabolism.

3.1.3.2. *Incompatibility between the concept of self-regulation and the aim of care*

These two concerns, despite their apparent opposition, are in fact only one. This is what Canguilhem very clearly highlighted, indirectly at least, in an article from the 1950s on the problem of regulation in the organism and in society (Canguilhem 2012).

Canguilhem compares two attitudes which are apparently contradictory but in reality complicit: the first attitude assimilates society to a machine, the second to an organism. The Erewhonians seem to favor the first option, fearing that the human is no more than a mechanical cog in a prosthetic super-machine, while Adorno, Horkheimer and Arendt see rather the danger of an extreme biologization of the human, a living being among others, in an organism-society whose structuring values would henceforth be dictated by adaptive necessities alone. Two opposing visions, says Canguilhem, but both complicit in that they also affirm that in human societies, regulation can be thought of as self-regulation. Machines (cybernetic at the time when Canguilhem wrote his article) have self-regulatory mechanisms thanks to which they can maintain the aim of a goal despite changes in their environment; organisms are of course based on self-regulation, the names of which may vary over time (homeostasis for example).

The machine-prosthesis society, as well as the adaptive-organism society composed of living entities, are two conceptions which are therefore based on the same main idea, namely that a human society may be conceived on the model of self-regulating dynamic systems. The Erewhonians as well as Adorno, Horkheimer and Arendt fear this prospect, they reject it; however, it seems to them to take shape implicitly in the developments of industrial machinery, and to result inexorably from these developments if nothing arises to thwart them.

We therefore understand the underlying reasons explaining the incompatibility between, on the one hand, the two approaches – evolutionary and in terms of organic projection – and, on the other hand, the requirements of a technologically caring and careful attitude with regard to the world and to living beings. This latter attitude requires, as we have already said, a continuous evaluation of the results of technological action with regard to the valuative activities of living beings, human as well as non-human – an attitude necessarily pragmatist and always revisable in its conclusions, incompatible with the very idea of self-regulation, that is to say with the idea of entrusting the regulation of action to mechanisms operating by themselves.

This is, moreover, the fundamental problem raised by care robots today. How to design robots capable of providing care, when, on the one hand, a robot is a machine which acts by applying rules of inference (whether a programmer initially provides them, or whether its algorithm leads to it to work out for itself the rules to be applied starting from a knowledge base), and, on the other hand, care consists essentially in doing something else, precisely, than applying rules – whether these are rules that the acting being gives itself (as an unsupervised learning robot does), or deterministic rules inherent in the structure that acts – in other words, self-rules in both cases?

Before these two paths of the biological philosophy of technology, namely technological evolutionism and organic projection, developed, Butler had therefore associated them in supreme anxiety, the consequence of which is ultimately the impossibility of a care which moves through technology.

3.1.3.3. *To take care, must we admit that there is an immutable nature of things?*

That being said, there is reason to doubt that by preventing technology from escaping human control, the Erewhonians made possible any form of caring and careful action. It should be noted that considerations on the evolution of machines and its dangers do not occur until the end of Butler's work. They are preceded by a long account, by the narrator, of the bizarre conceptions which are prevalent in *Erewhon.* These quirks are the result of a complete reversal of values, affecting in particular the definition of crime and disease. The Erewhonians in fact punish diseases of the body; the sick must appear before the courts and are sentenced. What we spontaneously call crime on the other hand (or at least an offense), for example theft or swindling, is there considered to be a disease of the soul which rather arouses lamentation and compassion, and falls within the competence not of the judge but of a kind of doctor of the soul: the "straightener"?

However, this inversion of values is explained by the Erewhonians' conception of social stability. Like any society, Erewhonian society seeks to preserve itself and, for that, it must fight against the dissolving forces which threaten it. Erewhonian quirks are explained by the particular conception that the Erewhonians have of the conditions of this stability – that is to say, first of all, by their conception of the human. "Man," writes Butler, "is the resultant and exponent of all the forces that have been brought to bear upon him, whether before his birth or afterwards. His action at any moment depends solely upon his constitution, and on the intensity and direction of the various agencies to which he is, and has been, subjected. Some of these will counteract each other; but as he is by nature, and as he has been acted on, and is now acted on from without, so will he do, as certainly and regularly as though he were a machine" (Butler 1970, p. 304). The very important word in this passage is constitution. No one can anticipate what these external influences will be which will act on a human being. The only fixed point which enables introducing a little predictability into the field of human affairs is precisely this constitution, the same in every human being. The existence of this fixed point is essential to moral and social life: "it is the foundation on which morality and science are built. The assurance that the future is no arbitrary and changeable thing, but that like futures will invariably follow like presents, is the groundwork on which we lay all our plans – the faith on which we do every conscious action of our lives" (Butler 1970, p. 306). No social life without this faith.

According to the Erewhonians, to behave like a thief or a crook is to be subjected to unfavorable influences that no one can predict and against which it is very difficult to fight. There is no "moral constitution"; there are only bad influences that affect us from the outside. On the other hand, there is indeed a physical constitution, which is nature within us. The integrity of this physical constitution is the only fixed point which can confer a certain stability to social life. Disease, which affects the physical constitution of individuals, which causes it to vary in all directions around this fixed point, is therefore deeply antisocial. It is therefore criminal.

The narrator attends a strange trial – strange in view of our current conceptions, but very logical if we think of Erewhonian ideas on the human being. The accused man suffers from pulmonary phthisis; he is therefore condemned very severely. To justify his verdict, the judge puts forward a remarkable argument: if the severity of the law with regard to the sick were to be relaxed, we would see those very suspect people, the doctors of the body, take considerable power, to such an extent that they would even become the sole holders of power in the state. "Their organization and their intimate acquaintance with all family secrets would give them a power, both social and political, which nothing could resist. The head of the household would become subordinate to the family doctor, who would interfere between man and wife, between master and servant, until the doctors should be the only depositaries of power in the nation, and have all that we hold precious at their mercy. A time of universal dephysicalization would ensue; medicine-vendors of all kinds would abound in our streets and advertise in all our newspapers. There is one remedy for this, and one only. It is that which the laws of this country have long received and acted upon, and consists in the sternest repression of all diseases whatsoever, as soon as their existence is made manifest to the eye of the law" (Butler 1970, p. 142). According to the Erewhonians, the supreme danger that threatens society is, in short, dephysicalization – that is, the loss of this fixed point that is the natural constitution of human bodies.

This anthropological conception explains the prohibition of machines, precisely because when left to their natural development, they represent a threat of the transformation of humans to the point of causing them to deviate from their native constitution. This posture, which could be described as radical bioconservatism, ultimately makes the preservation of an unchanging, biologically-based order prevail over any other consideration of any kind whatsoever – a posture certainly incompatible with the requirement of care, which implies, as we have seen, that value is attributed, not to positive and definite content, but to valuative activities.

At the environmental level, there is no doubt that the Erewhonians would defend a preservationist rather than a conservationist position[3]. However, current attempts to bring together environmental ethics with the ethics of care only make sense, explains Catherine Larrère (2012), on the condition of abandoning the classical preservationist posture: the concept of wilderness, that is to say of a wild nature constituting an unchanging order to be protected against any form of human activity, is incompatible with the very idea of a care for nature – which supposes rather, to use the terms of Tronto, an activity of maintenance, repair and even correction.

There is every reason to believe that Erewhonian philosophy is already based on a systemic and self-regulating conception of nature. This vision, theorized in the 1950s by the Odum brothers, stipulates that nature constitutes a self-regulating, cybernetic system, spontaneously tending towards equilibrium if nothing disturbs it.

3 Conservation refers to the attitude of making wise use of natural resources in order to protect them and avoid rapid depletion. Preservation means a more radical attitude that considers that human activity can only degrade nature: it is therefore necessary to preserve wilderness – nature without humans.

Human intervention, however small it may be, can only be disruptive: the concept of a healing technology is meaningless here. Taking care can only mean above all doing nothing, renouncing any form of technological action. Thus, a purely privative conception of care. The change in this theoretical framework in the 1980s (Larrère 1994, Larrère 2009), in the sense of a dynamic and processual conception of ecosystems, introduced the necessary (but not sufficient) condition for a more positive conception of care in technology. Therefore, since nature is no longer seen as a self-regulating system but as a set of entangled multi-scale processes, a system far from equilibrium whose fundamental law is disturbance, human beings can find in it a positive role by actively contributing, that is to say technologically, to its development.

The Erewhonian utopia then indicates, at least in the negative, that this new conception of nature, more appropriate to the idea of providing it with care, must necessarily be coupled with a renewed conception of the living being – more precisely, renewed with regard to its Kantian conception as a self-organizing organism, that is, establishing a self-organized and self-regulating order spontaneously ensuring its equilibrium if nothing disturbs it.

With regard to care, the Erewhonians are therefore right in what they say concerning prosthetic techniques: leaving these to their own dynamics, without being able to relate them to another measure than that of their endogenous logic of evolution, means sooner than later entering a situation of lack of care. On the other hand, they are wrong in what they claim concerning living things: opposing to this self-referential evolutionary dynamism of prostheses the absolute value of a self-referential biological constitution to be preserved will not lead us towards care.

3.1.3.4. *From Erewhon to Hans Jonas: an ambiguity in the conception of technology*

The stumbling block of the Erewhonian conception is its fundamental ambiguity with regard to technology. On one hand, evaluating technologies in terms of what they do to the human biophysical constitution implies sorting "good" technologies, which do not threaten the immutability of this native constitution, from "bad" technologies which on the contrary alter it and must therefore be pitilessly rejected. On the other hand, however, it is well attested that *all* technology, of whatever form, transforms and shapes human life down to its biophysical foundations – in other words, all technology is anthropobiogenetic. As anthropologists Margaret Lock and Vinh-Kim Nguyen maintain, the biological and social aspects of human life are co-constituted (Lock and Nguyen 2010). Social and cultural constructions are not, in humans, superimposed on a pre-constituted layer of biological life, identical in every individual at birth and constituting a universal basis for socio-cultural differentiation. Contrary to the idea that differences in cultures and societies stand out against a biological background common to all humans, Lock and Nguyen explain that these differences always stem from a process of differentiation that already operates at the biological level – differentiations which are inseparably biological and social (*biosocial differentiation*). The biological in humans is not the pre-social and the pre-cultural; it is always already reconfigured by what individuals are exposed to in their social and cultural environment, even before their birth. Humans are biologically shaped by their technologies. Under these conditions, where to draw the line between "good" and "bad" technologies, and how to justify the path of this line?

This is, moreover, the problem dealt with by Hans Jonas, defender of the concept of the intrinsic value of living beings. Jonas explains that what he calls modern technology, that is to say technology as it has developed in the context of the first industrial revolution and subsequently, presents quite singular and worrying characteristics, compared to pre-industrial technologies. This would entail "good" and "bad" technologies, even if Jonas does not use this vocabulary. The first chapter of *The Imperative of Responsibility*, as well as the texts of the 1980s on medicine (Jonas 2012), are very explicit on this point. Human technological action has undergone a profound transformation over the past 200 years, and this transformation – which concerns both the very nature of the power to act, its modalities, the scales of intervention and the nature of the operational chains enacted – makes us guilty of a serious moral error with regard to future generations. By acting as we do today to meet our needs, we jeopardize the possibilities that our children and our grandchildren will have tomorrow to satisfy their own.

That said, Jonas does not argue that only modern technology is transformative to the point of becoming problematic. The deployment of traditional, pre-industrial technology before the industrial era was in no way absolutely unproblematic, unequivocally in the service of life. The original human beings, explains Jonas (1982), cultivated a monist, pan-vitalist thought, insofar as they considered life as an obvious, omnipresent phenomenon, while death was seen on the contrary as the exception to be explained. This non-problematic posture with regard to life has been inverted into a thought according to which, on the contrary, life is an exception and a miracle awaiting explanation; the inert, that is, non-life is on the contrary the normal order of things. Jonas sums up this inversion in a concise and forceful formula: "our thinking today is under the ontological domination of death" (Jonas 1982, p. 12). However, although Jonas unsurprisingly situates the philosophical roots of this reversal of technology against life in the development of thought and science in the 17th century, the fact remains that pan-vitalist monism was already, he says, "disturbed ... by the experience of death and gradually eroded by technology" (Jonas 1982, p. 16). It is not only modern technologies that introduce a split into life. Any technology has in essence a relation of distance with life and comes to disturb it, even if this originally means a disturbance of life which unfolds within life itself – life is self-negation, technology is an alteration of life from the very interior of life – and not, as is the case with modern technology, an influence on life exerted from a point external to it.

According to Jonas, the technology of the Greeks is thus, as we have seen, the medium through which fragile, precarious living beings show cunning towards hostile nature, by building a liveable and durable habitat capable of resisting the power to sweep all things away that defines the *physis*. Technology bears witness to an opposition that arises within life, between the life that binds us to all living creatures (*zoe*) and the life that a subject wants to live by making sense of it and conforming to moral values (*bios*). Under these conditions, technology cannot be oriented towards care; it is constitutively struggle, cunning and an effort of mastery. Certainly, ancient technology is not marked by widespread carelessness, but this is explained simply by a factual reason: at the time, notes Jonas, the possibilities of technology were very limited, in no case were they a match for the power of nature. This factual observation does not in any way imply that technology can in its essence involve care.

Undoubtedly, on closer inspection, Jonas in no way defends a fixist conception of biological constitution in the manner of the Erewhonians. It is even the opposite since for him, all vital activity is the manifestation of a freedom: life is metabolism, that is to say a movement of liberation from the determinism of physico-chemical processes and transformation of non-self into self. That said, Jonas conceives the plant as the perfect living being insofar as, unlike the animal, it directly manufactures its own constituents by photosynthesis, transforming the inorganic into the organic. The plant signals towards an ideal of self-regulating autonomy, endowed with a high degree of immediacy. Unlike plants, animals do not themselves produce the ultimate constituents they need to live. They must draw them from elsewhere – precisely from plants, or from animals who have themselves assimilated plants – which leads them to the mediated behaviors of diversion and predation. Technology, already in animality, is the mark of an incompleteness, proof that the animal has moved away from the perfection of the plant. Any technology confirms a deficit in life. Because it appears in living creatures whose metabolism depends on what other living things (plants) produce, technology is fundamentally predatory. The very notion of an intrinsically caring technology, in Jonas, appears as a contradiction in terms. Care only appears at the stage where humans are no longer just making tools, where they also produce images (*homo pictor*) and where they bury their dead, thus testifying to their "trans-animal" vocation and to the unique meaning that life takes on in them (Jonas 1996).

Like the Erewhonians, Jonas therefore seems to think that technology can only disturb an ideal vital order (exemplified by the plant in this case). We understand the reason for this: if the living being remains conceived in a Kantian way, that is to say according to the categories of self-organization and self-regulation – and everything suggests that this is the case with Jonas – it cannot be otherwise; technology inevitably has the meaning of an alteration, a deviation introduced into the biological order.

3.1.3.5. *Caring for living beings: does this necessarily mean that we cannot modify them? A story of the goat-spider*

The first condition, therefore, of a possible conception of care in technology: giving up the search for a fixed point, an immutable self-organized and self-regulated order of the living being, identified as what should imperatively be preserved against the blows of technology. This point has very important practical implications for the evaluation of technologies of intervention upon living beings, human as well as non-human.

Indeed, it can be tempting to consider any manipulation of organisms, based in particular on technologies of genetic modification, as being in principle an example of technological *hybris*, of the guilty desire to master life for the purpose of exploiting it. A caring attitude towards nature and living beings in general would be *de facto* incompatible with these aims of modification. However, the situation is much more complex. As Simondon points out, possibly surprisingly, seeking to modify the living being may turn out to be much closer to an attitude of ethical generosity, than the posture of sacralizing one's native integrity and preserving it from any intervention of a technological nature.

"The gaze of sacrality," explains Simondon about the possibilities of intervention on human beings, "sees a being, the individual for example, as an inseparable and homogeneous unity, in which the part is inseparable from other parts, as if the whole was unanalyzable. Such a sacralization of the human individual gives birth to the concept of the person. On the contrary, the technological gaze grasps the latent plurality of the different parts of the being, because its operational approach makes the being, even the individual, a reality to be completed ... In the value judgements which technicality bears, there is... reference to the implicit assumption of a possible segmental intervention to repair, improve or optimize the being. An individual being seen technologically ... is to be repaired in a local manner; it is understood as that which calls for the restorative gesture" (Simondon 2014, p. 127) – which, in Simondon, is an indication of a very demanding ethical attitude since it involves evaluating technological interventions with regard to the benefit that they actually bring, concretely, for the individuals to whom they relate. On the contrary, as Haraway emphatically pointed out (Haraway 1991), conceiving the organism as a complete and untouchable whole, excluding in principle any type of corrective or ameliorative intervention, may lead to reactionary political positions.

This analysis can be further clarified by taking as an example non-human living beings which are effectively the object of modifying biotechnological interventions, in particular genetic ones. Let us return to the case mentioned above: the goat genetically modified by the Canadian firm Nexia Biotechnologies at the very beginning of the 2000s, in such a way that this goat would synthesize in its milk the proteins from which a spider weaves its web[4]. Spider silk thread indeed has remarkable mechanical properties (in particular, its solidity is six times greater than steel, at an equal diameter). Obtaining large quantities, on an industrial scale, is therefore very desirable; unfortunately, it is not possible to breed spiders for this purpose. The alternative imagined by Nexia Biotechnologies is thus to bring animals who thrive in gregarious breeding conditions (in this case, goats), to manufacture this silk thread. Raising goats carrying the transgene that codes for the protein of interest has become a technologically feasible solution.

So, let us compare a genetically modified goat for this purpose, with a standard goat whose phenotypic characteristics were subject simply to a classical process of artificial selection. Should the first approach, biotechnology, be condemned in principle on the grounds that it modifies the natural biological datum? If so, the selectionist approach of traditional breeders should be as well, since it similarly leads to a modification of the living being: artificial selection is indeed a technology of intervention that modifies the living being. We certainly sense, at least confusedly, that these two types of intervention call for different moral evaluations; however, reference to the representation of an original biological constitution, held to be immutable, is of no use here. In both cases, the organisms are technologically modified.

In general, artificial selection consists in intervening at the level of animal phenotypes, to intensify characteristics that are deemed interesting at a macroscopic level. These characteristics may relate to performance (racehorses for example) or to yield (dairy cows); they therefore have value only in the eyes of humans, but they do not fundamentally depart from what the philosopher Christopher Boorse calls the design of the species (Boorse 1975), defined by the set of distinctive properties and possibilities which define the "normal" type of the species in question, and which circumscribe its own biological goals. Admittedly, practices of animal husbandry promote goals which differ from these biological goals; however, they do not give rise to completely new biological properties and possibilities. High-performance or high-yielding animals intensify characteristics that are already present in all of their

4 The following analysis takes up the theme of an article published in *Cahiers Costech* (Guchet 2017).

congeners. This intensification may undoubtedly be taken to the extreme, when, for example, the technologies are no longer strictly speaking artificial selection but of intervention on the metabolism of animals via the administration of hormones. However, once again, if such an intensification can make a qualitative leap and pose very legitimate ethical questions (in connection with the suffering that these interventions may cause in animals), the fact remains that these are already-existing biological possibilities that are in a way exaggerated.

Transgenesis consists in intervening in intracellular mechanisms, and therefore at a completely different order of magnitude. This type of genetic manipulation aims to reconfigure living beings in the detail of their morphogenetic processes, at a molecular scale, by forcing them to develop new biological possibilities – such as, for example, for a goat to produce spider silk in its milk. This is undoubtedly a possibility that is not included in the design of species of goat, to speak in Boorse's terms.

It may be objected that this concept of "design" of the species, and the biostatistical conception of the norm which is its corollary, brings us back precisely to the idea of a native constitution of the living being which would serve as a reference term for the evaluation of technology. Does this not ultimately amount to remaining dependent on a concealed substantialist and naturalist presupposition, stipulating that there is an essential "nature" of living beings as species – summed up definitively in Boorse's concept of design of the species? The answer is no, since it is not purely and simply a matter of opposing the idea of a natural type to be respected by certain farming practices on the one hand, and technological interventions deemed to be deeply denaturing on the other (transgenesis), but rather distinguishing in both cases two radically different ways of articulating human goals with the normativity of living beings. Indeed, in both cases the processes of biological individuation are oriented by human ends, but this orientation operates according to distinct principles in each case.

In the first case – artificial selection – the breeder takes advantage of individual variations within the species. Such animals are selected for breeding because they express with particular intensity a characteristic that is of interest. Breeders direct a process which operates without them and which does not depend on them – namely, the production of this inter-individual variability. They become co-producers of this variability by orienting a process that started without them, towards their own needs. Breeders impose upon the animal a defined environment and control the reproduction process, which they can stimulate with the help of the veterinarian; but they do not intervene in the detail of the operations which produce inter-individual variability, at the molecular scale.

In the second case – the transgenic goat – breeders create a new process which would not have happened without them, namely the insertion of a transgene leading to a phenotype completely foreign to the design of the species. Transgenesis is of course a process present in Nature; however, it is not Nature operating here, but the human being. Human agency does not intensify existing inter-individual variability but creates a new type of variability between individuals. The inter-individual variation between a goat resulting from an artificial selection process whose milk is particularly abundant, and a goat with lower yield, is not of the same type as the inter-individual variation between a goat producing spider silk in its milk and a standard goat. The two types of technological interventions have different ways of affecting the ability of the living beings involved to exercise their vital normativity. In artificial selection, the breeder seeks to intensify particular characteristics in the goat (to continue the same example), but this goal is equivalent to making this intensification a component of the vital norms of the goat in question. For this to work, it is necessary that the human goal is somehow assimilated by the goat as an integral part of its path of life, in the words of Canguilhem. In other words, artificial selection consists in achieving a human goal (abundant milk) by taking advantage of the vital normativity of the goat, by arranging for this human goal to become a vital norm of the goat, a kind of goal for it – the term "goal" is to be understood here in the sense of the finalized, teleological dimension of living processes. More precisely, the human goal is not initially the goal of the goat, but it can become so because it is only a quantitative variation on what is already the goal of the goat: all goats produce milk, so it is just a matter of making them produce more.

In the case of the transgenic goat, it is different. The goals of the biotechnologist are completely foreign to the goals of the goat, that is to say to its vital normativity. Let us remember that vital normativity designates the capacity of a living being to forge new values of life when its environmental conditions change. Producing spider silk is certainly not one of the possible adaptations of goats in reaction to changes in the environment. Biotechnologists do not introduce their goals into a limited space of biological possibilities, within which living beings can be normative. They isolate within the goat a biological subsystem (the milk production apparatus) by inserting new possibilities, totally foreign to the normative possibilities of any existing goat. While breeders require the normativity of the living being to accomplish their goals, biotechnologists would in contrast aim for an ideal of modularity according to which the organism could be broken down into functional subsystems, emptied of any reference to vital values. While milk production has biological significance, a vital value for goats, the production of spider silk in milk has none. If the recovery and processing of silk proteins by appropriate technologies were to cease, strictly speaking there would no longer be any silk thread for the goat.

Finally, note that this modular and simplifying approach comes up against precisely the complexity of the normative organism; in this case, the inability of researchers to reproduce in goats all the operations that the spider implements to make its thread (polymerization of amino acids, production of beta keratin, regulation of the acidity of silk, weaving with the legs, etc.). Evacuated in the initial conception of the project, the vital normativity of the spider producing its silk thread reappears as the return of the repressed. Nexia Biotechnologies went bankrupt in 2009, but the work continues.

In summary, the mere fact of intervening technologically in vital morphogenesis does not necessarily imply a lack of care. The genetically modified goats from Nexia Biotechnologies, once born, must have been raised on a farm – that is, they must have been the object of appropriate care. We can well imagine that the staff dedicated to this care did not see these transgenic goats as simple protein factories on legs, but indeed valuating living beings who should be individually the object of the greatest attention – like any goat raised by a shepherd in the mountains. However, this consideration is inessential with regard to the biotechnological intervention itself. On the contrary, it is characterized by the most complete indifference to the individuality of each of the modified goats, which from the point of view of this intervention are no more than metabolisms with an economic interest. For the biotechnologist, milk production results from the functioning of a particular metabolic subsystem, considered to be part of the animal. Ideally, it would be interesting to be able to separate this part of the whole, in order to be able to focus upon it by getting rid of what is not of interest with regard to the desired end (in this case, the production of spider silk). There is no doubt that if we could produce goat milk industrially without goats, by finding a way to effectively isolate the subsystem which, in the goat organism, produces milk, Nexia Biotechnologies would not be burdened with having to raise goats. For the traditional breeder, on the other hand, milk production is by no means the specific functionality of an identifiable and ideally isolable subsystem: it is the result of a global arrangement which cannot be fragmented, and which requires for its effectuation an equally global care – care of the goat, not just care of its milk-producing metabolism. From the point of view of our traditional breeder, what produces milk in the goat organism is not identifiable as a part of this organism: milk production is rather an instrument of goat life, in this case what allows the goat to feed its young. Breeders divert for their own ends a vital instrument, not a biological subsystem.

The reference to the idea of native biological constitution, to the idea of a design of the species whose integrity should be defended at all costs against any modifying technical intervention, is in short not very satisfactory in the sense that it does not make it possible to distinguish, to continue our example, the activity of the traditional shepherd from that of the biotechnologist. On the grounds that in both cases it is a question of intervening technologically upon living beings, should we see in the two approaches one and the same attitude that the Erewhonians, just like Jonas, associate with the expression of unrepentant technological *hybris*? A more nuanced analysis emerges from the previous analysis, as well as a more precise criterion for the evaluation of technology: the criterion that distinguishes technological interventions which do not compromise the ability of the living being to make use of its instruments of life, from those which, on the contrary, fragment living beings into isolable and, if necessary, relocatable parts.

The first condition of care in technology – giving up the perspective of the fixed point – must therefore be supplemented by a second, more positive condition, echoing the perspective of the *natura instruens* introduced in the first chapter of this work: rediscovering the instrumental meaning of the organ for the living being whose organ it is, in opposition to its meaning as part of a larger whole. In a word, it is a question of rediscovering the meaning of the organ as *organon*.

3.1.3.6. *The "biological philosophy of technology": a plurality of approaches*

The two major vitalist approaches to technology seemed to point the way to a possible overcoming of the difficulties left to us by Kant. Does the problem not arise from the divorce of technology and life, and the rejection of the first by both sides of a dualism of intelligence and mechanics – final representations on one hand, the pure execution of activities that are indistinguishable from automatic movements on the other? *Nexus finalis* and *nexus effectivus*? Thus, paving a way for reintegrating technology into life, by proposing at the same time that technology emanates from the operations of the self-organized and self-regulating living being (the opposite of *nexus effectivus*), and that it becomes autonomous in relation to final representations in order to acquire its own life, a logic of evolution all its own (the opposite of *nexus finalis*): Does this herald a possible way out, going beyond the Kantian diagnosis?

Butler's genius is to have shown that this was not so, even before each of the two vitalist approaches was expressly theorized. In their respective foundations, these two approaches ultimately depend on the Kantian framework of thought. Once living things, including human beings, are defined as organized and self-organizing beings, that is, as a whole made up of parts, we can no longer understand at all how technology can inscribe itself and exercise its power of reconfiguration within these beings, without this inscription being *de facto* coupled with a lack of care, a frank carelessness. The temptation is then great to see in it a blind, autonomous evolutionary force, governed by an inexorable law of development, totally foreign to life.

The question was: How to overcome the divorce of technology and life? It is now clarified, and formulated as follows: what can be the alternative to the Kantian conception of a living being as an organism, in the sense of a being self-organized and self-regulated in all its parts? This question forms the common ground of what

Canguilhem called the “biological philosophy of technology”. What is it exactly? Can a conception of care in technology emerge from these new views?

In saluting the existence of a current of biological philosophy of technology, from Kapp to Leroi-Gourhan (Canguilhem 2008), Canguilhem perhaps made a confusion between approaches that should have been clearly distinguished. The biological philosophy of technology is indeed not one but four philosophical currents which, although all are unanimously attached to thinking of technology in connection with life, introduce in very unequal ways the possibility of thinking technology with regard to care.

The first current is that inaugurated by the Kappian thesis of organic projection. Introduced in France by Alfred Espinas at the end of the 19th century, it is found in Bergson and in Bergsonians like Edouard Le Roy (Le Roy 1931). It is this current that Canguilhem sees multiplying almost everywhere in the approaches in particular of ethnologists (he evokes Leroi-Gourhan), who seek to think of technology other than as a production of intelligence.

However, if the Leroi-Gourhanian conception of technology as the exteriorization of biological functions does indeed, at first sight, seem to be a reformulation of the concept of organic projection, it is not really so: Leroi-Gourhan defends a proposal for a biological philosophy of technology which cannot be assimilated to that of Kapp, thus opening the way to a second current of the biological philosophy of technology. As we will see, this second current is characterized by an intention to explain the originality of the human biological formula by mobilizing general categories applicable to all living things.

A completely different current of biological philosophy of technology, while starting from the same evolutionary premises as the previous one, brings together authors who on the contrary think that the human biological formula can only be interpreted and explained according to very specific categories, inapplicable to other living beings. This current, recognized under the name “philosophical anthropology”, is associated with the names of Max Scheler, Helmut Plessner, and Arnold Gehlen. Plessner and Gehlen will serve as a common thread for the analysis of this philosophical trend: Plessner because he offers a completely fascinating conception of the organic, capable of extension towards very current issues concerning care; Gehlen because of the very important place he gave to the question of technology in his philosophical anthropology.

A fourth and final current of biological philosophy of technology will be examined later, on the grounds that it involves an extremely different orientation from the other three – in particular, it is in no way concerned with the intention of accounting for the anthropological difference on an evolutionary basis. Its guiding question is quite different and, as we will see, it may lead to an absolutely original way of thinking about technology with reference to life and care.

3.2. Philosophical anthropology, a promising way to articulate care and technology?

3.2.1. *Plessner's biological anthropology: redefining the concepts of organ and organism*

3.2.1.1. *Positionality and excentricity of the human in Plessner*

Author in particular of a very voluminous work entitled *Levels of Organic Life and the Human: An Introduction to Philosophical Anthropology*, published at the end of the 1920s (Plessner 2019), Plessner proposes a conception of the human which rejects all dualism of life and the mind, of biological processes and intelligence. It is a question of grasping the human being in its tangible bodily materiality, of understanding the human from the viewpoint of the whole of the living world. Plessner thus intends to go beyond Kantian anthropology for which, to summarize, the human domain is that of *a priori* syntheses carried out by a subject which is from the beginning distinct from the living being. He is grateful to Scheler for having criticized this idealism, which he says is still very much alive in the Husserl of *Ideen*, who remained a prisoner of the choice between empiricism and apriorism, of human as nature and human as spirit, which presupposes that the categories according to which we access reality are given. For Plessner, however, it is precisely a matter of assisting their birth, from the basis of the direction fixed by human bodily nature and phylogenesis. A philosophical anthropology, emphasizes Plessner, necessarily rests on a philosophical biology, that is to say on "a theory of the essential laws or categories of life" (Plessner 2019, p. 71).

This philosophical biology cannot be a science in the sense of the natural sciences, since these sciences themselves are products of the form of human life. Therefore, the categories according to which they describe the natural world are not capable of describing the vital activity which gave birth both to these sciences and to the said categories. Philosophical biology must begin by putting the naturalistic thesis into parentheses: according to Plessner it can only be a phenomenology, in the sense of a science whose guiding concepts are not provided by the constituted sciences.

It is not the purpose of this work to present and discuss Plessner's thought in all its aspects, and in particular to explain in detail the concepts of his biological philosophy. It suffices to retain from it a few important considerations for the discussion of technology and of the value it embeds. On this point, moreover, and despite the scant space that Plessner expressly devotes to the question of technology in his book, several elements deserve to be noted.

First of all, an almost incidental remark which gives Plessner's work a decidedly political significance and scope: "Insight into new connections between physical and psychological functions, interest in the concrete human as the subject and object of education, medical aid, and political decisions are compelling the development of new anthropological concepts in response to a new way of looking at living things" (Plessner 2019, p. 69). In other words, philosophical anthropology is urgently required by the all-round extension of what Foucault later called "biopolitics". It is neither more nor less than technologies of direct intervention upon humans, and what is more upon humans as living beings, the effects of which it is absolutely necessary to evaluate. Let us understand that the underlying question is an axiological one: what exactly do these technologies do to human life? It is necessary to be able to judge their value and, for that, to have criteria of evaluation. Where may these criteria be found? Plessner's answer is unambiguous: only a solid theory of the human being as a living being – that is to say, philosophical anthropology – can make it possible to understand exactly what the values at stake in the deployment of biopolitical technologies are.

Philosophical anthropology is the only variant which assumes that "all of [a human being's] spiritual expressions of life depend on his physical characteristics" (Plessner 2019, p. 71), which means that any technological intervention operating at the level of said "physical characteristics" (in order to intensify its yield, improve its parameters with regard to determined political goals, etc.) must imperatively be evaluated in terms of what it implies, on the rebound so to speak, for the "spiritual expressions" characteristic of human life. Plessner insists on this point: "the essential original unity" of the human, "located this side of the separation into body and soul and for everyone a matter of certainty in active life, exists and can only be understood in the reference of all human characteristics to the goals, goods, and values of human behavior". And also: "It is only in the value-laden terms of a life laying claim to meaning that nature and spirit find themselves in that personal tension" (Plessner 2019, p. 69). This means that the human biological formula, going beyond all duality of body and mind, can only be adequately described according to axiological categories.

The non-human living being is also a being whose activity is oriented according to values, a valuative activity in short. Plessner indeed defines the living being as one for whom the difference between an interior and an exterior has an objective meaning, and which itself constructs the border between the two. In other words, the living being is the being that defines its own topology, insoluble in the metric universe where physico-chemical processes take place. The living being not only fills a given portion of space: it itself poses the border that constitutes it. It is therefore the posed-being, even autoposed. The first of all biological categories is thus "positionality", by which the living being does not entirely coincide with itself: by laying down the border which delimits it, it is already beyond itself, it transcends itself. Positionality consequently designates the fact that there exists, between the living being and the surrounding world, a relationship mediated by the living being itself: it is the border. We are already far from the definition of the living being as a self-organized and self-regulating body.

As proof, Plessner develops his thesis by analyzing the very concept of "organ". The organ has a double status: it is first of all what the living being *is*, as an organism, but it is also, then, what the living being *has*, insofar as it is not reducible to the complex formed by all the organs and all the systems which assure its biological functions. The living being is a center of activity; for it, its organs are means which it *has* and not only parts which constitute it, thus what it *is*.

This conception of living things (more precisely animal forms as opposed to plant forms) as identical *and* not identical to the organism is very important for the discussion of values. Indeed, to sum up the living being as an organism, as an organic *Gestalt* as Plessner more or less says[5], is *de facto* to see in it only the design of the species. The whole living thing can be interpreted as the result of a phylogenetic, evolutionary process, having resulted in the organic formula of the species to which the living thing in question belongs. However, this inappropriate way of seeing things, which ignores precisely that the living being (the animal) is a center of activity which *is* not just the organism, but which *has* it as well – that is to say, as Plessner says, which is not only its body but which is also *in* its body – according to him consists in importing into biology communitarian political theories which have nothing to do with it. These theories affirm the superiority of the group over the individual. Imported into biology, they lead to denial that the living individual is a center of evaluation, and reduce it to the *Gestalt* of its species (the equivalent of the norms of the group). According to Plessner, on the contrary, the point of view of the individual animal as a center of activity takes precedence over all memberships – taxonomic, phylogenetic, etc. – that may be assigned to it.

Very strong political choices therefore already underlie Plessner's conception of life. These choices are reflected in a coherent manner in his anthropology, that is to say in his conception of the human as a center of evaluation opposing its own values – as a total human being and therefore, also, as a living being – to those of the group equipped with its biopolitical technologies.

5 Plessner takes a position here in the Driesch/Köhler debate on the distinction between organic totality and *Gestalt*.

Plessner assumes to the very end the axiological orientation of his philosophical anthropology, firmly maintaining that there is no difference between the bodily organization of humans and that of other animals. The animal is "centered", that is to say that unlike the vegetable it is a center of activity and evaluation which is indissoluble in the organic *Gestalt*. Note that this distinction between plant and animal does not appear essential to the subject; it even seems somewhat arbitrary, even downright false, at the biological level. Hans Jonas is perhaps more correct when he defines all living beings by mediation, that is to say by a dimension of transcendence and freedom. A plant links together very disparate orders of magnitude, namely solar energy and the mineral substances present in the soil: it is also a "center" which does something with the means at its disposal.

In any case, between the centric living being and the non-centric living being, according to Plessner, there is indeed a difference in bodily organization: the plant does not, strictly speaking, have organs. As for the human being, it is not centric but "excentric", says Plessner. Now this "excentricity" does not consist of a positive difference in *Gestalt*; that is, it is not linked to particular characteristics of bodily organization. It consists of nothing more or less, for humans, than the introduction of a distance between themselves and their experiences. The human experiences itself no longer as an object of lived experience alone: it is in this sense that it exists as a "person". As an animal, the human is a center of positionality; however, unlike the animal, it also has the capacity to ex-center itself relative to this center: it never completely coincides with the content of its lived experience. However, excentricity, which according to Plessner circumscribes the personal sphere, has no specific counterpart on the side of living corporeality: nothing is required from this side. Strictly speaking, assumes Plessner, "Being human is not tied to any particular gestalt and… could just as well take on a variety of gestalts that do not correspond with our own" (Plessner 2019, p. 272). And also: the human being "is thus free to experiment with laying claim to the timelessness and placelessness of his own position (by virtue of which he is human), for himself and for every other being, even those that are entirely alien to him" (Plessner 2019, p. 279). The modernity of this position is incredible: that gorillas or chimpanzees might be considered as full-fledged persons – an idea that is gaining ground very gradually, and still with great difficulty today – raises no objections in principle for Plessner. The anthropological difference knows no taxonomic barriers: it is not the difference of the species *Homo sapiens* from all other living species.

Consequently, the reference to what the Erewhonians considered to be the human physiological constitution, that fixed point which they needed to make assessments of technology, loses all meaning here. Let us add that this conception of anthropological difference – excentricity – goes hand in hand with the introduction of the concept of the common world, defined by Plessner as that in which the spatio-temporal diversity of the places where humans inhabit is devalued. The animal stands in a place, as a centric being it stands before its place; it is frontally opposed to it, at the same time as it is in relation to it. This is its environment. Excentricity is the overcoming of any place or environment, as well as of the community organizations attached to it.

3.2.1.2. *The ambiguity of the question of technology in Plessner*

What exactly is the question of technology, in this theoretical framework? Plessner unfortunately leaves it a bit in suspense. More specifically, he reserves for it a somewhat ambiguous treatment.

Plessner interprets excentricity as a loss of balance. The animal is concentric, centric, in a balanced and (Plessner seems to say) unproblematic relationship with its environment – its being merges with the whole of its vital system. The excentric being is on the contrary in a precarious situation. It knows destitution: artifice is the correlate of human nudity; for the human being "the expression of his nature that corresponds to the human's essential dividedness, nakedness, and existential neediness is artificiality" (Plessner 2019, p. 294). Plessner therefore makes his own a classic *topos* of the philosophy of technology since the myth of *Protagoras*, often repeated in the last century: namely that the human is a *deficient being*, that is to say a being deprived of the securities with which nature has endowed all the other living beings in various ways (protection against cold for animals with fur, speed for felines, etc.).

Yet, this *topos* seemed at first sight to have no place in the context of Plessner's anthropology. Indeed, remember that for him, there is no specific bodily characteristic of humans: the human's mode of being, excentricity, can accommodate an indefinite range of possible *Gestalten* (the only condition is that they demonstrate the property of centration: the plant cannot claim this). Plessner logically concludes from this that the particular physical characteristics which physically distinguish *Homo sapiens* from animal species (for example the vertical posture, the short face, the hand freed from locomotion) have only empirical value: they are inessential. The assertion is bold, but consistent.

However, Plessner now maintains that the corporeality of *Homo sapiens* does indeed have characteristics, although all negative (absence of fur, claws, etc.), which are directly implicated by excentricity: they therefore appear, ultimately, to be essential. Must we therefore conclude that excentricity calls for a peculiar *Gestalt*, characterized by deprivation – the opposite of what Plessner previously asserted? Unless it is argued that deprivation is not a starting point but a consequence of the human situation in the world: that the human living being gradually lost the bodily characteristics which initially made it an animal just as well-adapted as any other. Deprivation would therefore not be an essential datum, but a secondary datum. This is the conception of Alsberg as we shall see, but not, however, that of Plessner.

In addition, it is hard to see in these conditions what could explain, at the very level of centric animal existence, the fact that certain animals somehow make the "choice" of excentricity. Would we not lose, in doing so, the prerequisite of Plessnerian philosophical anthropology, namely, not to introduce any explanatory factor for humans which cannot be explained itself in the terms of a situation where nothing that characterizes humans is presupposed?

Plessner also highlights another conception of technology, linked to the excentricity of specifically human positionality. He explains that excentricity cuts humans off from the places and times in which animals ensure, by means of various regularities, their security: by doing so, it makes them seriously insecure, which is eventually unbearable for them. The human wants to get out of this "unbearable excentricity of his being" (Plessner 2019, p. 289), and for that it seeks to rebuild a balance, a home (a place), through technology. The idea is probably consistent with the above, since in this case there is no longer any need to presuppose anything deficient in the human *Gestalt*. The deficit is constitutive of excentricity itself; there is no assumption of correspondences in terms of bodily characteristics. That said, it remains to be understood how this reading of technology as repairing an imbalance can be reconciled with the thesis of the *deficient being*. Doesn't Plessner say two incompatible things?

This ambiguity can be understood by the fact that, basically, Plessner seeks to hold together two incompatible requirements concerning the conception of technology. On the one hand, by relating it to the image of the *deficient being*, it retains a naturalistic meaning. Technology is a vital fact, proceeding from the construction of a biological formula capable of ensuring the viability of a being in a state of absolute nudity. On the other hand, however, by attaching it to the unbearable ontological suffering experienced by the excentric being, Plessner confers on it an extra-natural situation: to be human is to detach from the living being, and this detachment requires humans to create their own grounding. Plessner writes more exactly: "the excentric form of life creates its own ground" (Plessner 2019, p. 294). It is indeed a form of life, a separation of life from within life itself if we understand correctly – in other words, the opposite of an extra-naturalist position in the strict sense. And yet, Plessner does not really remove the ambiguity. It resurfaces, intact, in the combined formula by which Plessner sums up what he means: "man lives insofar as he leads a life" (Plessner 2019, p. 294), that is to say if he actively, in a reflective and voluntary way, does that which is part of the non-problematic functioning of the vital system in those animals that remain centric. An attractive formula, let us agree, but somewhat uncomfortable and, at worst, fraudulent: can Plessner not see that by using the same word, "life", he conceals that it is really two very different things? Life in the sense of organic life, biological processes, and life in the sense of leading one's existence according to value choices. It seems that the conception of technology remains trapped in the dilemma indicated by Kant, and that no progress has been made on this subject.

It is not surprising in these conditions that these few considerations on technology do not lead to any questioning concerning values. If technology remains caught in a duality of perspectives which seem irreconcilable, the values introduced into the world by life in the first sense of the term (what humans do to lead a good life) are not intended to be evaluated with regard to the values of life in the second sense of the term (being alive, of the living being that the human does not cease to be). Plessner finally arrives at a surprising formula which falsifies the whole analysis: "the human is naturally partial" (Plessner 2019, p. 298), man is half natural. The other half is quite extra-natural.

Another way of saying it is that, in the excentric form of life, excentricity has absorbed all vitality: it is pure creative activity of a second nature, of a new grounding, without there being a term of reference external to it with regard to what might be required to evaluate its productions, and its effects. Technology resolves an imbalance, a distortion in the excentric human form of life: but it is not considered able to cause new imbalances or conflicts of values by itself.

An encounter is therefore missed, and it is a shame. The fundamentally axiological orientation of the reflection at the beginning, motivated by concern about biopolitical technologies in full expansion, gave hope for a more acute treatment of the problem that a telescoping is always possible between the values that guide technological activity, which creates a fundamentally new situation in the living world, and vital values.

However, Plessner almost casually introduces a new point of view on technology, different from the two *topoi* of *deficient being* and ontological imbalance, which could have allowed him to push further in this direction. More precisely, this point of view is a corollary of the thesis of imbalance. For the second nature that humans create for themselves to effectively provide them with a place, a balance, it is necessary, says Plessner, that it have a certain weight: in other words, what humans make must detach itself from them and consist in itself as a world, objectively. It must have a solidity, a durability of its own. There is an objective side to instruments; their separability with respect to the process of technological invention is constitutive of their sense of being. In short, they are not only utilitarian means with a view to achieving defined goals, but also and above all beings endowed with objectivity, called to come and occupy a place in the world.

Plessner, however, takes absolutely no account of the fact that this second nature that we create for ourselves does not quietly overlap with our first nature. In reality, it colonizes it, it transforms it, it may degrade it; but of this Plessner makes no mention. Now the world that we make for ourselves to compensate for our unbearable excentricity is not a new place, completely isolated from the places and environments of other living things, animals and plants: it breaks open these existing places. Of this, Plessner does not say a word. He writes: "The myriad things with which we have to do every day, from a piece of soap to a mailbox, are only potential objects; as elements of our dealings with them, however, they are components of our surrounding field, members of our relationship with them" (Plessner 2019, p. 286). In other words, these petty things populate what remains, after all, our place, in that we are always at least "half" natural beings.

Once again, technology is split, torn apart: on the one hand there is, so to speak, "big" technology, that which relieves us from our ontological troubles, and then there is "little" technology, that by which we satisfy our most basic and everyday needs – which seem, let us say, philosophically uninteresting to Plessner.

Unfortunately, the actual small pieces of soap, and all other everyday products – all these little things that Plessner does not really care about – are now contributing to massive destruction of ecosystems by dumping all kinds of polluting chemicals into them. The rebalancing of our excentricity takes a heavy toll on all living things on Earth. Plessner's philosophical anthropology cannot say much about care. It is in any case rather curious that Plessner does not at all manifest the same preoccupation, of a moral and political character, depending on whether it is the biopolitical technologies of human management or technologies of the natural world.

In view of the problem that concerns us (the relationship between technology and care), Plessner's philosophy is ultimately characterized by a strong contrast between, on the one hand, a very original conception of the human which, in many respects, will prove to be decisive for thinking of care in technology (in particular for the concept of the organ-instrument), and on the other hand, a relatively poor conception of technology, far less in any case than what we might have hoped for from the analysis of the human as a positional and excentric living being.

We must be grateful to another great theoretician of German philosophical anthropology, Arnold Gehlen, for having better understood the threats that industrial technology might represent for what Plessner called the excentric situation of humans in the world. However, in doing so, Gehlen seems despite everything to retreat from Plessner's considerable advance, touching the anchoring of human behavior in the deepest strata of organic phenomena. With Gehlen, this anchoring seems indeed less clearly assumed.

3.2.2. *Industrialization, work, and life. Critique of modernity in Gehlen*

3.2.2.1. *The "loss of experience" of the* Handelndes Wesen *in the industrial era*

In the voluminous work he published in 1940 (Gehlen 1988), Gehlen expounds his ideas concerning the nature of humans and their place in the world, from a perspective that has many similarities to that of Plessner. Both seek to account for the originality and specificity of the human form of life, while carefully avoiding two symmetrical and inverse pitfalls: on the one hand, postulating from the start that which is in question (for example by explaining, like Scheler whom Gehlen criticizes, that humans owe their unique position to the fact that they have a mind); on the other hand, to consider the human as a living being among others, interpretable according to the same general categories applicable to all living beings. Both claim to sum up a biological approach to anthropology, specifying that it cannot in any case be an "animalizing" approach: humans, explains Gehlen, cannot be derived from animals.

Gehlen, following Plessner, therefore assumes indeed that all human activity must be related to their vital system, to their biological being. That said, he clearly distinguishes the two concepts of the biological and the organic, and relegates the organic to animality. The anthropological includes the biological, but human biology must be described in its own categories, which are not the descriptive categories of animal life. In humans, in short, the biological is supra-organic.

This was also Plessner's thesis, though expressed less directly: since no characteristic of the bodily *Gestalt* allows us to differentiate the human from the animal, it must be admitted that the human biological formula does not derive, by an evolutionary process, from animality. In other words, like Plessner also, Gehlen considers that the correct level of analysis is not that of organic structures, but that of action. Indeed, referring to Dewey's pragmatism, Plessner notes that "Gehlen conceives of the homo species exclusively in terms of its potential to act" (Plessner 2019, p. xxvii). For Gehlen too, the human is a *Handelndes Wesen* – a being defined by its disposition to action. In both cases, this disposition to action is related to the thesis of the *deficient being*: the human being, in a bodily sense, is characterized by its nudity, its lack of morphological specializations and biological means. In both cases too, this thesis of *deficient being* is not considered to be explanatory of humans: on the other hand, it explains that humans necessarily had to do something that animals did not have to do, in order to survive. The human, an unfinished being, could not be satisfied with simply living: it was forced to "lead" its life, that is to say, "to act" (Plessner and Gehlen have very similar formulations to describe this need). Humans are living beings to whom the regulatory mechanisms of their relationship with the environment are not given, not objectified in their bodily structure: they must invent them, they must transform the world to live there, they must make for themselves a second nature. Technology is their means.

So far, Gehlen is saying nothing very original, especially since like Plessner, he pays very little attention to technology in his great work. On the other hand, he pushed his reflection on technology much further in another work, published in the late 1950s (Gehlen 1980). It is a psychosocial study on the place and significance of technology in contemporary industrial societies, informed by more general anthropobiological considerations. Gehlen offers in this work a perspective which is completely absent from Plessner's book: he endeavors indeed to identify the deep meanings of the evolution of technology, and in particular of the industrial phase of this evolution. While technology remains a generic and unambiguous term in Plessner, Gehlen intends to examine to what extent the transition from artisanal to industrial technology has profoundly changed not only the content, but also the structure of human experience.

Gehlen takes up the idea that humans, *deficient beings*, have compensated by technological action for their impotence in the face of nature. The imperfection of human organs gave the first impetus to the *Handelndes Wesen*. However, Gehlen does not stop at this still general and now banal formulation. He introduces a finer distinction between different types of relationships between technologies and organs: these relationships may be substitution or replacement (when the tool makes up for a missing organ), enhancement (when the tool intensifies the possibilities of an existing organ), or delegation in the sense of relief (when the tool lightens the burden of humans in the task: it aims to reduce effort).

From the point of view of substitution, and jointly from that of enhancement and intensification of the bodily means of action, Gehlen underlines the shifting trend from the organic to the inorganic, especially in the materials used and in the energies mobilized[6]. This generalized substitution of the inorganic for the organic can be explained according to Gehlen by the fact that intelligence knows the first better than the second – the reference to Bergson is here explicit. It follows that the very modern conception of the world as an inorganic process accessible to reason, going hand in hand with the conception of a human cut off from nature and from life, finds its origin in an attitude inherent in technological practice when it emancipates itself from the organic.

Because the human is the *deficient being*, it is therefore not permissible to conclude that its separation from nature and life, made very manifest by industrialization, constitutes a basic anthropobiological datum. It is a historical

6 Gehlen no doubt refers to Mumford's evolutionary approach, which we will discuss further.

process which was in no way inescapable, and which involves choices of value – in this case, orienting the circuit of action by which humans make their place, their second nature, towards a project of unlimited mastery and domination of nature. The emancipation of technology from the organic, in favor of the inorganic, has *de facto* freed it from all limiting ethical frameworks: faced with inorganic nature, says Gehlen explicitly, there is no ethical attitude.

Unlike Plessner therefore (who says nothing about this), Gehlen clearly argues that the relation of the *deficient being* to the natural world originally had a deeply ethical character, which remained very much alive in agricultural morality but which was completely forgotten with the transition to industrial morality. Gehlen certainly emphasizes that humans have found themselves needing to extend and amplify their power over nature, because this is the law of their life. That said, extending and amplifying a power does not inevitably imply this attitude of predatory domination devoid of any ethical dimension, which, in Gehlen's view, ultimately comes down to the choice of industrialization. The difference between amplification of action and domination of nature is not justified according to the coordinates of Plessner's philosophical anthropology, it cannot find a place there. On this point, Gehlen undeniably goes further and introduces an interesting possibility for evaluating technological evolution with regard to the fundamental data of human life.

From the point of view of relief of effort, Gehlen also emphasizes that a crucial distinction must be made. The *Handelndes Wesen* in essence seeks to objectify its own material action by entrusting it to inanimate nature (tools and machines), and by letting the latter drive and potentiate it. Gehlen takes this up in mentioning a consideration of Hermann Schmidt concerning the process of exteriorization of human action in technology – from the tool to the automatic device, via the motor machine. This exteriorization consists for humans, explains Gehlen, of putting themselves in circuit with nature, that is to say of coordinating their work process with natural processes. Substitution, amplification, and relief of effort proceed from an extension and a deepening of this circuit of action. If the latter begins with the circuit of the eye, the hand, and the object, in elementary manipulative activities, it gradually frees itself from its dependence on living corporeality by objectifying itself more and more in machinic processes.

Now, if the exteriorization of material action in natural operations is indeed a fundamental datum of the human condition, it has taken on a disturbing appearance with industrialization in that it has ended up cutting the human being off from its own work process. Direct technological action allowed technicians to control their results and correct themselves. From now on, workers are inserted into indirect, complex, and specialized operations, over which they have no real grasp, either in terms of knowledge or in terms of the capacity for corrective and modifying intervention.

Industrialization has therefore completely changed the meaning of exteriorization, by making it pass from alienation in the sense of objectification, by the human, of its own work process, to alienation in the sense of loss, by the human being, of the possibility it had hitherto preserved (in direct confrontation with matter) of mastering its work process and results. The individual in industrial societies no longer controls the conditions of his or her own activity. As Gehlen says, industrialization has exacted a very high price, consisting of both a loss of sense of reality and a loss of experience. With a reduced range of experiential possibilities, the individual is more easily won over to second-hand experiences, group opinions and, eventually, to technologies of human management which aim to guide individual action and behavior according to group standards. On this point, Gehlen therefore repeats Plessner's warning, but he pushes the analysis further by detailing the psychosocial consequences of this group stranglehold over individuals who have become impoverished in experience – extension of the power of technology to non-technological domains; willingness to fully reprogram society; erection of consumption as the ultimate value; development of an extreme subjectivism which redirects the possibilities of experience to intra-psychic life, consequently emptying them of the capacity to transform society which only authentic experience possesses.

3.2.2.2. *The circuit of action and organic life*

However, Gehlen does not really seem to move forward compared to Plessner, at least on a key point for our proposal on the requirement of care. If his philosophical anthropology allows us to identify criteria for the evaluation of technological evolution (which are not found in Plessner), the fact remains that these criteria do not refer in any way to a stratum of life deep enough to include the living human being but also the non-human living being.

For Gehlen, objectification of the human being's material action via the circuit of action opens up the possibility of a rich and formative experience, for which it provides the conditions. According to him, this experience is the source of all value in humans: the loss of experience is therefore *de facto* an anti-value. However, given the fact that Gehlen, like Plessner, starts from the premise of *deficient being*, he is necessarily led to locate this origin of value, in humans, in an understanding of living experience which is foreign to animal living. Gehlen does not need (any more than Plessner does) to recognize valuation activities in animals: it is enough for him to recognize them in the specifically human modality of living, in order to draw perspectives for judging technological evolution in the industrial age.

However, it is still possible to find indications in Gehlen allowing us to overcome this foreclosure. In his book of 1940, Gehlen posits that the principal characteristic of the human biological formula is not the nudity of the *deficient being*, but what he calls the need for "self-interpretation". Humans are living beings who need to interpret their existence. For this, human beings seek a point of support in the world around them: they interpret themselves by inserting themselves into nature so as to be able to think in its image. In this rapprochement, they differentiate themselves. Human beings seek more precisely to introduce regularities into their self-interpretation, in order to compensate for their precariousness and congenital instability. They are fascinated by the regular processes of nature, but also find models of themselves in their own work process, which also has regularities. Consequently, by objectifying their material action in the circuit of action", humans not only secure their exchanges with the world: they may also draw models for their own self-interpretation from it.

Now, at this stage, Gehlen says something essential: the objectification of action consists in transferring to inanimate nature (that is to say to technology) those organizing principles which are already operating in several parts of the organism. There are indeed regular patterns – breathing, for example. This transfer is unconscious (the earliest technicians did not understand biological regulation). As we can see, Gehlen takes up the Kappian technique of organic projection fairly faithfully, but this leads him to considerations which do not appear in Kapp. More precisely, if the circuit of action consists in transferring to the outside, to inorganic nature, principles which are already at work in organic nature, then it is no longer possible to hold to the single premise of the *deficient being*, and of the human being as the inventor of a modality of living impossible to deduce from organic processes, that is to say from animality. In short, Gehlen seems to surreptitiously introduce into his anthropology that which he claimed to have excluded: the humanizing role played by organic positivities, in this case organic phenomena characterized by their regularity.

The premise of the *deficient being* fixed in a very negative fashion the conditions for the appearance of the human biological formula. From now on, it is clear that these conditions are not as negative as they seemed, and that the properly human modalities of life are not as disconnected from animal organicity as Gehlen says. In other words, the distinction between the biological and the organic is not as clear-cut as one might have thought at first.

Gehlen uses a remarkable term to designate this double movement of projection and reflection in the circuit of action: he speaks of instinctive "resonance". By objectifying its material action, the human being establishes a "resonance" between the processes of its organism and its own working process – between the organic conditions and the technological conditions of its existence. Gehlen would then have had to assume the logical conclusion of this conception of the human vital formula according to the category of resonance: to recognize that the organic, in its positivity, is part of the conditions from which an authentic human experience, the source of all value, may emerge. If industrialization is an anti-value, it is therefore not just because it cuts humans off from their work process; it is also because it expels organic life from the domain of values. Under these conditions, humans in industrial societies no longer have the possibility of "resonating" with the world.

Let us summarize. Plessner and Gehlen each seem to emphasize only one of the two dimensions of a possible thought of care in technology. While Plessner elaborates a biological anthropology in which the stratum of properly vital behaviors, in the human being as well as in the non-human living being, may provide a point of support for evaluating technology, he does not develop this evaluative perspective at all, and leaves largely undetermined the question of how to properly judge what technology does to living beings. Gehlen, on the contrary, is very sensitive to the consequences of industrialization for human life, but using a meaning of "life" which seems to reproduce the split identified in Kant: life disturbed by industrial technology is this stratum of living previously identified as the fact of living one's life – what technology does to the being-in-life in its properly biological dimension (in humans and even more so in non-human living beings) is not Gehlen's central concern – with the exception of some previous remarks, but which Gehlen does not develop further.

Can we overcome this situation by digging further in the direction indicated by Plessner himself, and which might also be articulated with Gehlen's perspective on instinctive resonance: that of an understanding of the organ, no longer as part of the organism, but as an instrument at the service of the living being? Might this understanding allow us to get closer to a caring conception of technology?

3.3. The organ-instrument. Second condition of care in technology

3.3.1. *From the organ as part of the organism to the organ as an instrument with use-value*

3.3.1.1. *Being one's organism, having one's organism*

That organic life is not only an area of pure factuality – that is, of what there is – but also of value, that is to say, of what should be, is not really considered in the end in German philosophical anthropology. However, Plessner at least implicitly recognizes this meaning of should-be.

The theory of the organ which he proposes is indeed particularly promising from this point of view. If the centric animal *is* not just an organism (i.e. a system of organs forming an organic singular *Gestalt*), if it also *has* its organism insofar as it is a center of activity which is not confused with it, then the organ is to be considered detachable in its very essence. The organ, says Plessner, is indeed a part which contributes to the whole "immediately" (as Kant had already said in his time), but also "mediately". The organ contributes immediately to everything conceived as an organism; it contributes to it mediately inasmuch as the animal is also an active subject which is distinguished from its organism, and which possesses it as a set of instruments. In short, it is not the tool which is a projected organ, as Kapp thought; it is conversely the organ which must be considered as the tool of the living being – it has the same meaning as the human tool.

Consequently, detaching an organ from the body is not an anti-biological gesture; in any case it is not necessarily so. This gesture may prove to be in full conformity with the very nature of biological existence. The practical implications of this idea are again very important. There is, in fact, enough here to cut down the peremptory assertions which assimilate organ transplants – the extra-corporeal replacement of organs or even transplantation of artificial organs – to a purely mechanical vision of the human body (Hacking 2006). Far from signifying that the organism is perceived as a set of replaceable spare parts in the event of malfunction, like the parts of a machine, these surgical practices on the contrary testify, if we follow Plessner, to a deep understanding of the very nature of the organ. Since this is not only a constituent part, but also an instrument, its substitutability (at least in principle, if not in fact) belongs to its very nature – the instrument is indeed a detachable and replaceable thing.

Furthermore, as not only a part but also an instrument, the organ cannot be reduced to a pure fact of the organism: as an instrument, it also has, one might say, a use-value. Therefore, to continue the same example, there is no scandal in placing it on the same level as transplanted grafts or artificial organs: they too are valuable for use.

The conception of the organ proposed by Plessner thus makes it possible to go beyond a sterile dichotomy when it comes to evaluating technologies of direct intervention upon the living being. If the organ is only conceived as a part of the organism, the technologies which take it as an object must be evaluated with regard to the whole of the organism, posed as the ultimate term of reference; this is the vicious circle in which the Erewhonians were entangled. If, on the contrary, the organ is conceived as an instrument, then technological practices may be evaluated on another level: that of use, that is to say by reference, not to the organism considered as a harmonious whole to be preserved in its integrity, but to the subject which is "centric" in that it has something to do with its organs and his organism.

There is one question, however: have we not long gone beyond evaluating technologies from the point of view of their use? Indeed, but we return to the distinction already introduced above between the two meanings of "means"; we should not confuse the conception of technology as a means in view of a defined *use* (in the objective sense of the possessive) and its definition as a means available to a *user* (in the subjective sense of the possessive).

The fact remains that Plessner did not himself draw out all the consequences of his conception of the organ. He did not conceive of the organic as the domain of a should-be, open to possibilities of correction and technological improvement whose criterion of evaluation is (since it is an instrument) a use-value. Can we, in spite of everything, go further in this direction?

3.3.1.2. *The two conceptions of the body in Descartes*

It should be emphasized that by rediscovering the concept of technology as an organ-instrument, Plessner is part of an ancient tradition of thought that may also include ancient Greece. The word *organon* originally designated the tool and not the biological organ. It is, for example, absent from the Hippocratic corpus. It was by transposition of the instrumental register to the biological register that it ended up designating the organ in the current sense of the term, attested in particular by Aristotle (Byl 1971). The latter nevertheless continues to use the term *morion* to designate the parts of animals, and describes organs as instruments available to living things to accomplish vital goals. In this sense, to use Canguilhem's words, Aristotelian biology is a general technology insofar as Aristotle conceives the organism as a convergence of highly specialized tool-organs (Canguilhem 1975). Nature does not proceed like a human craftsman who may use the same tool to perform a whole range of tasks: it has produced instruments much more perfect than ours, in the sense that each is strictly determined by a single use. Canguilhem therefore emphasizes that there is no need to oppose Aristotelian biology and the mechanistic biology of Descartes. The two are variants of the same technological conception of life; mechanism and organology are two species of which technology is the common genre. In short, there is no distinction between the organ-part and the organ-instrument in Aristotle: the organ is only a part of the organism insofar as it is understood as an instrument of the living being.

This distinction widens in Descartes; nevertheless, the mechanistic conception of the organism has not completely eliminated the sense of the organ-tool. In the sixth Metaphysical Meditation, Descartes examines the exact nature of the body which

I know to be mine and which can be described by analogy with an automaton. However, it turns out that my body is precisely not completely soluble in a description of the mechanistic type (Descartes 1911).

I form with my body an inseparable compound. However, this compound, explains Descartes, is subject to error: thus, people who are dropsical constantly feel a parchedness in the throat, which leads them to want to quench their thirst even when absorbing too much water is harmful. The difficulty is that from a mechanistic point of view, the very notion of an error in nature is devoid of meaning: just as a badly made clock observes all the laws of nature no less exactly than a clock manufactured according to the rules of the art, likewise a dropsical body is subjected to the same laws of mechanism as a healthy body. The clock which incorrectly tells the time does not serve its purpose well; nevertheless, it is indisputably a machine.

Likewise, from a body-machine point of view, there is no difference between health and disease. Consequently, the experience of those who are dropsical has meaning only in reference to a definition of the body which is not the definition of the automaton: no longer a body-machine subject to the universal mechanism, but a body-instrument at the service of life, which makes me seek out what is beneficial to the preservation of my existence and flee from what is harmful to it. It may happen that I am deceived, that I take for a desirable good that which in reality harms me: this is because my body, which is certainly a machine made up of parts, is also an instrument itself composed of instruments. With regard to the laws of the universal mechanism, there are no good or bad machines. As an instrument of living, however, my body can be a better or worse instrument.

More precisely, it is an instrument whose construction generally serves the interests of the living being very well, but which may in certain cases prove to be unsuitable. The feeling of thirst, which arises from the functioning of the body-machine according to the rules of the universal mechanism, is a useful feeling for most of us: it pushes us to hydrate ourselves and, in so doing, conserve our life. For those who are dropsical, of course, the feeling of thirst is wrong, in that it may cause them to act contrary to what the preservation of their life requires: in their case, the instrument is not a good instrument. The problem is that it is not possible to change bodily instruments depending on whether we are healthy or dropsical. We are not housed in our bodies like a pilot in a vessel, says Descartes; we are joined to it very closely, we are confounded with it, which means that we cannot opt for another body, when the one we have no longer properly serves the interests of our life. We certainly *have* a body, but we must immediately add that we *are* it also.

This is, moreover, the reason why the evaluation of what technology does to our body-instrument is deployed not only at the level of efficiency, but also at that of ethics. If we could change bodies at will, as we might get rid of one dysfunctional tool and use another, we would simply have to take note of the fact that our body is no longer suitable and that we must change it. However, this is not the case. The body, as the laws of bioethics rightly state, is part of the person.

3.3.1.3. *A liver that tolerates gin, or the risk of forgetting the individual*

It should be pointed out, however, that it would be wrong to simply oppose the two points of view of the organ-part and the organ-instrument. Already in Plessner, the excentricity characteristic of the mode of existence of human beings opens the way to the sciences. By ceasing to be merely centric animals, to occupy a position of excentricity with regard to ourselves and our environment, we were able to develop a relationship with the world distanced from our experiences – an objectifying relationship, that is to say science in the modern sense of the term, with its requirement of disjunction between subject and object. Now the scientific study of the organism, made possible by its analysis as a *partes extra partes* whole governed by the laws of the universal mechanism, must according to Descartes allow us to make great progress in medicine which is above all in the service of life, that is to say, precisely, in the service of the instrumental function of the body. Far from being mutually exclusive, the two points of view of the organ-part and the organ-instrument are closely articulated, provided however that the former remains oriented and so to speak enveloped by the latter. There is a risk, however, of seeing the approach to the organism based on the relationship of the whole and the parts becoming dissociated from the aim of a well-instrumented individual, and becoming autonomous with regard to the strictly ethical purpose of which technological activity, however, should never lose sight: that of ensuring that technological intervention upon the parts of the body means, above all, making the individual more capable, better equipped to live its life according to its own chosen standards. It is clear that the *partes extra partes* approach, when no longer finalized by the aim of the instrumentation of individuals, may lead to a double loss, both from the bottom and from the top, of what must never cease to be the ultimate goal: that of making individuals more capable. The organ may then appear as that which may be improved in its functionality, but without consideration for the fact that this organ, in the end, must be able to be a good instrument for the individual.

Thus, in the mid-1960s, Robert C. W. Ettinger (Ettinger 2005) suggested that organ engineering, made possible in particular by cell culture techniques and the conservation of parts of the body *ex corpore* (via cryopreservation techniques), should ultimately make it possible to add to the organs new functionalities deemed interesting, such as making a liver tolerant of gin[7]. Alcoholics would no longer have a problem with their liver. Even so, we know that an alcoholic is not only an individual whose liver malfunctions: he or she suffers from more global metabolic

7 The example is from Ettinger.

disorders and, beyond the strictly medical aspects of their situation, also very often experiences great difficulties in their social and personal life. In sum, a gin-tolerant liver may have a supplementary function, but chances are that by freeing people from the fear of getting sick if they drink too much, this liver will significantly reduce individuals' possibilities of life – in short, condemn them to repulsive rather than propulsive paths of life, to use Canguilhem's terms. Ettinger seems to have completely lost sight of the fact that the very notion of improving an organ only makes sense in reference to the life of an individual and as an instrument of this life. There is no doubt, however, that this gin-tolerant liver would greatly serve the interests of manufacturers of spirits, by considerably expanding their market.

The primacy of the individual is therefore well and truly forgotten, from the bottom up, since, to parody Canguilhem, the concepts of normal and pathological (that is to say, the instrumental value, good or bad, of organs and the organism) make no sense at a scale where the organism is broken down into isolable parts and treated separately, but also from above insofar as supra-individual interests – here, those of manufacturers – are always likely to take precedence over the interests of the individual. Forgetting the primacy of the individual ultimately risks depriving the conception of the *partes extra partes* organism, giving opportunities for technological intervention, of all ethical significance.

3.3.2. *Putting the individual point of view first in the ethical evaluation of technology*

3.3.2.1. *The two senses of the organ in Marx*

This point is well highlighted by Marx, in a domain which is not medicine but industrial productive activity. The analogy between natural technology and human technology allows him to introduce a duality of perspectives on the organ, which can be understood either as a multifunctional instrument made to serve the goals of the living being, or as a part of a larger whole – the organism – within which it must perform a well-defined function.

At the beginning of the third section of Book I of *Capital*, Marx evokes the particular work activity which consists simply in taking something from nature, gathering fruits for example, were "a man's own limbs serve as the instruments of his labor" (Marx 2019, p. 199). By taking hold of an external thing and making it a means of its work (for example, a stone for "throwing, grinding, pressing, cutting"), the human being does nothing but convert this thing into an organ "that he annexes to his own bodily organs, adding stature to himself" (Marx 2019, p. 199). Now, if the stone when added to the panoply of means available to humans becomes an organ, it is because the biological organs are themselves already instruments. Marx underlines this very clearly, referring again to Darwin: "Darwin has interested us in the history of Nature's Technology, i.e., in the formation of the organs of plants and animals, which organs serve as instruments of production for sustaining life" (Marx 2019, footnote 2 p. 406). Prior to being parts of an organic whole, the organs are means of living, in all living beings including humans. The lung is not only, or even primarily, a part of the body; it is that which humans need to breathe (Marx 2019, p. 422). It is an instrument.

However, in large industry, it is the worker him- or herself who becomes an organ in the sense that, "in order to labor productively, it is no longer necessary for you to do manual work yourself; enough, if you are an organ of the collective labourer, and perform one of its subordinate functions" (Marx 2019, p. 558). The word "organ" here changes its meaning – a monofunctional part of an automatism, and no longer an instrument of living.

Admittedly, in explaining, in the aforementioned passage of *Capital*, that the difference between the natural technology of animals and human technology lies in the fact that humans, unlike animals, always imagine what they do before doing it, Marx seems to remain faithful to an intellectualist conception of technology. However, his position is actually more complex and, if the faculty of representations is part of it, in Marx this is not in the sense of an intelligence separate from nature: it is always a living being grappling with its environment which acts technologically, by equipping itself with artificial organ-instruments that supplement its biological organs. Marx remains a thinker of the *organon.*

The essential point of the Marxian analysis concerning this duality of perspectives is that it is a translation of a duality of points of view on social reality itself. The point of view of the body-instrument leads us to give priority to the individual over any other consideration, in particular over the interests of constituted groups of whatever kind. However, to avoid any misunderstanding, let us clarify that the Marxian argument does not consist in opposing the individual and the group directly – a position suggesting that the individual can emancipate and flourish alone, the opposite of what Marx says. The opposition is rather between, on one hand, the group in the sense of what Simondon (using terms from American sociology) called the *out-group*: namely, the group which is experienced by individuals as being external to them, endowed with its own laws of operation as well as a desire to impose upon individuals standards of behavior which are foreign to their activity, and, on the other hand, the *in-group* or what Simondon also calls the transindividual: namely, the group as a collective within which the individual exercises its capacity to be the subject of its own norms, with others.

Any technological action must therefore be assessed in terms of what it does to the individual's ability to make the most of his or her body-instrument. The individual has native, organic instruments allowing it to deploy the vast field of possibilities open to it, which is also, of course, a function of certain social factors. The capacity of individuals to effectively deploy the whole range of their means of life, with a view to realizing some of these possibilities as they see fit, constitutes the ultimate reference term for moral and political assessments of technology.

The point of view of the body-organism made up of parts, on the other hand, gives priority to the point of view of the group, more precisely of certain groups, over the individual. The individual is made up of interacting parts and is itself a part of a larger whole, whose interests it must serve. Arthur Koestler very clearly identified this characteristic of the *holon*, that is to say of the whole made up of parts, which can itself only be understood as part of a larger whole: to reason in terms of all-parts is to situate oneself in an unsurpassable relativistic perspective. The whole which has parts is always likely to be part of a higher-level whole, to whose service it must put itself (Koestler 1967).

Marx had thus acutely identified the shift from a productive system in which the individual is in full possession of the whole range of his or her means of production (crafts), to a productive system in which individuals find themselves, on the contrary, dispossessed of their means of production for the benefit of capital, and find themselves reduced to the level of a simple cog. This analysis moreover acquires a more general scope in Ortega y Gasset, who makes it the matrix of a more global diagnosis of contemporary societies.

3.3.2.2. *Ortega y Gasset: the individual, the culture, and the species*

At first glance, Ortega y Gasset seems to remain attached to a split concept of human life, which suggests that he will not have much to tell us concerning the relationship between technology and care. In view of the famous text dated from the 1930s, *Man the Technician*, in which he recorded his reflections on technology (Ortega y Gasset 1941), Ortega y Gasset indeed adopts a classic dualist posture, distinguishing life in the sense of being-in-life, organic life, and life in the sense of living one's life in accordance with one's values, which transcends organic processes. Unlike Plessner, he in no way envisages basing anthropology upon the fundamental data of organic existence, quite the contrary: the human being is immediately assumed to be a being of two natures, the natural that is to say, the organic, which self-realizes (self-regulation, again and always), and the extra-natural which arises within what Ortega calls the "vital program". Human life transcends natural life. Humans do not just live, they want to live well. To do this, they must do something other than satisfy natural needs: they must give themselves the means to meet superfluous needs. Technology is precisely the set of all of these means. It is not at all intended to support the human being-in-life in the world. Technology aims to satisfy those human needs related to well-being: the superfluous. According to Ortega y Gasset, technology is the production of the superfluous; its purpose is to free man from the yoke of animality so that he can go about being himself. It is at the service of the vital program, which is essentially extra-natural.

However, Ortega y Gasset offers a completely different analysis, deeper in a sense, of the relationships between life and technology in a text prior to *Meditation on Technics*. This text, from a lesson given in 1921–1922, exposes what Ortega calls "the modern theme" (Ortega y Gasset 1961), namely the immense task imposed upon the young generation of the time. The situation of European societies at the time is indeed, according to Ortega y Gasset, marked by the realization that the opposition between reason and life has become deadly. This opposition, he emphasizes, has been established from the origins of philosophy by Socrates, who undertakes to dislodge spontaneous life in order to supplant it with pure reason (Ortega y Gasset 1961, p. 55). Reason has forgotten that it is an emanation of organic life, that it is its instrument, and it has ended up constituting itself in an autonomous domain, dedicated to submitting life in all its aspects to itself.

However, reason and all of its faculties, including thought, are indeed according to Ortega instruments of life. "Culture," he writes, "consists of certain biological activities, neither more nor less biological than digestion or locomotion" (Ortega y Gasset 1961, p. 41). Does this mean, however, that thinking, imagining, reasoning are activities that are perfectly immanent in the organism, like the processes of spontaneous life (secretion, digestion, etc.) which have precisely no meaning and no value outside of the organism? No, answers Ortega: human life is characterized by the fact that certain activities which are immanent in the organism transcend the organism. In humans, there is "a store of vital functions whose results or products possess a durability and importance independent of life" (Ortega y Gasset 1961, p. 41) and which constitute what is aptly called "culture" as "spiritual life". It seems that we see here again the duality of perspectives from *Man the Technician*: on one side there is organic life, and on the other side trans-vital life constituted by the functions of culture, which Ortega nevertheless describes as biological life. The biological and the organic do not overlap.

That said, trans-vitality does not escape the bounds of organic regulation. The trans-vital functions must remain attached to the life of the organism; they are its instruments (just like the functions of spontaneous life, moreover). Now, since talking about an instrument already signals towards use, it follows that the instrument must be evaluated with regard to the very thing it must serve: life in the sense of the life of the organism. The trans-vital functions of culture, "like all vital powers, [are] the servants of the whole organism of the living being and derives its regulation and modulation from that organism" (Ortega y Gasset 1961, p. 68).

Ortega y Gasset is very explicit: we must recognize the existence of vital values at the very level of organic life, in its own processes, and not only in what exceeds it, that is to say in the trans-vital. The creations of culture must be evaluated with regard to values which are strictly vital in the sense of organic life. Ortega does not however devalue the acquisitions of culture in favor of a suspect vitalism: life without culture is only barbarism, he writes. However, the instruments of culture have value only in proportion to their capacity to serve the requirements of life in its unity, including its organic dimension.

It must be admitted, however, that Ortega maintains a certain ambiguity. He considers for example that "culture is a biological instrument and nothing more" (Ortega y Gasset 1961, p. 58). The instrumental point of view is reaffirmed, but it is a question here of a biological, and not organic, instrumentality. The terminology is a bit vague: should we understand that the instrumentality of human functions (including the trans-vital ones) finds its term of reference, and of evaluation, in organic life, or else in biological life which is already a trans-vital life? Ortega does not seem very clear with this duality of perspectives of the biological and the organic. If it is less marked than in Plessner and Gehlen, it remains no less insistent.

Its major advance in comparison with Plessner and Gehlen, in any case, consists in positively assuming the existence of properly organic vital values. Indeed, he seems to anticipate Jonas somewhat in considering "the superiority of the act of living to all that is not life" (Ortega y Gasset 1961, p. 74): in Ortega, as in Jonas, there is a metaphysical postulate in the assertion that organic life – life as metabolism in Jonas – is a bearer of intrinsic value. "Life is valuable … for its own sake" (Ortega y Gasset 1961, p. 74).

What is more, the intrinsic value of life –in the sense of organic life – is in reality in Ortega a hierarchical plurality of values. This hierarchy is in turn understood by Ortega according to the category of vital potency. Different individuals of a living species express a more or less perfect state of realization of the power of the species in question. Vital values, positive or negative, are distributed along a continuum bounded, on one side, by the degradation of the species, and on the other by the full fulfilment of its own potency, which is its perfection (Ortega y Gasset 1961, p. 75). In short, the world of Ortega remains Aristotelian. It should even be stressed that concerning humans, this concept of vital power is to be understood in both Aristotelian senses of power.

In the first sense of the term, potency designates in Aristotle a possibility inscribed in a form of life, as opposed to forms of life in which this possibility does not appear at all – I am not a pianist but I can learn to play the piano, unlike a dog. Every human being is a potential pianist; we can all learn.

In the second sense of the word, potency means the capacity that I have to do something, but that I am not currently doing. I am a pianist in the sense that I learned to play the piano, but I am not always playing the piano – I may be reading a book or eating, in which case I am a pianist in action and no longer in power in the first sense of the term, but a pianist in power, and not in action, in the second sense of the term.

In every living species, biological potency characterizes the design of the species in contrast to other species, which means that it is always actualized in the first sense of the term: the lion can run fast, etc. The living being is only in potentiality in the second sense of the term: by the design of its species, it is everything it can be, it is fully actualized, but it is not always in the process of exercising its own potency – the lion does not always run.

The human has this very particular feature, Ortega y Gasset mentions, that in him or her biological potency in the first sense of the term, thus in the sense of design of the species, is not actualized: it must be "cultivated" says Ortega. "Culture is merely a special direction which we give to the cultivation of our animal potencies" (Ortega y Gasset 1961, p. 76). A remarkable idea: the human being is the living being who must itself fulfil, through passing over to a trans-vital regime of being, its life as a living being, its organic design. The organic is not for humans an initial datum; it is its responsibility.

At this precise point, however, Ortega y Gasset seems to hesitate between two paths.

The first consists in maintaining that this fulfilment of the organic consists in aiming for the perfect realization of what is fixed by the design of the species. Ortega y Gasset thus praises the "grace of an animal type" of certain humans who are particularly accomplished at this level –the grace "of the *genus* 'man' in its exclusively zoological aspects" (Ortega y Gasset 1961, p. 76). Evaluating trans-vital activities, for example technological ones, with regard to organic vital values therefore means, in the human domain, measuring the degree of fulfilment of the "*genus* man" which may be attained via these activities. However, reasoning in these terms raises the formidable problem of designating what are the organic characteristics supposed to constitute, precisely, the perfect "*genus* man". If the fulfilment of the design of the species *Homo sapiens* stems from a human activity oriented by cultural, trans-vital values, there is every reason to think that a large number of very different human groups will have very different notions of this organic perfection. A pluralism of values, again.

The major difficulty comes from the fact that Ortega y Gasset, while affirming the existence of an essential relationship of organic life to the concept of value, seems to maintain an ambiguity in the way of understanding this organic vital value. Indeed, this is conceived, on the one hand, as a categorical property of life, in the sense that it is fixed by the design of the species; but it is also conceived, on the other hand, as a property which is not categorical precisely, rather as a relational property, that is to say one which needs to enter into relation with something else to actualize itself. The human fulfils itself in the trans-vital regime, creating values other than the organic vital values of the species, while also thereby fulfilling the organic vital values of the species.

The essential point of Ortega y Gasset's analysis is that this double fulfilment of organic vital values and trans-vital values is not the act of the species as such, but of individuals united in societies, human groups scattered all over the Earth. Ortega y Gasset says it explicitly: in humans, it is not organic life itself, but culture, which has the capacity to fulfil the vital potency. However, culture is always situated, contextual; it passes through well-identified human groups and through the individuals who constitute these groups. It is therefore the human individual living in society, and not the species, who can be brought to experience a telescoping between trans-vital values and organic vital values. It is for the socialized individual, and not the species, that the problems arise; it is therefore he or she who must be able to resolve them, by being suitably instrumented.

Therefore – and this is the second path towards which Ortega y Gasset leads us, though timidly and as if hesitating – culture aims not so much to fulfil the design of the species, as to provide every individual with the instruments which will allow them to make use of the whole spectrum of their sensorimotor and intellectual possibilities, and to defend what is important to them. "The things to which we aspire, the things we believe in, the things we venerate and adore [that is to say, the things we 'value'], have been created in the environment of our individuality by our actual organic potency and constitute a kind of biological drapery in which we indissolubly swathed, body and soul" (Ortega y Gasset 1961, pp. 78–79). This remarkable formulation points towards a concept of organic potency of the species which it is in the power of individuals to actualize, at least when the latter are endowed with adequate instruments allowing them to carry out their valuative activities.

Ultimately, the diagnosis of Ortega y Gasset is that contemporary mass societies have progressively removed the possibility that every individual should have, provided that they are properly instrumented, to fulfil the vital potency attached to their species. Now, when the fulfilment of this vital potency of the species is no longer in the hands of each individual, but in those of a supra-individual group (the Nation, the Party, etc.), we know the barbarism towards which this can lead.

A key point therefore emerges from Ortega y Gasset's analysis: the value of a trans-vital activity must be assessed with regard to the individuals themselves, and not with that of a supra-individual entity such as the species, or even the human group taken as a whole. The theme is also very present in science fiction literature: if for example Humanity is the supreme value to defend, then an Artificial Intelligence to which is entrusted the task of defending Humanity may very well sacrifice an indefinite number of individuals to accomplish its goal[8].

3.3.2.3. *Is giving priority to the point of view of the individual compatible with the aim of caring for nature?*

All this analysis nevertheless raises a major difficulty. If care for nature passes through the recognition that every living being is a valuator in Dewey's sense, must we not admit that to take care, I will have to at one time or another give up exercising, individually, some of my instruments of life with which technology provides me? In fact, by using these instruments to make my own assessments, I may always be in a situation of harming the ability of other living beings to make their own. In short, is adopting a caring attitude towards *natura instruens* not, to parody a known formula, to accept that my own freedom of valuating ends where that of other living valuators begins? Is there not consequently an incompatibility between, on the one hand, the idea of a primacy of the individual endowed with their own instruments of life in the ethical evaluation of technological action, and the other side the very concept of care for nature understood as care for *natura instruens*? If the instruments which I individually possess to carry out my valuations must themselves be evaluated, not with regard to my own choice of values alone, but also with regard to the valuative capacity of all other living beings, it is necessary to conclude that the ultimate point of view for the ethical evaluation of technological action is not the valuating individual which each person is for themselves, but the system of all the valuations of all living beings, whose coherence and overall harmony are today seriously threatened by the valuations of certain human groups.

Ultimately, it seems that by emphasizing individual valuation and its own instruments, we are going a long way backwards to the question of "mastery of mastery" of our technologies and, therefore, to an externalist position on the ethical evaluation of technology. It will indeed be necessary that at one moment or another, I accept seeing myself limited from the outside in my valuative activities, and that a higher and overshadowing point of view prevails over these: that, precisely, of the harmonious system of all the valuations of all the living valuators. To use Leibniz's terms, the ethical evaluation of technology should be done, not at the individual monadic level, but at the global monadological level.

8 This is the theme of the film *I, Robot*, inspired by the eponymous novel by Asimov.

Must we conclude that we are at an impasse, or worse: that we are going around in circles? Despite appearances, this question must be answered in the negative in view of the progress made by Ortega y Gasset. We first diagnosed the risk of seeing an incompatibility grow between, on the one hand, the valuations which human individuals carry out by means of their technology, and, on the other hand, the requirements of their organic life and that of all other living things. However, Ortega y Gasset now explains to us that an individual in whom the organic biological potency of the species *Homo sapiens* does not flourish, for lack of adequate instruments, is an individual whose valuations are illusory. Far from being an incompatibility between the evaluative point of view of the individual on the one hand, and that of the species *Homo sapiens* inserted into a nature populated by other species on the other hand, it appears that the individual can only fulfil itself by fulfilling within itself it the biological potency of the species, and vice versa: the species is only fulfilled if individuals have at their disposal a very rich range of instruments of life, including technology, so that these instruments of life serve non-fallacious and non-illusory valuations – that is to say, such that these valuations actually contribute to the flourishing of the biological potency of the species.

Ortega y Gasset does not, however, give much more information on what exactly this very strong entanglement between the individual and the species means. How does this idea further advance us towards a possible implementation of care "in" technology? It turns out that these two points precisely constitute the common thread of French paleoanthropologist and ethnologist Leroi-Gourhan's reflection on the current situation of humans in the world. Leroi-Gourhan in particular has pushed the analysis of what it means for humans to imagine themselves disconnected from their own biological potency quite far. It is indeed in the light of paleoanthropology, that is to say the science of human evolution, that Leroi-Gourhan proposes to illuminate both the present of humanity and the perspectives of evolution that are open to it.

As we will see, it follows from the teachings of paleoanthropology that not being cut off from its own biological potency, for the human being, means above all that it recognizes itself as the result of a long evolutionary process. However, the terms in which Leroi-Gourhan describes this process make it clear why a human, which sees itself literally dispossessed of the free exercise of the whole range of instruments of life, both biomechanical and sensorimotor, with which the design of our species provides us (to which are also added the exosomatic, socio-technical instruments which a human derives from society and from its own activity) goes hand in hand with an evolution which will undoubtedly not stop, but which can only continue in the direction of an inexorable drying-up of human biological potency.

3.3.2.4. *From the human cut off from nature to the human cut off from its own evolutionary process: Leroi-Gourhan's analysis*

In his great account of hominization published in the mid-1960s, in two volumes, under the title *Gesture and Speech* (Leroi-Gourhan 1993), Leroi-Gourhan diagnoses that two centuries of industrialization are radically calling into question the foundations of human evolution from its beginnings.

According to Leroi-Gourhan, humans are living beings whose evolution has consisted in overcoming a succession of imbalances. However, it turns out that the current human is confronted with a new imbalance which, this time, does not seem able to be overcome. The reasons for this pessimistic diagnosis should be explained.

Leroi-Gourhan divides living beings into large functional series, which again, as in Plessner and already in Marx, opens the analysis to a conception of the organ as a means of life, that is to say as an instrument and not as a part. Each species is indeed defined by a unique way of coordinating three fundamental elements, which are found everywhere in the living world: the locomotory apparatus (forelimb), the organs of prehension and the organs of responsiveness (these last two elements forming the anterior field). Humans are no exception. They belong to a large functional series to which the primates also belong; however, Leroi-Gourhan does not consider the human as a primate among others. "In primates the actions of the hands are in balance with those of the face, and a monkey makes wonderful use of this balance. It even goes so far as to make its cheeks serve to carry food, which its hands, still required for walking, cannot do" (Leroi-Gourhan 1993, p. 210). Thus, for each species "a cycle is established between its technical ability (its body) and its ability to organize itself (its brain). Within this circle … a way opens up towards increasingly pertinent selective adaptation" (Leroi-Gourhan 1993, p. 60). "In early anthropoids, on the other hand, a kind of divorce takes place between the hand and the face. Thereafter the one contributes to the search for a new balance through gesticulation and tools, the other through phonation" (Leroi-Gourhan 1993, p. 210). The tool is not the product of intelligence, "the noble fruit of our thought" (Leroi-Gourhan 1993, p. 106), but a zoological reality.

The human is therefore this strange living being in whom a divorce has occurred between its technical ability and the ability to organize itself. With *Homo sapiens*, the balance is restored by bringing into play, on the one hand the individual, and on the other side "the body social which is the individual's indefinitely perfectible extension in action" (Leroi-Gourhan 1993, p. 20). The human remains in its body a mammal among others, almost identical in its anatomy and its cerebral apparatus to what it was when it hunted the mammoth; but it has displaced its evolutionary dynamism into a new body, completely exteriorized, "a collective organism with a practically unlimited potential for achievement" (Leroi-Gourhan 1993, p. 184).

Now this techno-economic based collective organism has entered a meteoric pace of progress, endowing humans with a total triumph over matter, to the point where, according to a striking and well-known formula of Leroi-Gourhan, "the last small oil deposit being emptied for the purpose of cooking the last handful of grass to accompany the last rat" (Leroi-Gourhan 1993, p. 185). So, the social form of humanity, which was established around 8,000 years ago on the basis of the dialogue between the individual and the social group, is perhaps ceasing to be ours, says Leroi-Gourhan. By dint of externalizing an increasingly broad spectrum of biological functions into artefactual devices, it has given rise to what Leroi-Gourhan calls a "dematerialized image" of humans (Leroi-Gourhan 1993, p. 402). This image is that of a human whose zoological determinations, which make it a mammal among others, have become covered by the constructions of civilization. "Born to hunt wild horses in the steppes and gradually adapted to locomotion in a seated position in an atmosphere of burnt petrol" (Leroi-Gourhan 1993, p. 402), *Homo sapiens* has forged itself in the image of a living being which can only fulfil its being by completely cutting itself off from nature.

However, contrary to what this false representation of the human suggests, civilization has not made the mammal within us disappear at all. "No one can deny the contradiction between a civilization whose power are almost unlimited and a civilizer whose aggressivity has remained unchanged since the time when to kill a reindeer was to survive" (Leroi-Gourhan 1993, p. 402). This obsolete mammal is now completely cut off from nature, yet it has retained the same physical and mental characteristics as when it had to hunt reindeer. Not only the human skeleton, but also human behavior has remained that of a social mammal "for which the constraints of territory, food acquisition and reproduction remained thinkable and interpretable in zoological terms" (Leroi-Gourhan 1993, p. 401).

This dematerialized image that the human has forged of itself has fueled a process whose disastrous consequences we are beginning to measure: "the denial of any connection between the human and the rest of the living world" (Leroi-Gourhan 1993, p. 402), as well as the loss of any form of direct link with the material world. The loss of contact with the material world no doubt appears as a culmination of the process of exteriorization of which human evolution has consisted – the process at the end of which human faculties (manipulative action, motor skills, intelligence, memory) found themselves displaced outside of human corporeality. However, this loss of material contact is not implied by this process as its necessary consequence: it was made possible by this conviction that the nature of man is the artifice, according to the well-known formula of French philosopher Emmanuel Mounier. Human beings seem today to have reached the end of this process of exteriorization. They are now on the way to being completely transposed into artificial organs. This goes as far as the imagination itself, which is transposed into the programs of the audio-visual industry.

The idea of an end of the "career" of humans on earth (Leroi-Gourhan 1993, p. 366) is also mainly motivated, in Leroi-Gourhan, by the observation of an exhaustion of the activity of the imagination. The role of imagination in human evolution is indeed essential. From an evolutionary point of view, humans are both zoological individuals and creators of the social memory in which the programs that govern their behavior are placed. What makes the human biological formula unique is the breaking of the link between species and memory, that is to say the possibility of an evolution escaping from paleontological drift. Operational memory is exteriorized in relation to the zoological underpinnings of the species, which has made it possible, on one hand, for the individual to leave the established tribal framework, and on the other hand for the tribal memory itself to progress. "The characteristic capacity of human societies to accumulate and preserve technical innovations is connected with the collective memory. Our role is to organize our operational sequences consciously toward the creation of new processes" (Leroi-Gourhan 1993, p. 233).

However, these possibilities of individual innovations introduced into operational programs imply the activity of the imagination. It is imagination which allows individuals to free themselves from the operating chains deposited in social memory and to make them evolve. Progress depends on the vitality of the creative imagination. If this should end up being completely controlled and transposed into an already codified program, leaving less and less margin for personal variants – the human will inevitably be doomed to play the role of a "specialized cell" (Leroi-Gourhan 1993, p. 247), in what looks more and more like a beehive or an anthill. Therefore, "freed from tools, gestures, muscles, freed from programming actions, from memory, freed from imagination by the perfection of the broadcasting media, freed from the animal world, the plant world, from cold, from microbes, from the unknown world of mountains and seas, zoological *Homo sapiens* is probably nearing the end of his career" (Leroi-Gourhan 1993, p. 407).

The question is basically the following, and this is how Leroi-Gourhan poses it: how can this being which remains, more or less, the mammoth-hunting mammal that it was thirty or forty thousand years ago adapt to the entirely artificial living conditions it has made for itself? For this, "*Homo sapiens* would have to enter upon a new biological stage where perhaps he would be able to control his aggressivity" (Leroi-Gourhan 1993, p. 183). Without losing its fundamental values, that is to say without ceasing to be itself, *Homo sapiens* can still evolve in its skeleton and/or in its neuro-motor apparatus. On this precise point, however, Leroi-Gourhan is quite skeptical: "we have to acknowledge that not a great deal of scope is left if we are to remain human beings in the physiological and mental sense accepted today" (Leroi-Gourhan 1993, p. 129). Physically, *Homo sapiens* is a species that probably has a certain future, but the possibilities are still limited. Furthermore, evolution at the osteological or neuromotor level can only happen at a slow pace: if it is possible, it will take millennia. Yet there is an emergency; we do not have much time ahead of us. In any case, there is reason to think that a major transformation cannot occur in terms of an increase in the volume of the brain, if this quantitative criterion is really important, says Leroi-Gourhan. Certainly, operations of artificial selection might achieve this result, adds Leroi-Gourhan. Nevertheless, this is surely not the main thing, and what no one has seen so far is that "no change is possible without the loss of the hand, the teeth, and consequently of erect posture" (Leroi-Gourhan 1993, p. 129).

Leroi-Gourhan delivers here a remarkable analysis in counterpoint to current debates on augmented humanity. Indeed, these debates mainly focus on the possibilities of technologically increasing the possibilities of the brain, of controlling the emotions by neuropharmacological means, of lengthening the lifespan indefinitely. Nothing is envisaged, however, at the level of the bone structure; no biotechnology aiming towards a regression of the hand or intervention affecting the erect posture is promoted as likely to bring us into post-humanity. However, if we stick to the strict teachings of paleontology, it is clear, explains Leroi-Gourhan, that the possibility of a post-*sapiens* humanity is constitutively linked to this regression of the hand, that is to say an in-depth reorganization of the bony framework, going so far as to question the upright position. A new evolution of the human being can only be envisaged today at the cost of a reorganization at the level of the most important element of all human evolution: bipedalism. In other words – and this is the contradiction with which we are confronted today – shouldn't humans today consider to "break away from [themselves]", something which they have never had to do since the beginning of their ascent? Let us understand: is *Homo sapiens* not obliged to consider questioning the foundations that made humanity what it is (standing, short face, free hand, detachable tool) to be able to continue "pushing its rock up the hillside" (Leroi-Gourhan 1993, p. 407)?

The analysis thus dramatically underlines the fact that technology, a zoological fact and a structuring element of the human biological formula, has ended up turning against this very formula – which means that in humans, life has turned against itself. Leroi-Gourhan is here poles apart from a definition of technology as an extension of the body, inscribed in the logic of action of this body – a simple exosomatic extension of the possibilities inherent in the fundamental data of human biomechanics and neuro-motor apparatus. On the contrary, technology testifies to the tensions within life when it becomes human life. In this way, Leroi-Gourhan is much less attached to Kapp than to Bergson.

Leroi-Gourhan may not be entirely pessimistic in *Gesture and Speech.* There is still some room for maneuvering, however narrow, to find a balance in our relationships with nature, on the basis of fundamental biomechanical data which define the human biological formula. In addition, the double image of the "human machine", an osteo-muscular and neuro-motor device inherited from paleontological evolution, and of "its improved artificial copy" (by transposing his faculties into the social super-organism; Leroi-Gourhan 1993, p. 269) does not exhaust all of human reality. Not everything has been exteriorized. While technological and social behaviors are lived according to collective norms which imply compliant execution, aesthetic behaviors liberate the human from established programs and enable it to invent new programs. In the dialogue between the individual and society, the exosomatic organs are intended to constitute a repertoire of means at the service of the individual. They must allow the individual to amplify his or her knowledge and skills. The individual is a creative dynamism, a process of individuation rather than a form that is fixed and enclosed within existing social structures.

"The dilemma", writes Leroi-Gourhan in a formula that could have been that of Marx, "is between the individual as the motive force of a social microcosm built to human scale, a microcosm in which the individual commands the full gamut of aesthetic and technical possibilities, and the individual as a cog in the infinitely perfectible machine of a completely socialized society [that is to say of a society in which individuals no longer have any room for their own invention: their behavior is entirely subject to existing social norms]" (Leroi-Gourhan 1993, p. 358). In other words, the assertion of the individual endowed with "the full gamut of aesthetic and technical possibilities", that is to say its instruments of life, both sensorimotor and technical, does not imply in no way that said individual is freed from all its dependencies and all its attachments with regard to the natural world. This face-to-face confrontation between the individual and society undoubtedly implies that the former can free itself from the rigid determinations that the latter imposes. However, this liberation is in no way a departure for the individual from the natural world. Individual affirmation in the face of society has an evolutionary meaning; it is a mode of balancing between the technical ability and the ability to organize itself of the species *Homo sapiens*.

In Leroi-Gourhan's eyes, it therefore appears imperative that humans abandon the dematerialized image that they have forged of themselves, and that they see themselves as what they really are: a being whose range of technical and social possibilities remain dependent on the fundamental coordinates of its mammalian existence. Leroi-Gourhan suggests that, fundamentally, without this preliminary turn of the human gaze towards its own reality, no satisfactory solution can be found to the current disastrous situation. However, this turn of the gaze cannot be driven either by existing social and political institutions or by the evolutionary process.

Regarding the former, there is indeed every reason to think that they have already resolved the dilemma of which Leroi-Gourhan speaks – that is to say, their functioning is based on a collective organization in which humans are actually only considered as “a cog in the infinitely perfectible machine of a completely socialized society”. This fear echoes that of Gehlen, who saw the imposition of a vision of society as a fully programmable and reconfigurable reality, according to an engineer’s vision. In the mid-1960s, Leroi-Gourhan said he was optimistic; he still believed that the institutions would take the gravity of the situation head-on. At the end of his life, in the 1980s, he no longer believed this at all.

Regarding the evolutionary process, Leroi-Gourhan does not see an adequate solution either, given that its temporality is incompatible with the urgency of our situation. However, there is undoubtedly a more fundamental reason to suggest that the solution to generalized carelessness can only come from a human initiative, and not from a process of natural evolution (which would lead for example to the regression of our predatory instincts).

3.3.2.5. *A facetious little robot, more human than “humans”*

This point can be illustrated by an animated film released on screens several years ago. This film, *Wall-E*, depicts a small robot whose function is to clean up all the waste from the planet Earth which has become uninhabitable.

Humanity has indeed departed, to quote Michel Serres, that is to say that it has (apparently) freed itself from all its old dependences with regard to nature. It takes refuge in a sort of gigantic spaceship where everyone’s needs can be met, pending an improvement in the situation on Earth and the possibility of being able to return there. These new conditions of existence made for humans, in a completely artificial environment, have manifestly triggered an adaptation mechanism, an evolutionary process: the human biomechanical formula has undergone very significant changes, namely, precisely, the regression of the hand, weakening of the teeth and loss of erect posture.

The film therefore shows, in its main characteristics, the human evolution of which Leroi-Gourhan envisaged the possibility, at the end of ten or twenty thousand years. However, by itself, the evolutionary perspective does not provide any criteria allowing to judge the value of the regression of the hand and the teeth, or even the loss of erect posture: it is simply a question of taking note of a possible form of human evolution, which may lead in the long term to a new biomechanical formula adapted to particular, artificial, life conditions. Humanity in the film *Wall-E* has perfectly restored the harmony between the biological formula and human biomechanics on one side, and the technological conditions of existence on the other side.

Leroi-Gourhan, as we have seen, considers it true that efforts to technologically steer the human evolutionary process, in the sense of better adaptation, cannot lead to a satisfactory solution. Saying this, however, he does not reject all value of this aim of technological steering of humanity: it is only a question of recognizing that this type of solution is not commensurate with the challenges we face. The solution is ineffective or even inoperative, but not morally indefensible. Strictly speaking, if we managed to effectively manage our biomechanical evolution to adapt to an artificial environment, there would be no problem: the purpose of the human evolutionary process has always been to restore the balance between the technological and organic conditions of existence, whenever this balance is broken. If an evolutionary process restores equilibrium, it is sufficient and there is no call for special consideration regarding the modalities of the process. It is apparently a problem of fact, not of value. The question of the criterion for evaluating the way in which human life is prolonged in "technological plasticity" remains unresolved.

In addition, to consider that humanity has the possibility of adapting, by evolving towards a living form which will no longer have any of the biomechanical attributes that have defined the human since its beginnings, may be destabilizing and vertiginous, but for what moral reasons should it be opposed at all costs? In sum, and to put it trivially, the adaptive point of view of equilibrium does not, by itself, allow us to consider the humanity described in *Wall-E* (and envisaged by Leroi-Gourhan) as an antivalue. The fact that this departed humanity no longer rubs shoulders with living beings (which have totally disappeared) is no more justifiable as an antivalue from an evolutionary point of view. We are once again clearly at odds with the demands of caring for nature and living beings.

However, there is every reason to believe that for Leroi-Gourhan, this evolutionary perspective in the sense of a regression of the fundamental data of human biomechanics is indeed an antivalue in itself. Based in fact upon his functionalist approach, as we have suggested, here again he derives an instrumental conception of the structural data of the living being. Animal structures, that is to say, at least in advanced mammals, both the bone structure and the neuro-motor apparatus, are not only parts of the body; they are above all means used by the living being to be able to maintain a balance between its physical possibilities and its means of expression. The living being *is* not its biomechanical framework and its nervous system, it *has* them to do something with: Plessner's formula is perfectly appropriate here.

In humans, the almost complete exteriorization of the means of life from individual organisms may lead to two types of balance between skeleton and brain. The first consists in seeing in society a new organism having its own means, that is to say its own skeleton (techno-economic) and its own brain (externalized social memory). In this case, the individual is no longer very much at all, a simple cog necessary for the proper functioning of the techno-economic skeleton (the productive individual) or social memory (the employee of bureaucracies). The second type of balancing consists in saying that society is not an organism but a set of means-instruments available to individuals. The group no longer takes precedence over the individual; it is the opposite. The individual possesses its instruments of life, including the exosomatic instruments of technology.

The film *Wall-E* presents what might be an exit from the impasse by adaptive evolution, that is to say by deep changes in human biomechanics (the first type of equilibration). According to this orientation, however, the social super-organism would find itself possessing all the instruments of life, with individuals having absolutely nothing more to do. They would not possess their body as an instrument; the structures and operations of their organism would not be the means of any vital activity.

The problem is therefore not so much that an evolutionary process deprives us of what has made us human so far; the problem is that to allow us to keep pushing our rock up the hillside (that is to say, for humanity to not disappear but to be transformed by adapting), this process could lead us to a situation where "living" will no longer consist of an activity at all. If *Wall-E* is an anti-value, it is not because the human is no longer the human (a concern which, moreover, would be very difficult to explain); it is because the notion of "life", conceived as an activity of using an appropriate range of instruments – organic as well as extracorporeal or technological – to accomplish goals, has been completely lost. The problem is not the regression of the hand and the erect position in themselves: it is that by losing these two biomechanical characteristics which have hitherto structured human life, there is ultimately no longer any rock to push, or even a hillside.

A second condition, therefore, for devising technologies through which care may be taken for living beings, human and non-human: to jointly rediscover the meaning of the organism and the organs, as well as the meaning of technology, as instruments which the individual may use to proceed to its valuations.

In summary, an individual exteriorized to the point of being completely dispossessed of all its instruments of life, now transposed to society, leaves a feeble margin for an evolutionary future for the human species. Adaptive logic indeed wants this future to go in the direction of an almost complete disappearance of the organ-instruments (erect posture, removable hand) which made the evolutionary process of exosomatization possible. Having become unemployed in a completely artificial context, where the individual no longer needs to mobilize any instrument whatsoever, natural or artificial, to live (as illustrated by *Wall-E*), these organ-instruments are inexorably brought to regress to the point of disappearing altogether.

A remarkable idea emerges from this analysis: the biomechanical and neuro-motor characteristics that define the species *Homo sapiens* are no longer just a simple zoological fact; they now appear as values to defend. More precisely, if the dialogue between the individual and society is, at the outset, the evolutionary response to an imbalance that has arisen between technical ability and the ability to organized itself in humans – a zoological fact, in other words – this dialogue appears now as the framework outside of which the very notion of a *human* life seems to lose all meaning. This idea echoes the analysis, albeit barely sketched out, of Ortega y Gasset: the human individual is fulfilled only by fulfilling nature in itself, and not by opposing it. We fulfil ourselves as individuals, not by freeing ourselves from our belonging to the species to make ourselves into pure individuals, but on the contrary by bringing the vital potency implied by our belonging to the species *Homo sapiens* to a high level of expression. Ortega y Gasset does not say: the individual, to flourish in its valuations, must consider the requirements of nature within it, that is to say, the requirements of organic life. He is more radical in maintaining that nature within us, that is to say the design of the species *Homo sapiens*, does not have the meaning of a limiting frame for our technical action: it is a task to be fulfilled, it demands our responsibility. Nature within us is not what is given and which limits us by the constraints which it sets up against us; it imposes itself upon our moral attention as what must be done.

Dewey and Simondon particularly developed this idea of a sort of naturation of nature by the well-equipped human individual – of nature within us but also, more generally, of nature outside of us, thus opening up to a concept of care for nature more positive than the simple act of not compromising the ability of non-human living beings to make their valuations. Taking care of nature is no longer just refraining from doing what may harm living beings; it is now, more positively, to act in a way that fulfils nature.

What does the expression "fulfil nature" mean, which seems to presuppose that nature is incomplete and incapable of actualizing its potentials by itself? How does this expression lead to care for nature, when assuming the right to reconfigure nature, on the pretext of "fulfilling it", seems on the contrary to go hand in hand with the most complete technological *hybris*, the opposite of a careful and caring attitude?

By way of illustration, a discourse in full swing today, nourishing the promises of transhumanism, stipulates that thanks to their technological means, humans are on the verge of being able to control and orient their own evolutionary process. They could ideally design it. Does Leroi-Gourhan recommend that we go in this direction? We must immediately answer in the negative. Leroi-Gourhan is very clear on this point: it is not by tinkering with humans, for example by modifying their genetics, that we will resolve the very serious situation of imbalance that we must confront. How can we understand that humans have to "fulfil nature", including their own, if there is no question of intensifying the project of technological mastery of nature?

Dewey brings to this point very precious elements of understanding, which Simondon also developed by taking them further, as we will see, in the sense of care.

3.3.2.6. *Technology creating objective connections in nature: Dewey's analysis*

According to Dewey, technology is undoubtedly the means by which the human being manages to master the succession of events and transformations of which nature consists, which means that it serves the desire of humans to materialize their goals, and above all to stabilize them by making natural mechanisms contribute to the accomplishment of these goals, rather than being an obstacle to them. Technology thus transforms natural events, marked by a strong level of contingency and hazard, into instruments of life, that is to say into means of objectively bringing into existence that which humans value, subjectively. In short, Dewey always seems to depend, apparently, on a conception of technology according to a duality of ends and means.

However, the goals here are not set by an intelligence cut off from nature and from life and, as Dewey explains, "the foundation for value and the striving to realize it is found in nature" (Dewey 1929, Preface p. iv). Values are not added to a supposedly objective nature; they are part of nature or rather, they must be established in nature through the instrumental control of natural events that a very particular living being, the human, may acquire (in particular through science). Correlatively, technological means are not simple stratagems stemming from subjectivity and intended to disappear when the goal, itself purely subjective, is achieved: they are the means by which values, resulting from valuating acts, can acquire a consistency and a sort of objective existence in nature.

The tool is therefore not the means that a pure subjectivity, separated from objective nature, gives itself in order to trigger exploitable effects within nature; it is in its essence a connection between things: "a tool is a particular thing, but it is more than a particular thing, since it is a thing in which a connection, a sequential bond of nature is embodied" (Dewey 1929, p. 122). The tool has a power of revelation of natural things, in the sense that it grasps them according to their objective connections to something other than themselves – to use an extension of Dewey's example, the plough establishes an objective connection between the wood it is made of and the growth of plants. As Dewey writes, in a summary formula: "a genuine instrumentality for is always an organ *of* an end" (Dewey 1929, p. 368).

Technology thus introduces a metaphysics of relationships; natural existences appear as ways towards other ways. Instrumentality is not external to the aim and values; these acquire an objective existence only in the natural connections established by means of the instruments used. This is why, explains Dewey, "the notion that means are menial, instrumentalities servile" (Dewey 1929, p. 367) prevents measuring everything with which the sought-after end is in objective connection. It is therefore, inevitably, to doom it to precariousness and the risk of inconsistency.

Ends and values, in short, only apply through the capacity to perceive, as exhaustively as possible, all the real connections which link them to the world. These connections are, precisely, the very fact of instrumentality. Using one instrument rather than another is indeed to trigger a natural sequence that will fully participate in the objective reality of the intended end. There is therefore reason, concludes Dewey, to be very careful and caring with regard to instrumentality: "to entertain, choose or accomplish anything as an end or consequence is to be committed to a like love and care for whatever events and acts are its means" (Dewey 1929, pp. 366–367).

There is something else here which is more than a simply technological meaning of care: instrumentality calls for moral vigilance. Let us also note that Dewey brings out, implicitly in his analysis, a new meaning of caring for nature. Until now, the expression had mainly had a negative meaning: it meant not compromising the ability of living beings to use the full range of their instruments of life. A more positive sense now seems to be emerging: as soon as humans reveal the objective connections between natural things through their technical action, it must be admitted that care takes on a productive meaning. Taking care of nature is in a sense fulfilling nature.

3.4. From anthropology to aesthetics

3.4.1. *Is it enough to recognize that humans "belong to nature" to orient technology towards care for nature?*

3.4.1.1. *Beyond the theme of the deficient being. Third condition of care in technology*

The foregoing analyses ultimately seem to boil down to a fairly simple prerequisite: doing away with the understanding of the human as a *deficient being*, shared by Plessner, Gehlen and Ortega y Gasset.

Indeed, if the original human being is the naked being, then its own domain can only result from its action (particularly technological). As such, though it may be recognized as a living being among living beings, it eventually apprehends itself as a self-referential being, detached from the living world as a whole. Technologies appear to it as having to be evaluated, morally and politically, only in terms of what they do for its own domain of action. The human separated from the animal is both the only source and the sole reference term for this evaluation. There is no sense in morally evaluating techniques with regard to the organic, which is not a value at all. Plessner and Gehlen consider the organic as unessential in terms of human self-production, which is supposed to be the only source of ethical values.

Admittedly, Gehlen himself recognizes the existence of a kind of positive feedback loop between action and sensorimotor apparatus, that is to say the nervous system: technological action upon the world shapes the sensorimotor apparatus that underlies this action. In spite of everything, he remains Kantian (according to his own admission) assuming that to penetrate the secret of intra-organic processes, it would be necessary to have a different understanding than ours – an intuitive understanding capable, precisely, of knowing the type of causality to which the organic processes belong. This is not the case with human understanding, which is discursive and not intuitive.

Let us therefore conjecture that the abandonment of the premise of the *deficient being* – explaining in part at least the fact that the organic, relegated to the animal, is immediately excluded from any questioning on ethical values – is a *sine qua non* for being able to progress to reflection on the relationships between technology and care. It must be noted in any case that the major advances towards a possible implementation of care in technology, including the aforementioned authors but also Dewey and Leroi-Gourhan, went hand in hand with going beyond this understanding of the human as a *deficient being*. The conceptualization of the organ as an instrument at the service of life, or even the idea of a human fulfilment of nature, precisely allowed us to escape from the interpretative framework of *deficient being*.

A third condition of care in technology, therefore: to do away with this old theme, already in Plato, of the human as a *deficient being*.

Is this yet sufficient? Do we now hold all the conditions for care in technology? It seems not. Indeed, it is very possible to subscribe to the three conditions already identified (the absence of a fixed point, the organ and technology conceived as instruments, the abandonment of the theme of *deficient being*), while defending an analysis of the human situation in the world which places us at the antipodes of care. This is in any case the general impression that emerges from the position of Paul Alsberg.

3.4.1.2. *From body-adaptation to body-liberation: the analysis of Paul Alsberg*

Paul Alsberg was a German doctor who fled Nazism and emigrated to the United States where he restated, in a book published in 1970 posthumously and entitled *In Quest of Man* (Alsberg 1970), ideas on human evolution that he had already aired in writing in the early 1920s (this latter work was burned by the Nazis, so it disappeared). Plessner as well as Gehlen cite him; he was, in short, not at all an unknown.

Alsberg's theory has the premise that humans have an animal origin, that is to say, they must be derived from a precursor which is completely animal. However, deriving the human from the animal does not mean that the human has remained an animal, but that it comes from a being who has been an animal and who, for reasons which must be explained, has ceased to be an animal. The human is not at all an animal, even a very special one: it represents something else, something totally different – a new category of living being. Alsberg says it unambiguously, and so to speak counter-intuitively: humans are not primates. On the other hand, from the fact that the human is located outside of animality, Alsberg does not conclude, however, that it is outside of nature: it remains caught within nature, biologically related to the conditions of life on Earth. It remains "heavily afflicted with animal heritage". In short, if we understand correctly, humans are not animals, but they share with animals a fundamental condition which is that of every living thing on earth. Concretely, humans are undoubtedly not a primate, but like primates they must provide energy for their metabolism, they need oxygen, etc. Organic processes and their unconditional requirements transcend the distinction between human and animal; they are imposed upon all living beings.

Alsberg's reasoning for deriving humans from animals is original. It starts from the idea, inherent in Darwinian theory, that natural selection favored those most adapted to their environment. This adaptation of animals is controlled by the principle of *body-adaptation*, or of *body-compulsion* to use Alsberg's terms: it is the principle according to which evolution consists of a selection of bodily criteria. However, Alsberg notes without great originality that the human as we know it today is naked. It lacks all the adaptive bodily characteristics of which animals dispose. Assuming that humans have an animal origin implies in these conditions, very logically, admitting that the animal precursor of humans was, like all animals, adapted: it must not have been naked, it most certainly had to have bodily characteristics allowing it to effectively ensure its survival. Otherwise, it would not have survived; it would have been eliminated by the process of natural selection.

Alsberg concludes from this – and this is the originality of his thesis – that humans are the result of a loss. The human precursor had bodily characteristics that were gradually lost, to give this *deficient being* that which is human. The *deficient being* is not at the origin of the process of evolution; it is a result, an emanation. The human is not the animal plus something, but the animal minus something. Alsberg does not say, however, that humans are a failed animal, as Ortega y Gasset maintains. According to Alsberg, the human was not born, as a human, completely destitute, effectively failed, which would have obliged it to compensate for this deficit by an effort of labor. Rather, humans are deflated animals, the result of deflation in an animal.

According to Alsberg, artificial tools are therefore not projections of organs: far from extending the body and intensifying its adaptive characteristics, or creating new ones for it, the tool, on the contrary, has caused those characteristics which the organs possess to regress, then wholly disappear. Technology cannot be explained by the body; on the contrary, it is the very particular structure of the human body which is explained by technology. Tools are not appendages of bodily functions. No doubt the body is involved in the use of tools, but bodily function does not create the tool, which has its own existence. The intervention of the body is limited to raising the power of the tool from potentiality to actuality.

How to explain that a living being possessing all the adaptive resources necessary for its survival ended up losing them? Alsberg's hypothesis is as follows: the *Primeval Man*, as he puts it, this animal precursor of humans, began to make use of tools external to its body, of stones found in its immediate environment. In the use he made of them, it happened that some of these stones broke. The cutting edges thus obtained proved to be particularly useful and effective. The *Primeval Man* finally decided to produce them himself, this time deliberately cutting the stone. He discovered that he could vary the size and shape of the cutting edges, which allowed him to diversify the range of his tools by adapting them to the various uses he made of them. In other words, Alsberg has humanity originate from a perfectly adapted animal which simply discovered, by chance, the benefit of substituting for those services which its bodily provisions provided for it, the much more efficient and extensive services that could be provided, first by pebbles found along the way and accidentally broken, then by fabricated tools.

Note that the human precursor animal must therefore have already developed cognitive skills, allowing it to make associations of ideas – between the repetition of the occurrence "the stone breaks" and the observation of material action accomplished more effectively. The idea is already found in Pitt-Rivers, but unlike the latter, Alsberg never explicitly mentions these cognitive skills of the association of ideas. However, there is nothing to prevent the view that the *Primeval Man* possessed them. This assumption does not in any case contradict Alsberg's premises.

A process is therefore set in motion, which Alsberg calls *body-liberation*: from now on, the human-in-becoming seeks its adaptive resources less and less within its own body, but begins to manufacture them. It gradually integrated tools into its evolutionary scheme, to the point of making them the very engine of its evolutionary process. Thus, losing its usefulness, the adaptive bodily equipment of *Primeval Man* ended up deteriorating and disappearing. The evolution is therefore a natural evolution; there is no need for Alsberg to involve an extra-natural datum in the explanation. This evolution consisted, for this living being of a new kind which is the human, of freeing itself from the "natural bonds which connect life intrinsically with the exigent conditions on earth" (Alsberg 1970, p. 69). Humans are therefore opposed to nature while remaining interior to nature.

We can interpret this apparently paradoxical formula by supposing that human evolution, controlled by the principle of *body-liberation*, gradually freed the human being from its initial bodily determinations, but that the human being remained despite everything dependent on the elementary, organic conditions of all life on earth. In short, the human freed itself from its animal body to forge its human body, but invariably remained an organism. This duality of the body and the organism would make it possible to maintain that the human being is the living being that self-produces its conditions of existence, by means of tools, by causing a regression of its initial bodily determinations, but that it remains nevertheless inserted in a broader totality, which defines for it a range of elementary needs to be satisfied. By the expression "natural bonds which connect life intrinsically with the exigent conditions on earth", Alsberg understands the natural limitations of the body, but not the basic conditions of all life on earth.

In short, humans are not purely self-referential; they are not the sole source of their own being. Do we not therefore find in Alsberg an essential prerequisite for thinking of care in the technological relationship to the world, as we have seen previously? In any case, an essential point of Alsberg's thesis is the fully assumed abandonment of the concept of the living being as a self-regulating organism. Indeed, since the adaptive attributes that *Primeval Man* possessed have disappeared, replaced by more advantageous technical artefacts, both must be considered at the same level: that of the instrumental value of the means of life.

3.4.1.3. *Overconfidence in science*

Alsberg specifies that the human evolutionary process, transferred to technology, proceeded in an unconscious way but that today, through science, it becomes conscious or rather it must become so. Human beings must now evolve by following their reason, convinced of their responsibility towards themselves, the entire human community and nature. They are called to take control of their own evolutionary process and to take responsibility for it – a moral responsibility, if we understand it correctly. Let us note in passing, in the scientist-like optimism of Alsberg, the questionable (to say the least) ethnocentrism which he seems to display. To affirm in the 1960s, and perhaps already in the 1920s, that a human being provided with science (western and modern, if we follow him correctly) is considered to be the only one who can become fully aware of the concrete modalities of its terrestrial existence and its evolution, and that it is therefore the only one who can embody a relationship of responsibility for nature, may come as a surprise. Must we understand that every being other than human beings in advanced industrial societies could have only had an unconscious relationship, one therefore devoid of any sense of responsibility, with regard to its way of living in Earth ?

It is clear that we do not find a line in Alsberg concerning the devastation to which industrialization is subjecting this nature to which we do not cease to belong, and towards which we therefore have a responsibility. Alsberg even explains, on the contrary, that industrialization and automation complete the process of *body-liberation*, that they are in short the culmination of an evolutionary process initiated by *Primeval Man*. From the first flint cutter to the fully automated factory, there is no break in continuity. Technology, whatever it is, is always part of the process of liberation from bodily limits, and, as such it seems, it does not suffer any criticism. This is moreover a direct consequence of Alsberg's conception of technology. From the first cut pebble, let us remember, technology has no essential dependence on the body of the technician, which it does not extend; it is only defined by its intrinsic potentials, by its instrumental value. The manufacturing gesture only updates these potentials, which means that nothing fundamentally changes, in the sense of being technological, when the tool passes from the hand to the machine tool: industrialization in no way interrupts the process of *body-liberation*, which only continues in other ways.

Alsberg is undoubtedly worried about the major dangers which hang over humanity, like the atomic bomb which threatens to explode at any time (we were at the end of the sixties). That being said, when he speaks of the climate for example, Alsberg notes that we are in an interglacial period and that in the more or less short term, humanity will be confronted with a complete change of its environment: the northern and southern regions will be covered with ice. Nevertheless, let us have confidence, he urges, in the capacity of future science to solve all these problems, including climate control.... In short, science and technology are not a problem, but always a solution.

Science may take on two meanings in its relation to technology. If it is in a dominant position, technology is then only applied science, that is to say, it consists in acting in the natural world according to the general laws identified by scientific investigation. This is the intellectualist conception of technology that Canguilhem denounces. Alternatively, science is at the service of reflective judgement; it sheds light on technological action that is contextual, singular, and therefore not identically reproducible everywhere. Alsberg's conception of science seems rather to be oriented towards the first definition, foreign to the aim of care.

The idea of responsibility towards nature, due to the fact that we belong to it and that we depend on it for the maintenance of organic processes, might appear sufficient to found an attitude of care. We must, by conscious effort, organize our existence on earth "in harmony with nature", says Alsberg. That said, if it is indeed care that is in question, this can only be a non-moral conception of care. Organic processes are necessary processes; we cannot not satisfy them, except by letting ourselves die. There is no question of value in this area, but of unsurpassable material constraints. We humans have gone beyond the bodily constraints to which animals have remained subject; but we were not able to free ourselves from certain necessities, such as the supply of oxygen, etc. That said, nothing in Alsberg would allow us to morally disqualify the quest for ever greater liberation from constraints – no longer only bodily, but organic. For it to be possible one day to oxygenate our organism without breathing terrestrial air, to provide energy to our metabolism without eating, and so on: Alsberg would probably have no complaints. The process of body-liberation would intensify into the process of *organic metabolisms-liberation*. In the meantime, it is better to take care of nature: it alone can provide us with what we urgently need. At no time, however, does Alsberg say that nature, as well as the non-human living beings who inhabit it, must be the objects of care of a moral nature. The universality of organic needs, to which humans remain subject like animals, does not introduce any moral obligation with regard to the *sine qua non* conditions which must be fulfilled for living beings to be able to carry out their valuative activities. If we understand Alsberg, we must be careful, for example, not to ransack the rainforest: not however because by doing so, we inexorably condemn a considerable number of living beings (including humans) to live according to repulsive paths of life, but because we need them for the production of oxygen in the atmosphere. A calculation of interest, not moral consideration.

Consequently, the fact that Alsberg unambiguously assumes the idea that humans, although they are unique living beings, share their organic condition with all animals, in no way guarantees that their technical action must be penetrated by values of care. Alsberg, who was a doctor, must thus have thought, even in the 1960s, that medicine was the only technological activity that was genuinely concerned by the moral requirement of care.

Alsberg's analysis therefore makes us think, at least negatively, that in order to penetrate care into the technological relationships of humans with nature, at least one additional condition is missing in addition to the previous three. An indication is given to us by the way in which Leroi-Gourhan formulates the very terms of the problem, in the passage already quoted: it is a question, it should be remembered, of giving all rights to "the individual as the motive force of a social microcosm built to human scale, a microcosm in which the individual commands the full gamut of aesthetic and technical possibilities" (Leroi-Gourhan 1993, p. 358). The solution to the general carelessness of which industrial societies have been guilty seems to lie – as Leroi-Gourhan seems to say, in any case – in the adoption of new technological choices guided by new structures and new contents of sensory experience; that is to say by a reconfiguration of the *aisthesis*.

As prisoners of the dematerialized image that we have made of ourselves, we certainly know that we are living beings, but this knowledge is purely intellectual: we do not experience in a sensory way, except negatively, the fact that we are starting to seriously suffer from the environmental upheavals for which we are responsible. Yet such seems to be, according to Leroi-Gourhan, the fundamental fact of the problem: on the one hand, learning not only to know ourselves abstractly, but also and above all to perceive ourselves and to feel ourselves very concretely as belonging, very deeply, to the immense community of living beings populating the Earth; on the other hand, learning to perceive and feel that this community of living beings is not just an evolutionary fact, but a moral issue.

If we understand correctly, this is the sense of hope that Leroi-Gourhan places in the political domain, at the very end of *Gesture and Speech*: it is about fashioning for ourselves a new sensitivity. How? Leroi-Gourhan does not respond and leaves us with the problem. Nevertheless, he points out here the fourth condition of care in technology: reconfiguring our perception.

3.4.2. *Creating a new perception. Fourth condition of care in technology*

3.4.2.1. *The Bergsonian "bigger soul": learning to perceive differently*

It is in Bergson that we find the most acute formulation of this problem. At first glance, Bergson's position on technology seems unoriginal. He faithfully takes up Kapp's theory in the famous passage of *Two Sources of Morality and Religion* where he introduced his famous consideration on the need for a "bigger soul": "if our organs are natural instruments, our instruments must then be artificial organs. The workman's tool is the continuation of his arm, the tool-equipment of humanity is therefore the continuation of its body. Nature, in endowing us with a tool-making intelligence, prepared for us in this way a certain extension. But machines which run on oil or coal or "white coal" [that is to say electricity], and which convert into motion a potential energy stored up for millions of years, have actually imparted to our organism an extension so vast, and endowed it with a power so mighty, so out of proportion to the size and strength of that organism, that surely none of all this was foreseen in this structural plan of our species... We must add that the body, now larger, calls for a bigger soul" (Bergson 1977, pp. 267–268). The suggestion seems clear: humans are living beings who have exteriorized their functions into artefacts. The human being has by nature enlarged, that is to say equipped itself with exosomatic prostheses which enrich the range of its means of action, but until quite recently this enlargement of the human being through tools had remained within the limits of possibilities of individual action without tools. I could do more and something else with a hammer than without a hammer, but the handling of the hammer remained commensurate with my bodily and cognitive possibilities. The crudest tool was already an enlargement, but an enlargement which did not go beyond the framework fixed by the data of the species – the accent is still that of Rousseau.

In the context of industrial machinery, the rupture is consummated: we have made an artificial body whose possibilities are incomparable with those of our simply equipped action. Industrial machinery no longer extends the body; the possibilities of action that it gives us greatly exceed the limited framework of that which nature intended for us.

Our morality, said Bergson, was once commensurate with our possibilities of action; as these latter have experienced an enlargement of unprecedented scale, we need an according moral reform. Here we are tempted to understand: we need a body of new rules, commensurate with what we can do – a body of rules which govern our actions. Bergson would therefore already have a top-down type of normative approach to the regulation of technology, consisting in enacting binding rules in order to provide a framework for human technological action.

However, another reading of the Bergsonian invocation of a "bigger soul" is possible. Our body, says Bergson in an enigmatic way at first, "comprises everything we perceive, it reaches as far as the stars" (Bergson 1977, p. 221). However, this surprising formula is in reality perfectly consistent with the Bergsonian conception of the relationships between the soul and the body expounded in *Matter and Memory* (Bergson 1912).

Contrary to a whole metaphysical tradition which considered perception as an activity of the mind coming to add something to pure sensation, Bergson indeed defines perception as an activity which comes, not to add, but to take something from reality. More precisely, if we classically conceive action as that which follows representation, itself being the product of an activity of our mind which organizes the variety of sensations through which the real affects us, Bergson completely reverses the terms of the problem and affirms the primacy of action over perception: perception is only the reverse side of a possible action upon things. Every living being deploys a range of possible actions which depend on the design of its species, as well as environmental opportunities. The perception that this living being has of reality is an exact reflection of its possibility of acting: the living being perceives only those aspects of things which are of interest to its possible action upon them.

Bergson's formula, that our body goes "as far as the stars", then becomes clear: it is not a metaphor; our body is truly, as a power to act, considerably enlarged by the unprecedented possibilities of industrial machinery and scientific instrumentation. We might even add that today our body goes as far as molecules and even atoms, given the possibilities of acting upon them given to us by certain instruments like the scanning tunneling microscope.

Our range of possible actions upon things has been profoundly transformed: the consequence is that we have to create a new perception, in connection with these new possibilities of acting technologically upon the beings of the world. Is this not the meaning of the bigger soul that Bergson demands – not therefore a corpus of rules, but a new perception proportionate to our possibilities for action? Bergson's genius is to suggest that in humans, perception is no longer proportional to action according to a datum of the species. The price to be paid for the considerable enlargement of the body by machinery, far exceeding the possibilities inscribed in the species (what Bergson calls nature), is that from now on, perception can become proportional to action again only by a free act, a decision which is not in line with nature's programming. We must *want to* perceive things and the world in accordance with the possibilities of our action upon them.

Bergson is not limited to suggesting that industrial technologies participate in the construction of new ways of perceiving and feeling, which is an obvious point already mentioned in *Creative Evolution*: "a century has elapsed since the invention of the steam-engine," he writes, "and we are only just beginning to feel the depth of the shock it gave us. But the revolution it has effected in industry has nevertheless upset human relations altogether. New ideas are arising, new feelings are on the way to flower" (Bergson 1944, p. 153). Bergson argues more radically that these new ways of perceiving and feeling are not immediately proportionate to our possibilities of action: it is therefore necessary, not only to understand how technology shapes our perception, but literally to make for ourselves a new perception.

We make use of all kinds of artefacts. These are produced within the framework of socio-technological systems on a planetary scale, of which it is useless to point out the extent to which their deployment and their functioning may wreak havoc in the domain of life, on non-human living beings but also on the living beings that we are. The production of our commonly used artefacts most often requires the exploitation of natural resources in such a way that ecosystems are weakened, threatened, and destroyed. Through our artifacts, we are linked to polluted rivers, degraded environments, the extinction of species. However, we do not see these links. You have to take the words literally: we literally do not *see* them. Our perception is not proportionate to our action upon things, upon living beings and upon the world. The moral reform that Bergson demands consists precisely in forging for ourselves a new perceptual equipment commensurate with our technological action – an equipment that allows us to see all the processes and beings that, literally, are in direct connection with the manufacturing, the structure and the functioning of our technological artifacts (degraded environments, vulnerable populations, etc.).

What however does it mean, in concrete terms: to make for ourselves a new perception? We must measure the enormity of the task: how can we make emerge, of our own free will, a new perceptual equipment? In what sense could we *want to* perceive otherwise than we do by virtue of the sensorimotor arrangements with which we are equipped? It seems dizzying and impossible, even devoid of meaning, and yet such is the meaning of this bigger soul that Bergson calls for. He also underlines how difficult the task is, even impossible, by admitting the extreme difficulty of a moral reform that is so complete – that is to say, a creation of novelty in the field of *aisthesis*. Bergson did not believe in the providential personality, in the exceptional individual who would arise and initiate humanity into another perception by mere force of example. We cannot hope that our will to learn to perceive will translate *de facto* into the effective creation of a new sensorimotor device, and therefore of a new sensory perception.

The Industrial Revolution undoubtedly gave birth to new ideas and feelings, and ultimately a new sensitivity, says Bergson. He specifies however that this transformative action of our technologies upon the categories of our thought and the forms of our sensitivity takes place in a subterranean manner. It is only afterwards, long after technology has exercised its transformative power, that the effects of this transformation are felt: we measure change when it has already changed. In other words, it makes no sense for Bergson to argue that we can voluntarily drive this transformative process. The techniques of industrial machinery profoundly transform the sensory and cognitive structures of our experience; but the direction of this transformation is not within our power. However, the bigger soul indeed designates the hope of a conscious, piloted change in our perception in the necessary direction by means of our power to act.

The difficulty is therefore extreme, and can be summarized as follows: we must find a way to trigger within the moral and political field certain processes whose effects must be equivalent to those that would produce a directed transformation in the biological and sensorimotor bases of our perceptual apparatus. We must produce in the moral and political field the equivalent of a creation of novelty in the species. "Failing so thoroughgoing a spiritual reform," Bergson says, "we must be content with shifts and submit to more and more numerous and vexatious regulations, intended to provide a means of circumventing each successive obstacle that our nature sets up against our civilization" (Bergson 1977, p. 274). In other words, lacking the ability to create new perceptual equipment, we have no other choice than to oppose new values to the old values: to the values of unrestrained consumption and artificial needs, we must oppose values of frugality inspiring a regulation capable of containing our consumptive and destructive frenzy. Moral, social, and cultural values versus moral, social, and cultural values: we are in the familiar domain of conflicts of values which are resolved by pluralist deliberation.

3.4.2.2. *Supplementing our perception with technological devices: a possible way to direct our action towards care?*

It is not impossible to push Bergson's analysis to the point of considering that this ideal of an invention in perception might benefit from current technological progress, particularly in the field of augmented reality. Could our perception be technologically instrumented in such a way that now we can effectively *see* the real effects of our technological action on the natural world?

Everything seems to be done now so that our technological devices are perceived within the strict limits of their individuality here and now, cut off from the processes of their production and their effects. The whole socio-technological system which supports consumption is constructed in such a way that by buying a good, we do not have to (morally and politically) buy the whole process which led to this good. We live in a regime of metaphysical dualism, of dissociation between the being and the process, the thing and all of the operations that led to it. We buy a tray of meat at the supermarket, but we do not see the whole agro-industrial sector which led to this tray, especially the most unbearable parts of it. This latter requires that activists film in hiding places inside slaughterhouses, and denounce what happens there through the diffusion of unbearable images. More perception, then. Activists who speak out against animal suffering force us to see the processes. They oblige us in a sense to assume that by buying a tray of beef from the supermarket, we are also buying this suffering.

In short, Bergson suggests an in-depth reform of our perception, a meaning that would precisely oppose this division of being and processes. The possibilities of technology might then lead us to consider things, not from the point of view of outright substitution of a new perceptual device for the old, but from that of the replacement of our perceptual apparatus by devices of augmented reality. One detail, however is that instrumenting our perception with such devices will only add a few more gadgets to our equipment, if this instrumentation is not accompanied by a critique of the dualistic metaphysics that underlies our way of perceiving reality.

Once again, care for nature is revealed to have a primarily regulatory value: in our efforts to contest a distribution of the sensible that produces carelessness, we rely on technologies which may also produce a lack of care elsewhere. This observation is not an indication of a failure in principle; it beckons towards the moral injunction to never consider ourselves free from our obligation of care with regard to what is affected by our technological action.

Bergson's genius lies in having closely associated the two aspects of the problem, or at least having indicated the double direction of this moral reform which is today essential. On the one hand, a reflection on our implicit metaphysics, but also on our values and the needs to which we give value; and on the other hand, a reflection on the design of our technologies considered as processes and not as individualized things. Certainly, Bergson himself put the accent on the first part, less on the second. This task is more specifically that of design, or rather of a particular current of thought in design, whose remarkable characteristic is that it has proposed, precisely, a conception of technology satisfying the four conditions for care of living beings and of nature as we have identified them.

3.5. Transition

Let us summarize what these four conditions are.

Firstly, a technological activity caring for living beings and nature in general only makes sense by reference to a conception of the living being as a valuator, and not as a self-organized being whose constitution is an unchanging fact to be preserved.

Secondly, care requires ensuring that the living beings to whom we attribute moral consideration indeed have the instruments necessary for the exercise of their valuative activities. Care, as Cynthia Fleury likes to say, is to make capable – in any case, it is to exercise vigilance with regard to the instruments of capacitation. Instrument here has an extended meaning; it may be specific biological arrangements, appropriate environmental conditions, and not only technologies themselves. If valuation defines vital activity, then the living being *is* not only its organism as "a society of organs" (the expression is from Simondon): it also *has* its organism, as a panoply of instruments which should allow it to act on and in its environment. Taking care of nature therefore means ensuring that these instruments, as well as the environmental conditions that are necessary for their proper functioning, are not degraded.

Second condition of care, therefore: regaining the meaning of *organon*, both organ and tool, as an instrument of life, and not as part of a whole.

The last two conditions of care are in a way corollaries of the previous one. The third – rejecting the theme of the *deficient being* – stipulates that this instrumental approach to life does not separate the human from all non-human living beings.

These first three conditions ultimately boil down to a fairly simple imperative: ruthlessly hunting down any dualistic conception of the human. Indeed, asserting that humans "belong to nature" does not take a step towards care, except to challenge any residue of dualism: a dualism implied by the idea that the living being has an intrinsic value conceived as that which must be preserved with regard to all human technological action (a dualism of technology and life, in short); a dualism implied by the idea that human instruments of life are not of the same nature as those of non-human living beings (a split introduced into the concepts of instrumentality and of *organon*, consequently); a dualism involved in the conception of the human as a *deficient being*, whose vocation would be to produce itself in an order of reality irreducible to animal life (a dualism, therefore, of the human and the non-human living being, even if we recognize the dependence of humans with regard to demanding environmental conditions).

The fourth condition of care is challenging the distribution of the sensible that structures our perception. Seeing and not seeing are political constructs, not starting points. Far from being purely and simply determined by the way in which our sensory-motor devices are biologically constituted, the division between what we see and what we do not see is fixed by the interplay of certain power relations. A caring technology is necessarily accompanied, for any individual, by the means of challenging this distribution of the sensible – and these means may also be technological. A technology that helps me to see differently – to see what is made invisible by the distribution of the sensible that structures the social space in which I find myself – is a technology that may potentially help promote a more caring and careful relationship with nature and living beings. Care, before being ethical, is of an aesthetic nature.

Now, it is well known that one particular human production teaches us precisely to look at the world differently, to see it as it has never been seen, that is to say in a way which, precisely, cannot be learned other than by the very special type of education of the gaze that this human production makes possible: that is, a work of art. The work of art of genius is not only a visible thing among visible things, it reconfigures visibility itself, it dimensionalizes the visible according to new coordinates. It disturbs the distribution of the sensible.

From this point of view, it is possible to identify, in the second half of the 19th century and the first half of the following century, a whole current of thought which presents no recognized real unity, which brings together authors of very different philosophical styles, but who have all asked the question of knowing how this fundamental property of a work of art might be transposed to the design of technological devices. It will moreover appear that these very diverse thinkers have all, in one way or another, undertaken to reassess technology based upon the requirements of organic life.

This is what we may consider as a fourth current of the biological philosophy of technology, undoubtedly the least known as such but the most promising from the point of view of a caring conception of technology, which the last chapter of this book intends to examine.

4

Design, Technology and Life

4.1. At the sources of design for life

4.1.1. *The premises of design*

4.1.1.1. *Art and industry in the 19th century: a conflict still in Kantian terms*

In an article dated from the end of the 1940s (Francastel 1996), French art historian Pierre Francastel emphasizes that the idea of an opposition between art and technology takes on meaning in a context that is both dated and situated, linked to industrialization. The guiding values of art no longer seemed able to play their traditional regulatory role in this new productive context. As explained by Auguste Comte in volume VI of his *Course in of Positive Philosophy*[1], aesthetic education might up till then have been a corrective for the "lamentable restriction, mental and moral, which is the attendant danger of industrial activity" (Comte 2000, p. 200). Industrial development had been, according to Comte, accompanied since the Middle Ages by a parallel aesthetic development. "The aesthetic movement followed close upon the industrial, tempering its danger by inciting a more general and disinterested mental activity than was required by daily tasks" (Comte 2000, p. 200). However, this regulatory function of aesthetics with regard to technology now seems to be weakened by industrialism and the rise of machinery which accompanies it.

In pre-industrial worlds, art and technology are in no way opposed, but are even inseparable, says Francastel. The tool extends the hand, the beautiful and the practical are two complementary points of view on the same reality, and any object has a practical aspect as well as an aesthetic aspect. It is no longer the same in the industrial era: machinery undoes the immemorial link between the hand and the

1 Translated in English.

making of objects. As Gérard Wajcman points out on this subject, more often than not it is most difficult to imagine behind the mechanical perfection of the mass-produced object, not only another who designed it, but someone's hand or even some kind of physical work (Wajcman 2012, p. 73). Nothing is more elusive in manufacturing than the trace of a hand, its concern being rather to erase all traces of the subject in the object. Once with industrialization, technology no longer extends the human gesture but is constituted in a separate domain, the question of the relationship between technology and aesthetics arises anew – a question which, if we follow Francastel, has not received a satisfactory response.

This reading of the conflicting relationships of art and technology in the industrial era therefore points, in counterpoint, towards the classic figure of the pre-industrial artisan, who embodies *de facto* as unproblematic this alliance of the operational gesture and overall human existence – an alliance that industry is said to have made disappear. The tutelary figure of the artisan undoubtedly dominates the debates of the time. In some respects, it would for a time yet obstruct the emergence of real reflection on design, and on the possibility of inscribing the values of care within it.

It is in Great Britain that the debate on the opposition of arts and industry was launched. John Ruskin, attached to the aesthetic and social values of craft production, was opposed by Henry Cole, one of the organizers of the London World Fair in 1851, for whom industrial machinery might lead to the dissemination in society of objects with new forms, made from new materials, capable of changing aesthetic values. Far from trying to simply add a dimension of purely external beauty to industrial objects, according to Cole, it is a question of providing industry with forms adapted to mass production. Cole is above all concerned with compromise; he does not fundamentally question the concepts of art, beauty, or taste. Nevertheless, his conviction that industry is capable of giving birth to new values in the appreciation of forms and materials fully participates in the emergence of design.

In France as well, the development of industry aroused concern among artists of the first half of the century, among writers and painters in particular. The general tone which emerges from these reflections is that art and industry seem in this era to be in opposition without possible conciliation. Romantics are very concerned by the profound transformations to which industry is subjecting landscapes (Laforgue 2003), as shown for example by the mobilization of painters from the Barbizon school against planned logging at Fontainebleau[2]. The movement to protect nature

2 The painter Théodore Rousseau launched a campaign at the end of the 1830s against the exploitation of the Fontainebleau forest. France, resolutely committed to the path of industrialization, needed wood. Rousseau was effective; in 1853 the movement obtained the

seems to have been driven in France by artists. Care for nature is on the side of art, carelessness on the side of industry.

Between art and industry, it seems that no agreement can be found and that a protective barrier must preserve the first from the second. Baudelaire could thus write in 1859 about the way in which photography claims to impose itself, in the eyes of some, as a substitute for painting: it makes sense that industry, making an irruption into art, becomes its deadliest enemy, and that the confusion of functions prevents any of them from being fulfilled. Poetry and progress are two ambitious beings who hate each other with instinctive hatred, and when they meet on the same path, one must serve the other (Baudelaire 1868, p. 261).

This attitude of opposition from artists towards industry and technology certainly aroused fierce criticism from a few who, like the writer and photographer Maxime du Camp, towards the middle of the century castigated those arts which had become decadent, shriveled, unable to grasp industrialization and its effects as new material. Science works wonders, industry performs miracles, Du Camp writes, and we remain impassive, insensitive, despicable, strumming the warped strings of our lyres, closing our eyes so as not to see, or stubbornly looking towards a past that nothing can make us regret. They discover steam power, we sing Venus, daughter of the bitter wave; they discover electricity, we sing Bacchus, friend of the vermilion cluster. It's absurd! (Du Camp 1855, p. 5)

At best, it seems, the relationships of art and industry are limited to concealing the practical utility of objects with ornaments (Georg Maag quotes for example iron beds decorated with swan heads, or gas lamps in full bloom [Maag 1983]). For du Camp, the gap between modern technologies for developing objects and traditional forms of expression should thus give rise to a new aesthetic reflection highlighting the identity of form and function (Du Camp 1855, pp. 41–42).

4.1.1.2. *Towards a more nuanced perception of the relationship between art and industry*

Relations between art and industry were, however, at the same time the subject of reflections and feelings that were more nuanced, more complex in a way. Architecture is undoubtedly the area where the integration of new technologies into creative activity was initiated very early. In "Paris, Capital of the 19th Century" (Benjamin 2002), Walter Benjamin thus underlines how the increasing use of iron, then later of glass, contributed to deeply transforming the thought of architects. While initially the new flows into the old, with architects "designing supports

protection of several hundred hectares of forest and, in 1861, the publication of a decree creating sites for artistic use, protecting part of the forest of Fontainebleau from any form of development.

resembling Pompeian columns, and factories that imitate residential houses, just as later the first railroad station will be modeled on chalets" (Benjamin 2000, p. 16), the functional nature of iron is gradually recognized and with it "the constructive principle begins its domination of architecture" – a domination evidenced by the constructions of engineer Gustave Eiffel.

The penetration of industrial technologies into art becomes even deeper, insofar as it is the very structures and contents of artistic sensitivity which are transformed by assimilating them. As professor of literature Pierre Laforgue points out, the emergence of the theme of the machine and machinery in French literature, in the years 1830–1840[3], has profoundly transformed the consciousness and imagination of romanticism, which Laforgue examines through the mutations of a sentiment dear to romantics: melancholy. Historian Nicolas Pierrot (2010) shows that a certain interest in industry manifests itself in art long before the impressionist moment, and that those who shared in it did not immediately designate industry as cultural otherness. Historian John Tresch (2012) also questions the reading which opposes the artistic currents of romanticism and industry. On the contrary, it shows that from the beginning of the 19th century a "mechanical romanticism" develops, notably through the role of technology in the invention of new artistic practices which escape the codes of academic art. Panoramas, circular paintings that cover 360 degrees inside rotundas and give the viewer the impression of immersion in the landscape, illustrate this phenomenon well. Benjamin has, moreover, left us a description of panoramas and of the way in which they overturned the relationship between art and technology (Benjamin 2000) or even dioramas (panoramic paintings projected in a dark room, made to give the illusion of movement thanks to the play of light). Tresch uses the term *technaesthetics* to describe this way of technically producing aesthetic effects, taking up a neologism which he found in Ampère – a neologism of which Simondon is therefore not the inventor.

In the second half of the century, it was in particular impressionist painters, such as Gustave Caillebotte and his factories in Argenteuil, Camille Pissarro and his industrial buildings on the banks of the Oise, Claude Monet and his factory smokestacks in Rouen or his paintings in the Gare Saint-Lazare, or even Vincent Van Gogh and his factories in Asnières, who took a fresh look at industry by treating industrial landscapes as a source of inspiration, and even a source of a new sensibility. It was not just a matter of noting the profound changes to which industry was submitting landscapes and society, and of simply making industry and everything surrounding it a new subject for painters; in a deeper sense it meant, through pictorial work, inaugurating a new sensory experience, a new type of gaze, a new perception of the changing world.

3 An irruption preceded, it is true, by a few mentions of the machine by authors like Stendhal, in the 1820s, but in a sporadic fashion.

Authors were not left behind, and the works of Victor Hugo (Charles 1997), Emile Zola or Jules Verne bear witness to an essential link to industrialization and technology.

Lewis Mumford considers that it was the Cubists, in the next century, who truly overcame the association of the ugly and the mechanical (Mumford 1950, p. 288). "In this perception of the machine as a source of art, the new painters and sculptors clarified the whole issue and delivered art from the romantic prejudice against the machine as necessarily hostile to the world of feeling" (Mumford 1934, p. 337).

4.1.1.3. *Birth of arts applied to industry*

The question of the relationship between the arts and industry was not only a matter for the literary and artistic communities. It also interested political circles. The French politician Léon de Laborde, who was sent to London in 1851 for the first World Fair, thus reported the conviction that France must affirm its originality in relation to its rival Britain by developing strong links between art and industry. At the 1855 World Fair in Paris, devoted to both industry and fine arts, the first retrospectives of French painters (Delacroix and Ingres) were organized in a dedicated building. Laborde's idea, however, was not limited to bringing machines and works of art together in the spatially and temporally limited context of the World Fair. He had a broader vision of leveraging the available industrial methods to reproduce in mass quantities, and in miniature, great works of art inherited from the past. His ambition was to promote the widest possible distribution, in French homes, on all mantelpieces, of these miniatures whose daily contemplation should, he thought, permit the development of taste and general culture among French people.

In 1856, Laborde published a voluminous text on *The Union of Arts and Industry*, in which he develops his views (de Laborde 1856), and which was immediately criticized by believers in the definitive incompatibility of art and industry. For example, Gustave Planche, a renowned literary critic at the time, wrote a long review of Laborde's work (Planche 1857), in which he opposed his ideas by recalling the opposition of the beautiful and the useful, disinterestedness and money, finality and means: the opposition established by Kant between art and technology still guides the positions taken.

The situation changed, however, in the second half of the century, with the creation of many schools of arts applied to industry. These schools aimed to train a new generation of artists capable of meeting the needs of industrialists. At the time, terms such as "industrial arts" or "arts applied to industry" were used to designate this expected encounter.

In spite of everything, in these initiatives there remains in principle an exteriority of art and technology. Arts applied to industry had the major function of adding an aesthetic dimension to technological reality, with an economic goal above all: to make objects which, beyond their usefulness, were designed to arouse a feeling of aesthetic pleasure among people, in the aim of greater sales. Art is applied to industry, but from this rapprochement, in some respects opportunistic, there arises no in-depth re-evaluation of the very concept of technology, or even of the concept of aesthetics.

4.1.1.4. *William Morris's Arts & Crafts movement*

This is precisely the point criticized by William Morris, founder of the *Arts & Crafts* movement, which proposed an ambitious program of reform of the decorative arts. Morris in fact criticized the applied arts conceived as simple additions of ornamentation to industrial objects. According to him, the industrial era was characterized above all by the loss of what he called "tradition", which consisted of the transmission of conditions in which human production was spontaneously, and therefore unconsciously, oriented by the "instinct for beauty" possessed by every human being (Morris 2012b). For Morris, if nothing consciously arose to contradict it, the spontaneous tendency was instead to make ugly objects. The era was thus marked by the repression of the instinct for beauty, which it was therefore a question of reviving. It should be understood that here beauty designates not only a conformity of human productions to certain aesthetic canons fixed by the era but rather a fundamental need of human beings: the capacity that they have, through objects that they manufacture and which come to take a lasting place in the world, of giving meaning to this world and making it habitable as a common project. What undermined the era, according to Morris, was the loss of the sense of the beautiful object, that is to say of the object that manifested within the exteriority of the world the interior harmony – bio-psycho-social, one would say today – which in pre-industrial times was experienced in daily life by the artisan at work.

In Morris's terms, the artistic concept of production, which prioritized the object produced, its characteristics and even its history, became obscured behind the commercial concept of production, for which the object was nothing: only profit counted. Art and craft were disjointed. The former was in a way isolated in an ivory tower; the latter disappeared, erased by the rise of industrial machinery which prioritized the commercial concept of the manufacture of objects.

Ultimately, according to Morris, what disappeared was the pleasure taken in production, in technical activity, in particular "that mysterious bodily pleasure" which goes hand in hand with skillful exercise of the physical faculties (Morris 2012a). Most human beings now work without pleasure; yet without this "unreasoning, sensuous pleasure", that is to say, exceeding the total planning of the activity, we are only machines. Preserving the memory of extinct craftsmanship, art must restore pleasure to the producer and the user of everyday objects; it must suffuse the entirety of daily life and inhabit the very least of our objects.

The pleasure of which Morris spoke here is consequently that same one of which Kant spoke: it is less a question of a psychological sensation which pleasantly colors the course of life than of a fundamental experience, enabling anyone to "feel his or her life" in a meaningful community. Artisans experience in their body this harmony between being and doing; moreover, by this very test, their existence is articulated with the existence of others, within a community supported by tradition (Morris took as a model the association of artisans in the Middle Ages, in the construction of cathedrals for example – a reference that can also be found in Gropius, the founder of the Bauhaus). Pleasure at work is the essential dimension according to which a community of subjects may be formed, oriented towards building a common world. The loss of pleasure taken in technical activity is not only a purely psychological manifestation of the arduousness of work; it more fundamentally indicates our inability to do common work, to bring the "works of co-operative arts" to life, and in the end to build a sustainable world.

The *Arts & Crafts* movement was therefore driven by a strong political intention, that of recreating, within the industrial era and in a sense against it, the individual and collective harmony characteristic of craft, through deeply renewing the concept of decorative art by the rejection of ornamentalist designs. "Against it" does not mean that Morris rejected the machine on principle. His position was not quite comparable to that of Ruskin. According to Morris, it was rather a question of rescuing the machine from capture by the purely commercial concept of production, in order to put it at the service of a production penetrated by artistic concern – which meant, according to Morris, reviving within the heart of industrial production the values of a craftsmanship not dissociated from the arts, in particular by restoring respect for the intrinsic properties of materials.

In understanding this task from the point of view of the need to recreate the conditions for pleasure at work and in the use of objects, Morris thus demonstrated a remarkable understanding of what was at stake in the era's debate on the relationships between art and industry, strongly marked by the Kantian heritage of an opposition between the beautiful and the useful. As we have seen, this opposition was probably not the essential issue; it stemmed from a more fundamental divorce between technology and life, which Morris fully recognized.

That said, Morris did not really advance in relation to the terms in which Kant posed the problem. He did not conceive of what a reunification of technology and of life (including in its biological dimension) might be, other than by reference to the figure of the craftsman who realizes the whole range of human possibilities, individually and collectively, by coordinating his manufacturing activities through his body. Undoubtedly, in evoking an instinct for beauty, Morris clearly underlined the need to reconnect technical activity to the biological foundations of human existence, denouncing the repression of this instinct in industrial societies. However, he did not envisage that this attachment of technology to life can be done other than by reference to a tooled body which can, by itself, support the unity of the artisanal manufacturing process – which, however, does not prevent this body at work from being usefully assisted by machines, whose function is to relieve it from painful tasks.

Despite everything, on one essential point, Morris achieved a decisive advance compared to the Kantian theoretical framework: his conception of the produced object. The opposition of the artistic and commercial concepts of production covers a distinction between two concepts of the object produced. On the one hand, a concept according to which the object is a fixed reality, indifferent to the process which produced it, a kind of ontological nullity whose sole purpose is to satisfy the financial interests of the producer. On the other hand, a reality which is on the contrary inseparable from its history and the process of its production, endowed with a kind of independence with regard to the producer, and taking value from its contribution to the construction of a sustainable world where the instinctual demand for beauty can be satisfied. There is already in Morris this fundamental idea, which we will find later in Dewey and Simondon, that a unified human experience, of which pleasure is the standard, can only flourish in a world of objects which is conducive to this development; that is, in a world where objects are themselves the location of this unification of experience.

This equally means that experience has organological conditions of effectiveness, according to Stiegler's expression (Stiegler 1998, 2008 and 2010). Objects which are not alienated as commodities, that is, not cut off from the process of their production and their becoming in the world, constitute the condition of possibility of a non-fragmentary experience. There is consequently in Morris a conception of design which was very far removed from the mercantile aims of the arts applied to industry.

The ideas of William Morris had a considerable echo in Europe at the end of the 19th century, among all defenders of a conception of the decorative arts aiming to overcome the mediocrity of the applied arts. The ideas underlying the development of the arts applied to industry testify to a very conservative posture, fixated on the opposition between beauty and utility, incapable of returning to the fundamental gesture that had previously consisted in expelling technology from life – that is to say, in Kantian terms, cutting technology off from the feeling of pleasure and pain. On the whole, the reflection of the era on the relations between the arts and industry ultimately struggles to escape the theme of industrial arts and their social as well as economic function.

4.1.2. *Overcoming the conflict of arts and industry: the rational aesthetics of Paul Souriau*

4.1.2.1. *Beyond the Kantian division between the beautiful and the useful, the idea of "rational beauty"*

In France, this reflection gained a new and decisive impetus with the publication in 1904 of the book by the philosopher Paul Souriau, *La Beauté rationnelle (Rational Beauty)* (Souriau 1904). This voluminous work is entirely animated by the intention of going beyond the opposition between the beautiful and the useful, and to draw the outlines of what Souriau calls a rational aesthetic. It seems, indeed, says Souriau, that we are currently witnessing the reconciliation of industry and art; the two ideas of the useful and the beautiful, unduly separated, tend to come together again in our minds (Souriau 1904, p. 200). An industry without any concern for art and beauty, and reciprocally an art disdainful of all the realities of life, must be surpassed.

Souriau therefore seems to stick to a relatively trivial, even simplistic, formulation of the problem: there is a need for industry to open itself to the search for a beauty which, if necessary, can be very well satisfied with being only surface, and for art to be concerned with the material existence of individuals (a program which remains rather vague).

However, Souriau goes deeper and, like Morris, his program leads him to redefine the very concept of the technical object. By taking the opposite view of the dominant ideas at the time, which were still in conformity with Kantianism, Souriau explains that technologies have an intrinsically aesthetic value: they have a kind of internal finality, like works of art, to the extent that, paradoxically, they have a utility and more precisely an operation. As Souriau writes, there is a mechanical beauty in a watch (Souriau 1904, p. 389), which we can only appreciate if we know how a watch is made, and how it works. This beauty of the watch is not therefore in addition to the mechanics of watchmaking; it emanates from these very mechanics.

This concept of mechanical beauty is also valid for those technical objects which seem to lend themselves the least to this idea of an inherent beauty; namely, industrial machines. Souriau exclaims, admiringly: we do not always appreciate the beauty of machines, which are a truly wonderful product of our art. A locomotive, an electric coach, a steamship, and soon the aircraft, is human genius passing before us. In this heavy mass despised by aesthetes, the apparent triumph of brute force, there is as much thought, intelligence, finality, or in a word, true art as in a masterpiece of painting or in a statue (Souriau 1904, p. 392 note 1).

Souriau's idea is therefore that we must liberate objects, including those of technology, from the implicit valuations by which we usually judge them. The idea that motivates the program of a rational aesthetics is that our feelings are not really our own. It is very doubtful that Nature gave us the instincts to judge artificial things which she did not produce herself, and which did not enter into her plan. Our aesthetic feelings, says Souriau, in essence, are products of culture. We must therefore submit these feelings to rational inspection, so that we may rectify them if they are wrong. Our aesthetic feelings may be deceptive; we need rules to guide them. It is a question of being able to magnify those objects which deserve it according to the criteria provided by reason, independently of the social criteria of the moment which often push us to consider objects very poorly (by considering them in a binary way, either as exceptional or as nullities).

Rational aesthetics must thus increase the field of our aesthetic evaluations, show us the beauty in a large number of things that we did not think could be beautiful, in particular technical objects. It is a kind of justice done to the object, to all objects, and not just to works of art. By avoiding the prejudice of the rare or exceptional work, we will learn to admire beauty where it is found, without prejudice. Not only will we admire more objects, but in each of them we will admire more things.

However, this appreciation of the beauty of machines at their fair value requires careful attention to operations, to comprehension of the role of each mechanical part. Complication and gigantism are not in themselves criteria of beauty: making the complicated and the grand when one could make much more simply and modestly is a source of deep dissatisfaction in terms of aesthetic judgement. For there to be a judgement of mechanical beauty, every element of the structure and functioning of the machine must be necessary to the accomplishment of what it was built for, that is, of its end. The most aesthetic form that an object constructed with a human hand can have is that which responds most simply and directly to its destination (Souriau 1904, p. 364).

Souriau's rational aesthetic is entirely governed by the criterion of the object's conformity to its purpose. It is in the field of technology that the correspondence between the object and its end is most strict and obvious. This correspondence is indeed more relaxed in works of art; greater freedom is left to the artist. If the essential task of rational aesthetics is to determine what the end of things is, it is therefore, paradoxically as it may seem, in the field of technology that it is most at home.

The first two principles of this aesthetic, the most important, are thus formulated as follows: first, beauty lies in evident perfection; second, everything is perfect of its kind when it conforms to its end.

4.1.2.2. *Rational aesthetics at the service of broader human experience*

Unlike Kant, to whom he explicitly says he is opposed, Souriau therefore intends to identify an objective basis for aesthetic judgements. This foundation consists in the conformity of the thing to its end, which only reason can make manifest. Souriau is in short much more Aristotelian than Platonic: if reason must provide an objective criterion for appreciating the beauty of things, it is not in the sense of giving access to a single ideal of beauty, a Beauty in itself. There is no ideal of beauty that is an absolute term of reference. A thing, including in the technological domain, can only be said to be beautiful by virtue of its own end; it is in itself its own term of reference.

However, this message may surprise and leave people perplexed.

On the one hand, is it not paradoxical to relate the technical object to its end as a criterion of aesthetic judgement? What, in fact, can the "end" be for a technical object, if not its utilitarian destination, that which it is used for? But then, how do we include in this category all that is implied by the aesthetic dimension of experience, as theorized by Kant: a constitutive dimension of living subjectivity, engaged manifold in the uncertainty of existence and, moreover, capable of a relation to reality that cannot be directed by the faculty of concepts?

Souriau's answer to this question is penetrating: the end of the technical object has aesthetic value insofar as it fits into a hierarchy of ends and, moreover, in the idea that there is a supreme end which is that of all human existence. Not only is there no absolute idea of Beauty, there is also no beauty in things independently from us. Souriau thus seems to support two antagonistic positions, but he is not inconsistent. On one hand, he explains that the beautiful has objective criteria; on the other hand, he also explains that the beautiful is not independent of the subject. How can beauty be both objective and subjective? The answer here again lies in a certain proximity to Aristotle. Human beings must seek to achieve their own perfection, which resides in the life of the soul, the most elevated life. Physical life is a means; the development of psychic faculties is the real end. Appreciating the beauty of things, especially technological things, at their true value, is necessary for the improvement of the soul in which lies the ultimate purpose of human life. Souriau thus maintains that this aim of improvement passes through the ability to perceive the intrinsic richness of objects, including those held in bad esteem by constituted cultures – Souriau is obviously thinking of the culture of industrial societies which is based on the division of the beautiful and the useful, and which furthermore excludes the entire, abundant field of objects and technological practices from what has human value and significance Thus, the contemplation of a wheat field in Flanders gives Souriau a feeling of beauty, which includes admiration for the work of men, the joy of seeing the earth respond, by its fertility, to their labor, and the emotion felt that this wheat brings wealth and life from the soil.

In short, aesthetic feeling as Souriau conceives it is not so far removed from what Heidegger expresses in "The Origin of the Work of Art" (Heidegger 1971), concerning the peasant shoes painted by Van Gogh: here too, it is about unveiling a world. According to Heidegger, however, the utilitarian aspect of shoes, their being produced with a view to an immediate technical finality, must be understood as an inessential aspect, a secondary derivation from this fundamental modality of "revealing" hidden within the beings of the world, including those related to technology[4]. For Souriau, on the contrary, the utilitarian, functional aspect of technical objects is not secondary to their revealing aspect: on the contrary, it is in their properly utilitarian and functional dimension that technical objects give access to the richness of a reality that in every way exceeds what concepts of understanding can put in order.

The divide between the beautiful and the useful expresses a considerable impoverishment of our experience, which has become incapable of using the support of things in the environment, including the most banal and everyday – technical objects – to gain access to the beauty and richness of the world. Each object conceals a possibility of total experience, indivisibly sensory, emotional, and cognitive, which society represses by imposing onto our sensitivity and our perception a fragmentation according to partial categories – the beautiful on the one hand, the useful on the other.

4.1.2.3. *Liberation of aesthetic judgement*

On the other hand – a second perplexity aroused by the idea that aesthetics has an objective foundation which reason must be able to identify – it seems to follow from this perspective that for Souriau, the relation of the human subject to a reality that cannot be grasped through rational categories and principles, must ultimately, despite everything, be guided by reason. Souriau indeed maintains that aesthetic feeling must fall under the jurisdiction of reason. Should we therefore conclude that from now on, nothing in our relation to reality can escape the grip of reason? It seems only one step from there to argue that the artisan must abandon their lived intimacy with the material and their work to, having become a worker, submit to the prescriptions of the design offices and become amenable to the scientific organization of work.

---

4 At least when they belong to the category of "tools", unlike the manifestations of *Gestell* which are characteristic of what Heidegger calls modern technology.

Souriau does not take this direction, however. If he maintains that aesthetics must become rational, it is not in the sense that feeling, which is irreducible within it, would no longer have any part. Souriau rather postulates the existence of a harmony between reason and feelings. Everything that transports us is undoubtedly beautiful. When a work gives the feeling of beauty, there is nothing to reason about; no rational evidence can prevail against this evidence of feeling. Reason is not intended to dissolve feeling and ensure that it counts for nothing; on the contrary, it must rather provide it with criteria of greater accuracy, having regard to the objectively aesthetic value of things. If feeling sometimes leads us to find things to be beautiful which do not satisfy the rational criterion of beauty, it is because society has perverted the natural harmony of our faculties; it has made it so that our experience of the world can only be fragmentary.

The program of a rational aesthetic does not therefore mean that reason must reclaim aesthetic judgement for itself, which would thus be called to merge with logical judgement. It means rather that a feeling which is neither redressed nor guided by reason is a feeling whose true aesthetic significance has withered. Reason must not come to exhaust the aesthetic dimension of the judgement of taste; on the contrary, it must magnify it by freeing it from the categorizations imposed by society.

Rational aesthetics therefore orientates towards a patient educational process aimed at gradually reconfiguring our feeling by detaching itself from sedimented psychosocial valuations, in order to conform to objective criteria of beauty spontaneously and naturally. It is not a question of forcing feeling, of exerting immediate pressure on it. If my reason shows me that a particular work is admirable, and I do not admire it, it is not a question of constraining me. Let us not rush nature. Harmony between our reason and our feeling, the unification of all our tendencies, will be established over time, through education. Souriau's position is therefore much closer to a posture characteristic of ancient Greek philosophies than to Kantianism. While Kant radically separated the realms of the true, the good, and the beautiful, Souriau's rational aesthetic reconciles them: there is an aesthetic truth as well as an essential relation of aesthetics to moral perfection.

In this sense, Souriau's philosophy is incomparably more consistent and liberating than the position of Léon de Laborde. The latter advocated a rapprochement of the arts and industries by exploiting the capacity to industrially reproduce in miniature the great works of the past, to distribute them in homes and forge the taste of the French nation. The aim of moral pedagogy is obvious, but unlike Souriau for whom rational aesthetics is above all a questioning of the social categorizations of the moment, de Laborde puts the merger of the arts and industry at the service of a better dissemination of official codes of taste.

According to Souriau, on the contrary, taste should not be oriented according to these accepted codes: it must be proportionate to the intrinsic value of the objects, and oriented in the direction indicated to us by our representations of the supreme goal of life.

From this point of view, his message echoes that of Morris. Both already maintain that technical objects – due to the very fact that they do not allow themselves to be completely mastered by the socio-cognitive categorizations according to which the real is ordered, in a given society and at a given time – introduce a critique of these categorizations and the effort to restore to human experience the full range of its sensory, emotional, cognitive and even motive possibilities. (Morris, more than Souriau, moreover, insists on the importance of the active relationship of confrontation with matter.)

4.1.2.4. *A yet-insufficient integration of the two meanings of life in the rational aesthetic*

That said, Souriau lacks one essential idea: namely, that unified experience should necessarily integrate the needs of the organism into its equation. Admittedly, Souriau evokes what he calls a practical wisdom consisting in orienting our tastes in the direction of our interests, of what is useful to us. This formulation may, again, surprise us: did Souriau not insist heavily on the fact that it is necessary to provide objective criteria for the judgment of taste? Does not the reference to our interests contradict this point? Let us remember, however, that reason is not intended to become hegemonic; it has a much more regulatory function in the task of recreating the conditions for the unity of experience.

That said, our interests do not point towards an anthropology of needs in the same way as Rousseau. Souriau does not go so far as to maintain that there can be no unity of experience without consideration for the essential needs of life, in the sense of the needs of the organism. Souriau therefore, in a sense, does not go as far as Morris. He remains undoubtedly, after all, still too Kantian, in the sense that he does not really question the split between the two meanings of life: on the one hand the feeling that the subject has of its own life (the unity of experience of which Souriau speaks), and organic life on the other hand. In this line of thinkers who have endeavored to reconsider the problem of the relationship between technology and life, Souriau therefore appears, as we shall see, less radical than others like Mumford, Moholy-Nagy, Dewey or Simondon.

However, a passage from *La Beauté rationnelle* introduces consideration for the needs of living things; but in a very indirect way, and ultimately without penetrating the overall philosophical vision defended in the work. Souriau evokes the art of gardening. It happens that certain gardens, says Souriau, give the impression of being pure products of nature, without human intervention, whereas of course they are the object of all the gardener's care (a memory of Julie's garden in Rousseau, perhaps?). However, he criticizes this attitude: this illusion of non-finality in what is the product of human intention serves neither nature nor art. Better to assume human intervention, artificiality, than to hide it under the appearance of natural spontaneity. Finality disguised as non-finality is absurd.

On the other hand, specifies Souriau, if technology does not need to hide behind a false valorization of natural spontaneity, this latter must nevertheless be scrupulously respected in the individuality of each plant. The garden may appear as a human construction; but the gardener chooses to sow seeds in those places where, he or she has made sure beforehand, the plants will find the conditions favorable to their full development. In other words, the guiding values of technology must be affirmed and brought out into the open, but these values must themselves be evaluated in relation to the valuations of individual living beings. The argument is remarkable; it already points towards the idea of an artistic technological intervention suffused by the values of attention to the needs of living things, and therefore to the requirements of care for nature. But it ultimately remains unpursued, and does not give rise, in the rest of Souriau's work, to further philosophical elaboration.

This philosophical elaboration, on the other hand, is found in Dewey and Mumford.

4.1.3. *Industry and the fragmentation of experience: anthropology and instrumentality in Dewey's work*

4.1.3.1. *Perceiving things according to their constructive processes and their objective connections to other things*

In his introductory work to the philosophy of technology (Mitcham 1994), Carl Mitcham designates Dewey as one of the two precursors of a discipline not yet established at the time, that is, precisely, the philosophy of technology – the other precursor being, according to him, Heidegger. Dewey therefore figures very prominently in the historiography of the discipline and, indeed, it is undeniable that the question of technology occupies a central position in his thought, particularly with regard to the problem of experience and its unity.

As we have seen, one of Dewey's major contributions consists in renewing the concept of instrumentality, by refusing to make it a simple product of the subjective imagination and of calculation, to give it an objective value. The instrument is defined by the natural connections it establishes – in short, in Dewey as in

Heidegger's *Being and Time*, instrumentality is defined by the structure of the "in-order-to", that is to say by the network of real connections that constitute the world.

Instrumentality is therefore not only the means of an activity that creates values but is also the organ of perception. It allows us to perceive the real connections of natural things, at least when not conceived of according to the classic and impoverishing dichotomies of ends and means, subject and object, human and nature. However, Dewey argues, the prevalence of these dichotomies in industrial societies has in a way degraded these perceptual possibilities, by deteriorating their organs – worse, by deteriorating any capacity for humans to forge new instruments for perceiving the connections of things.

As for art, it has retained this sense of being an artificial organ of perception, but precisely, it has deserted the world of everyday experience. Dewey takes up a very classic theme of the time (the 1920s) and also underlines, like others, the necessary reunification of art and everyday experience.

According to him, this reunification can only be envisaged on the condition of discarding a tenacious metaphysics, which has its roots in distant antiquity, and which introduced a dissociation into our perception: we perceive things as static realities, independent of the processes that brought these things into existence. The split between art and industry echoes this split in perception: the aesthetic experience provided by art gives us access to a unified perception of those particular things called works of art, while in normal processes of existence, we always encounter things (especially those resulting from human industry) in such a way that they appear totally cut off from their processes of production, as well as from the relations and connections which nevertheless constitute them.

What characterizes the work of art according to Dewey is that the work itself, the final result, is never dissociated from the technological operations by which it was produced. Technology is perceived as immanent in the work, which also means that in the aesthetic experience of art, there is no hiatus between perception and action: it is not a static thing that is perceived, but a creative dynamism. Artistic experience is therefore unique in that it allows us to see, not separate and fixed things, but processes and relationships.

4.1.3.2. *Recreating, through design, the conditions of undivided life and experience*

Now, if this metaphysics which separates Being from becoming has been able to flourish and impose itself as the fundamental structure of our perception, leading to the isolation of artistic experience from the normal processes of existence, this is, according to Dewey, because it is based on an original split in life itself: the split between the organism and the environment. This division is itself the consequence of a fragmentation in the conception of instrumentality (the end is opposed to the means), which fragmentation echoes a social organization that has degraded work into labor and shattered the unity of experience. As Dewey writes, "the traditional separation between of some things as mere means and others as mere ends is the reflection of the insulated existence of working and leisure classes, of production that is not also consummatory, and consummation that is not productive" (Dewey 1929, p. 368).

Conversely, authentic work introduces the possibility of unified experience at the sensory, perceptual, and motor levels. It finds its first impulse in an essential need of the organism: to create for itself an environment allowing it to deploy the whole range of its sensorimotor possibilities. Indeed, the organism should not be conceived as enclosed within its bodily limits, strictly speaking; it is not entirely contained in the envelope of the epidermis. It overflows the latter to extend to the objects surrounding it, in particular tools which are not external to its being but are part of it. Before Leroi-Gourhan, therefore, Dewey assumes that the tool is fundamentally a biological fact, a living reality.

In the process of work, the living human being creates "a more extensive balance of the energies of the organism and those of the conditions under which it lives" (Dewey 1980, p. 14). From this perspective, the things produced by a process of work are not realities enclosed within their own limits, defined by their utility alone: they are above all physical beings destined to come and take their place in the world and to produce effects in this world. "[The machine] took its place in nature as an object that has the same physical effects as those belonging to any other physical object" (Dewey 1980, p. 285).

The productive action of the organism which works is immediately perceived, with the consequences of this action. "The action and its consequence must be joined in perception. This relationship is what gives meaning… The scope and content of the relations measure the significant content of an experience," writes Dewey (Dewey 1980, p. 46). Aesthetic experience is precisely the perception of the links between my action and its effects in the world; between the things that I make and the relationships that these things, once they have taken place in the world, will form with other beings in the world. This experience knows nothing of the split between the technological operation and the product, action and its consequences, the thing and the relationships in which it is engaged. The whole difficulty therefore lies in finding an adequate metrology for these relationships: how to measure them?

Aesthetic experience plunges deep roots into animal life, which precisely knows nothing of the separation of the organism and the world. The distinction between art and industry is the consequence of an organization of production in which work no longer opens onto the unified experience of the living being, but quite the contrary.

The organism is no longer seen here as a center of activity that creates values, but as a machine that must be maintained, and whose function is to make other machines run. It too is now seen as something cut off from its environment.

As for the products issuing from technology, they are no longer seen as beings of the world but as utilitarian supports for finalized human action. "In products that strike us as merely useful, our only concern is with something beyond the thing, and if we are not interested in that ulterior product then we are indifferent to the object itself" (Dewey 1980, p. 183), and "we are accustomed to think of physical objects as having bounded edges" (Dewey 1980, p. 200).

In modern industry, the worker-turned-laborer only experiences an impoverished and fragmented experience. They can no longer perceive the relations which constitute things, nor the consequences of their action. Their conditions of existence atrophy their sensory possibilities, forcing their senses to remain an "excitation on the surface" (Dewey 1980, p. 21) – that is, to be nothing but simple receptors of punctual excitations, without the possibility for the individual of integrating them into a meaningful experience. The laborer may be efficient in the field of action without, however, raising this action to the rank of experience: they may accomplish it mechanically, depriving it of all aesthetic character.

Like Kant therefore, Dewey seems to consider that technical activity carried out as an automatism of the body does not have the character of real experience. Dewey expresses, in still-Kantian terms, this loss of any possibility to live an unified experience in modern industry: "Wherever conditions are such as to prevent the act of production from being an experience in which the whole creature is alive and in which he possesses his living through enjoyment, the product will lack something of being aesthetic" (Dewey 1980, p. 27). Dewey does not say that an aesthetic relation to industrial objects is absolutely impossible. Perhaps a mechanic in large-scale industry may experience this relationship. That said, the industrial organization of production has the consequence that the act of producing provides less experience for the human being, which results in a loss of pleasure.

As with Kant, therefore, for Dewey pleasure is not an isolated and almost anecdotal psychological feeling. It is the experience undergone by the organism when its concrete, economic and social conditions of existence do not hinder but, on the contrary, encourage the development of the whole range of its sensorimotor possibilities, including when these possibilities are supplemented, extended by technology. Unlike Kant therefore, Dewey does not exclude technology from the feeling of pleasure: it is not technology itself which is in question, but the social conditions in which it is deployed. Dewey's criticism does not target machinery, but industry.

For Dewey, therefore, design and production must serve the unified vital experience of the living being that we are, both producers and users. Indeed, it often happens, notes Dewey, that "the product may not appeal to the esthetic sense of those who use the product" (Dewey 1980, p. 5). In industrial societies, the conditions for this unification are lacking, which implies that our experiences can only be fragmentary.

This analysis has remarkable implications for technological design. On the one hand, Dewey also asks that we create a new perception, that we learn to see things other than within this metaphysical framework where they always appear cut off from their processual and relational reality – which implies an organological conception of the sensorimotor assemblages of living beings. "To know what to look for and and how to see it is an affair of readiness in the part of motor equipment" (Dewey 1980, p. 98), which equally means that this "readiness" may be organologically oriented – that is, that is, it may be technologically supplemented.

On the other hand, although we can read Dewey from this organological perspective, it would be totally wrong to conclude that for him, creating a new perception purely and simply means augmenting our perception technologically in a direction that we consider relevant: Dewey is not a forerunner of transhumanism. An organological reading of human existence does not lead to militating for a purely technological approach to human problems. Dewey is very clear: the task ahead is above all that of a social and political critique of the conditions of existence which are imposed on the overwhelming majority of human beings in industrial societies. Learning to see differently cannot mean (at least not only, nor even as a priority) acquiring new technological equipment. It means above all freeing oneself from a certain type of social and political organization that has operated a distribution of the sensible in which technological processes and their consequences have become invisible. The proximity to Bergson is very strong on this point.

In short, individuals in industrial societies have lost the power to equip themselves with instruments that would render them capable of experiencing the real connections of things and, based on this experience, of objectively creating values in the world. The individual is nothing more than part of a larger whole, that is to say of an organization which itself possesses the real instrumental power – namely the power, precisely, to decide what the objective connections of things are, what is linked and what is not.

The instrumental meaning of the organ has given way to the organ conceived as a part: the individual has become the organ-part of a productive system depriving individuals of the faculty of freely enjoying the full range of "the full gamut of [their] aesthetic and technical possibilities". To restore to individuals their instrumental power is, therefore, to undermine the foundations of the social and political organization which dispossessed them of it. It is also, inseparably, to unravel the different dualisms resulting from a conception of the human being separated from nature.

The instrumental conception of nature certainly does not mean, in Dewey, that a subject cut off from nature, eager for domination and a creator of values whose reality would be purely subjective, would have a vocation to become a predator upon natural environments. On the contrary, this means that this particular living being that is the human, belonging to nature like all other living things, is dedicated through its technical activity to revealing nature according to its potentials for objective connections. There are objectively causal relationships and relationships of succession in nature, and humans are living beings capable of "converting physical and brute relationships into connections of meanings characteristic of the possibilities of nature" (Dewey 1929, p. 370). By converting "actual immediate objects into better, into more secure and significant, objects" (Dewey 1929, p. 158), fitting better into their real connections with other objects, the instrumental power of the human being, as we have seen, accomplishes nature by objectively developing its potentialities.

The human being is, in short, an instrument for improving the natural world. Such might be a definition of care resulting from Dewey's analysis: care consists in acting technologically in such a way that nature is accomplished, improved, optimized according to its own potentialities, and not by reference to the interests of a subjectivity cut off from the world.

4.1.4. *Life as judge of technology. Lewis Mumford*

4.1.4.1. *At the origin of the separation of the mechanical and the organic*

In many ways, Mumford extends Dewey's reflection on the complex inter-determinative relationships between a metaphysics separating being and processes, a dual anthropology introducing a split into human life, a fragmented and shriveled sensory experience, a narrow conception of technology and the rise of industrialization.

Mumford considers technology to be an essential manifestation of life: "Man's technical contrivances have their parallel in organic activities exhibited by other living creatures" (Mumford 1952, p. 17). He writes unambiguously, in a text dated 1972: "my own analysis of technology begins… with organisms" (Mumford 1972) and "the point I am stressing here is that every form of technics has its seat in the human organism" (Mumford 1972). Technology therefore finds its origin in the behavior of living beings.

Art also responds to a fundamental need of human existence, that of going beyond activities devoted to satisfying the bare needs of the organism, with a view to the development of the personality. Art and technology are therefore two essential dimensions of human life, inherent in the organism. They both "represent formative aspects of the human organism" (Mumford 1952, p. 31). Their link is profound: the full development of the possibilities of the organism, and consequently the non-repression of its own needs, are conditions of personal life. Human beings have indeed found the way "[to fulfil] his own nature in cooperation with the environment, he creates a third realm, the realm of the arts, in which the two are harmonized and ordered and made significant" (Mumford 1934, p. 318).

Mumford therefore makes his own, at least at first analysis, a conception of aesthetics which remains in certain respects classical or Kantian in its general inspiration. Aesthetic experience concerns, on the one hand, the lived experience of a harmony of all our faculties, and, on the other hand, the lived experience of a harmony between our aspirations and material nature, between mechanism and finality. In short, internal harmony and external harmony, in a relationship of mutual coordination.

Because he has remained "in command of technical operations, working slowly and pleasurably on the job" (Mumford 1952, p. 48), the handicraft worker appears as the model of this harmonious relationship to oneself gained and maintained via a harmonious relationship with the outside world. The goal of effectiveness in transformative material action implies that the body becomes an instrument, by developing the necessary *habitus* for human gestures to handle highly organized materials, whose multi-scale structural properties restrict the spectrum of possible actions and transformations. In this successful test of the compatibility of the operative gesture and the intrinsic properties of the materials, the artisan also experiences a harmony between the personalist goal and the demands of their organism.

We are not very surprised to see Mumford associate this harmony experienced by the artisan with pleasure, and to rediscover here Kant's feeling of pleasure and pain. But for Mumford, precisely, this pleasure can only arise from an aesthetic experience connecting the respectively personal and biological dimensions of existence. In Kant, technology only reveals itself to be able to be situated against the background of a split experience of life. In Mumford, on the contrary, technology introduces a unified experience of life, in the double sense of being-in-life and of personal life. More precisely, technology, in the era of industrial machines also, *should* open the way to this unified experience, and be at the service of life. Historically, this has not been the case and the development of machines has proceeded, on the contrary, in the direction of a divorce between technology and the life of the organism.

Mumford locates the origin of this divorce of technology and life in the Middle Ages, and more precisely in the adoption in monasteries of a life regulated by the clock (Mumford 1934, pp. 12–18). The clock brought into human existence a demand for quantification and standardization, thus preparing the conception of a world "with mathematically measurable sequences" (Mumford 1934, p. 15). This mechanical time, accurately marking the moments of human existence, is completely foreign to the human organism. "Human life has regularities of its own, the beat of the pulse, the breathing of the lungs" (Mumford 1934, p. 15) – and these do not readily submit to this rigorous metric of time. Organic time is opposed to mechanical time. This life regulated by the clock, including in its organic dimension, spread to all of society from the 13th century onwards. It has resulted in the subordination of the needs of the organism to requirements of a different nature, summed up by the aim of putting human activities in order. The encyclopedist project of bringing the arts and crafts into discipline is thus the culmination of a long process that began around the 10th century.

Far from sticking to classic invectives against industrial machinery, Mumford proceeds to an archaeology of the industrial era for which the fundamental fact is indeed the divorce of technology and life. We must at this stage recall that for Mumford, technology is not in itself the Other of life, including when it imposes on life the appearance of a mechanism, or at least of regularities indifferent to the requirements of the organism. As paradoxical as it may seem, in beginning to organize its existence according to a metric foreign to the organism, the human being of Western societies of the Middle Ages certainly atrophied its sensorimotor faculties, but at the same time, it continued to serve the needs of life. There is something of the *pharmakon* in Mumford's conception of technology – technology is both poison and remedy.

The mechanization of life, and the resulting eviction of the organic, are in fact not only manifestations of a desire for power and domination of nature foreign to the needs of life: they are also the means by which the living human being has gained at least relative security, by channeling the contingency of external phenomena. "The submission to orderly processes of work" (Mumford 1952, p. 44) is not in itself characteristic of industrial machinery, nor even of the regulated organization of life that was enacted in monasteries of the Middle Ages; it is inherent in all technical activity. This submission is reflected at the level of the organism by the adoption of sensorimotor assemblies which can be qualified as mechanisms. This form, already, of the mechanization of life, expresses the need which living human beings have found to take account of the constraints imposed by matter. In order to avoid the vagaries of nature and to ensure a minimum of security, human beings have had to regulate their operations upon the intrinsic organization of materials and natural processes.

Technology is inscribed in the very movement of life; nevertheless, its deployment passes through a certain negation of life, through a mechanization of life in the form of an incorporation of regularity, of the standardization of operative gestures. In echo of Bergson, Mumford maintains that human life has only been able to flourish through technology, but also that this life supplemented by technology is always in danger of turning against itself and of intensifying its own mechanization, to the point of completely losing sight of the organism's needs. According to Mumford, this is thus what happened from the Middle Ages, and especially from the modern era (from the 16th century onwards) up to the 19th century, in which the repression of the organic by the mechanization of life was brought to its peak. The development of social organizations aimed at minimizing the needs of the organism, and devoting human beings exclusively to the nonorganic, became intensified in the 16th century with mining and modern warfare – the two scourges which accentuated the rupture between technology and life, according to Mumford – and reached its climax in the industrial era.

Undoubtedly, "these habits of mechanization, this submission to routine and drill, this willingness to deal in familiar terms with the nonorganic and with that, from an organic and human standpoint, looks deformed if not frightful, were not achieved without considerable resistance" (Mumford 1952, p. 47). More precisely, until the second half of the 18th century according to Mumford, a balance had been successfully maintained between the mechanical and the vital. So-called "eotechnic phase" of the civilization (10th century–mid-18th century), except in the mines and on the battlefield, had all in all remained primarily oriented towards the intensification of life (Mumford 1934, pp. 148–149), particularly – and this is decisive for our purpose – at the sensory level: "one by one the senses were refined" (Mumford 1934, p. 149).

The separation of the mechanical and the organic had certainly begun in the Middle Ages, but "not until our own age [that is to say in the industrial era] did man prove willing to throw his whole subjective and qualitative life overboard" (Mumford 1952, p. 47). The so-called "paleotechnic phase" of the development of the machine, from the middle of the 18th century, was at first devoted, says Mumford, to pushing to the extreme the "quantification of life" (Mumford 1934, p. 151), by tracking the smallest residues of a qualitative life. Unsurprisingly, Mumford quotes Ure: it is primarily a question of "training human beings to renounce their desultory habits of work and to identify themselves to the unvarying regularity of the complex automaton" (Mumford 1934, p. 173).

In writing the article "Art", Diderot could probably not perceive the overall logic in which his injunction to subject the arts and crafts to external formalization took place. This logic nonetheless turned out to be dominated by the aim of a generalized exhaustion of life: not only to expel workers from all artisanal know-how, but as much as possible to snuff out everything in them relating to the satisfaction of the needs of the organism, not immediately oriented towards machine production.

A more positive relationship between the machinic and the organic, however, seems to have become established in a third phase, around the 1870s suggests Mumford, motivated by the return in force of life and its own demands. Far from seeking to replace the living organism at all costs with mechanical substitutes, the emerging "neotechnic phase" seeks rather to establish the conditions for an environment favorable to the flourishing of life, not by an illusory return to some pre-machinic golden age, but in the very heart of the civilization of machines – exit, then, the nostalgia for craft.

We are only at the beginnings of this, says Mumford. We must even admit that this new phase of the machine, which is signaled by greater attention to the demands of life, particularly in the workplace, remains under the domination of the previous phase. The direction is however clearly indicated. Mumford's fundamental idea is that the machine is not only a utilitarian object defined by its capacity to produce quickly and in great quantities: it is above all a fact of civilization; it is charged with cultural values. The machine indeed has a fundamentally anti-organic tendency, but this can be put at the service of life on condition that it is inserted into a new system of values. This system must be able to accommodate pluralism of values and the always-possible confrontation between different value orientations among individuals.

Mumford is a fairly faithful Bergsonian on this point, but it is also possible to bring him closer to Dewey: the most urgent thing is to invent in the moral field, to forge new values for ourselves. Technology, including mechanics, can be maintained in the prolongation of life, provided that an adequate system of values, prioritizing life and its own requirements, constitutes the framework from which the design of technology proceeds.

Mumford therefore insists on an essential point, which Souriau had only glimpsed: technological development has value only with regard to the values of life in the biological sense of the term. Harmony between the technological and organic conditions of existence is a cultural creation; it is, first of all, a choice of values. Even more fundamentally, by maintaining that in humans, technology has an essential relationship to both organic and personal life, Mumford seems to defend the idea that technology fully participates in the configuration of forms and manifestations, both biological and personal, taken on by human life at a given time.

The organism is therefore a historical reality, the result of techno-social modeling; it is constructed by its technologies even in its modalities of interaction with the environment, but also in its ways of feeling and experiencing. However, for Mumford the human organism is also, moreover, the result of a long process of natural evolution. As such, it is controlled by requirements and by needs that no society can ignore, lest it pay the price for a fundamental absurdity in existence. No society can thus discount "the meanings and values and purposes of the organism, as established … in the general plan of life of the species" (Mumford 1952, p. 21).

The very thorny point, extremely difficult to clarify, is that in the case of the human species, evolution has, as we know, included technology as an essential element. Tools are attested to in the oldest representatives of the species – today, it has even been shown that the first flint cutters were Australopithecines. How then to understand, on the one hand, that tools participated in the constitution of the human biological formula, even in its bone structure as Leroi-Gourhan for example shows; and on the other hand, that evolution gave birth to an activity oriented by values – the activity of the organism itself – intended to serve as a term of reference for the evaluation of that which, precisely, contributed to the constitution of this activity: technology? How can technology be judged from the point of view of an instance (the life of the living being, in its own valuative operations) in whose constitution it has always been involved? If the possibilities of human existence, including sensorimotor and perceptual possibilities, are technologically constituted, how can we justify evaluating technology from the point of view of these very possibilities? How can the constituted be both judge and assessor of the constituent?

4.1.4.2. *Bioaesthetics: constitution versus institution of experience*

Also faced with this difficulty, Italian philosopher Pietro Montani is at first glance consistent (Montani 2014). Close to Stiegler in many respects, Montani embraces the organological conception of perception and sensitivity – that is, he sees them as technologically constituted faculties. However, sensory experience today is technologically informed by devices that provide it with sensory contents in abundance, at such a rate that the individual retains no possibility of symbolically elaborating them. *Aisthesis*, the sensitive faculty, is thus oriented towards the purely passive reception of these contents, that is to say towards consumption.

According to Montani therefore, the organism as a support for being-in-life cannot serve as a fulcrum for a liberating critique of the sensory faculty. This body is indeed entirely subject to technical prostheses; these entirely orient and plan the experiences that it can make. The organic body can always be objectified by technological devices which we may fear to be toxic for experience; that is, that they are at the service of an integral normalization of life. The life that can serve as a term of reference for the critical evaluation of technology is therefore, necessarily, a life that exceeds the life of the organism, which is always stuck in the positivity of the sensorimotor assemblies entirely accessible to the influence of technological devices and to planning. The living being is literally grabbed and subjected to indefinitely extensible possibilities for technological intervention and economic development.

Conversely, according to Montani, the flesh as opposed to the organism escapes this total grasp. It opens onto unplanned territories from which the reconfiguration of sensory experience in an emancipatory sense is possible. The concept of flesh is associated with the idea of institution (rather than constitution); that is, the process by which external conditions, provided in particular by technology, can become fundamental structures of experience but only by undergoing significant elaboration, and therefore a critique.

The concept of institution signals both the idea that the subject is not self-founded (it is not the origin and source of its own being, it is not constituent of itself) and the idea that what makes it exist is not a set of structures imposed on it by a foreign authority. The subject does not belong to itself, but this non-belonging is not a heteronomy.

The flesh thus introduces the idea of an always-possible abandonment, of an escape from the devices of capture, that is to say, the idea of the subject's autonomy without the need for the subject to be postulated as self-referential. Between the hetero-constitution of the subject and the forms of its experience (through technology) and the absolute self-constitution of the subject, there is a third path that the notion of institution aims to capture.

Still, according to Montani, it is not the life of the organism which, in its positivity, can take the place of such an instituting body. It is aesthetic life, that is, a sensory life whose forms are precisely instituted and not constituted, and which consequently is not entirely channeled and planned by the technological devices.

We must therefore admit that there is something non-constituted in the constituted, a charge of pre-individual reality that is not channeled into the processes of somato-psycho-technological individuation. How to account for this?

Stiegler and Montani certainly invoke the ambivalence borne by any technology – both poison and remedy, *pharmakon*. That said, it is not technology itself that contains, in itself, emancipatory potentials. What is potentially emancipating is the system of reality formed by the human being with its technologies, a system containing potentials which may be actualized in the sense of an effectively emancipating individuation. Montani, moreover, very rightly invokes Leroi-Gourhan and Simondon, finding in them the idea that our technological prostheses must retain an essential link to human sensitivity and contingency, and not constitute an autonomous, self-referential domain, dedicated to indefinite optimization.

From Montani's organological perspective, however, what is this system of pre-individual reality in which the pharmacological potentials of technology can be actualized, both in the sense of poison and in the sense of remedy? He provides a very interesting answer: it involves the imagination in its "interactive" dimension (this term refers to interaction with the outside world). The imagination is this system of pre-individual reality.

In his interpretation of Kant, Montani indeed reassesses the schematizing imagination from an organological point of view, in the sense that the synthesis without concepts which defines is activity always already operates – such is the hypothesis of Montani's reading – based on externalized technical objects. In short, Montani embraces the perplexity felt by Kant when faced with these syntheses which seem to be enacted by the body without reference to representations (similar to the activity of the organ player). Perhaps, Montani suggests, we should consider a sort of somatic and perceptual-motor schematism implemented by a body grappling with the world.

This is decisive, but supposes something essential: that the somatic and the perceptual-motor be precisely no longer reduced to the pure positivity of a body whose underlying structures, sensory and motor, are supposed to be constituted by a totally external instance, thus delivering this body to a situation of radical heteronomy. Montani's reading seems to us to grant too much to Kant in the interpretation he proposes of schematism, precisely because the *Anthropology* finally excludes the functioning organism from being anything other than this pure positivity, a living machine and not an instrument.

Beyond this inessential exegetical question, it remains that Montani puts his finger on a point of capital importance: defending an organological perspective is not the same thing as defending the thesis of the anthropological constitutiveness of technology. The organological perspective calls more for an understanding of technology as an instituting rather than constituting exteriority – failing which, again, we have difficulty seeing how the constituted might be able to turn towards the constituent to criticize it.

In short, admitting that technology participates in the institution within us of historically dated forms of experience does not mean that we are a *tabula rasa* and that everything is possible. Human life is not technologically constituted in all its determinations, which would leave no outside and therefore no possibility of criticism whatsoever; rather, it is instituted in new experiential possibilities, through the process of individuation. But these, as Simondon has strongly observed, always retain a charge of pre-individual reality, of non-individuation which has not been structured and may provide resources and guidance for further individuations.

Mumford precisely considers the life of the organism both as that which is historically shaped, in its experiential forms, by technology, and as that which is not exhausted by it. The being-in-life is not completely individualized by technology; it retains potentials for elaboration which, according to Mumford, manifest themselves in the pressure of a vital activity which may serve as a term of reference for the evaluation of technological developments. This evaluation does not, moreover, consist in comparing technologies to established norms which would define living things. Rather than norms, it would be better here, already, to use the Canguilhemian concept of vital value: the norm still designates the stratified, the positive; the value is rather transductive; it is the pre-individual of the norm.

Going beyond the paleotechnic phase of the development of machines, towards a stage that Mumford calls biotechnic, is consequently to create conditions so that the human being may technically enrich the range of its sensory and motor possibilities (to intensify rather than to atrophy life), without losing sight of the fact that the value of this enrichment is measured by the yardstick of the vital values of the organism, on condition of conceiving the latter not as a fixed organization, governed by established and immutable norms, but as a creative activity enjoying the whole range of its instruments, organic as well as technological. “Before man could take even the first timid steps toward conquering nature, he first had the job of discovering, controlling and utilizing more effectively his own organic capacities” (Mumford 1972). Here again, the body-instrument is opposed to the body-machine, a simple cog in the great paleotechnic industry.

It follows that for Mumford, connecting the potentialities of machines and the requirements of being-in-life means above all creating the conditions for a new sensory and perceptual-motor experience to be constituted, capable of integrating machinic potentials, rather than against them as Morris advocated. Mumford thus welcomes attempts “to assimilate the meanings of technics, to take in, as part of one’s conscious perceptions and feelings the new area of human experience opened up by the machines” (Mumford 1952, pp. 53–54). He adds that this transformative power of perception and sensitivity in the context of industrial machinery must lead us to see “things as they are” (Mumford 1952, p. 55) – let us understand by this, to see them according to their actual mode of existence, and not according to the dichotomous categories through which we usually relate to them (being/process, nature/artifice, living being/technology).

The fundamental structures of perception and sensitivity are not givens for our species. They consist of a historical process in which technology plays an essential role. This process is in no way deterministic, and it is not an evolutionary process in the sense of natural evolution. Our perceptual-motor assemblies are intertwined with choices of values; that is, through the design of objects, we choose not only the world in which we want to live, but also the categories through which we want to perceive this world. The design of objects always establishes, or reinforces, a determined distribution of the sensible.

The paleotechnic phase, characteristic of the industrial era, distributed the sensible in such a way that life was divided into mutually exclusive components – respectively, being-in-life and personal life. Between the two lies technology, operating in a single direction: from the person who sets goals and gives themselves the means (technology) to attain them, to living beings and to nature which purely constitute material for technology, interpretable in purely mechanistic terms. Dynamism of values is rejected on the side of the subject cut off from the world and from life. The world only gains its objectivity by emptying itself of all subjective, qualitative, and processual life.

To question the distribution of the sensible in the paleotechnic era, it is therefore necessary, according to Mumford, to begin by unravelling its fundamental organizing value – which consists in devaluing the life of the living being, in favor of a technological life devoid of any consideration for the requirements of the organism. It is in short necessary – unsurprisingly, for Mumford as a Bergsonian – to rediscover the original meaning of technology as the deployment of a unitary life which, in the same movement, establishes both its world and the sensorimotor assemblies which perceive this world. This is what Mumford calls the biotechnic era – which he says he sees "over the edge of the horizon" in the 1930s (Mumford 1934, p. 353). Yet one may sadly wonder if, seventy years or so after Mumford prophesied it, it is not still a very distant horizon.

Still, the direction is given: to forge a new sensibility and a new perception, by bringing to their full expression the aesthetic potential of industrial machines. The task is to bring these potentials to actuality by making them the organizing principles of a new sensorimotor experience. For Mumford, art still has an essential role to play in the accomplishment of this task; this is why he speaks of reforging the unity of art and technology at the level of their common origin, the life of the organism.

4.1.4.3. *Technology, nature, person*

Mumford adds that surpassing all division between the technological and aesthetic domains of human experience is also to consider the person and his or her values as a domain of objectivity. If it is a question of seeing things as they are, reciprocally it is also a question of bringing into existence, through industrial design, things which "also see us, so to speak, as *we* are", objectively (Mumford 2015, p. 67). A curious expression, at first glance: are things to have a perception, and are we to be objects that things would perceive? This is incomprehensible if we stick to the classic dualisms – of the human and the world, of subjects and objects. In this framework of thought, in fact, only subjects are endowed with perception; objects can only be perceived. *Esse est percipi aut percipere.*

Now, Mumford's strange formula – things see us as *we* are – only has meaning on the condition, on the contrary, of refusing any fundamental split between subject and object. Indeed, according to Mumford, it is impossible to promote a personal world if one begins by excluding from the equation the properly objectual component of this world. Objects (in particular technological ones, including machines) are not the Other of the person; technology is not a domain of axiological neutrality. Conversely, values are values only if they acquire objectivity, which supposes a conception of objectivity which is broader and more comprehensive than the current conception according to which the objective is opposed to the subjective. Mumford writes: "true objectivity must include every aspect of an experience, and therefore one of the most important side, the subject, himself, must not be left out" (Mumford 1952, p. 55).

The design of industrial objects is therefore an essential element of the world of the person. Mumford explains that "respect for the object", "this conscious interest in the actual go of things", is "a valuable contribution to the whole personality" (Mumford 1952, p. 54). The person can only be fulfilled by objectifying itself in the order and the regularities brought about by technology, including mechanics. "How we think, how we feel, what our purposes and values are, all enter into the final equation" (Mumford 1952, p. 55), that is, in what counts as an objective fact.

Undeniably, this anthropology which stipulates that the human being fulfils its nature only "in cooperation with the environment", and not against it, goes in the direction of care for nature. Mumford has moreover, in *Technics and Civilization*, left us unambiguous pages on the destruction of the environment in the paleotechnic age (Mumford 1934, p. 167). Pollution of air and water, but also the etiolation of entire regions whose multiple potentialities, in terms of productive activities and human development, are negated for the benefit of industrial specialization, in such a way that "a variety regional possibilities were neglected" (Mumford 1934, p. 171).

The regional specialization of productive activities is not a bad thing in itself, since it is modeled on the reality of climatic, geological, and topological differences of the different environments. However, the new specialization (in the paleotechnic era) was not done in accordance with regional opportunities: it consisted rather in imposing upon living environments an organization totally indifferent to their specific characteristics.

In short, Mumford seems to introduce the idea that for humans "[to fulfil] their nature in cooperation with the environment" is also to fulfil external nature according to its potentials and its own organizational ways. A remarkable idea indeed, already present in Dewey as we have seen – that of a living being which is only accomplished by fulfilling, through technology, the nature from which it arises, and to which it does not cease to belong.

A whole philosophy of care as an activity of maintenance, repair and improvement of the world thus emerges from Mumfordian anthropology as from that of Dewey, in support of the definition of care by Tronto and Fischer.

4.1.5. *Towards a design for life: László Moholy-Nagy*

4.1.5.1. *The Bauhaus experience*

The task of bringing together the arts, the potentialities of industrial machinery and life down to its biological foundations, cannot be limited, of course, to just theoretical remarks – at risk of losing sight of the ultimate goal: the creation of a new sensitivity and new values. This task must therefore be materialized through initiatives in the fields of decorative arts, industrial design, and construction. Morris himself was the founder and manager of a decorative arts company intended to create the conditions for the reunification of arts and craft in the industrial age; Mumford calls for a commitment from designers in the direction of this convergence.

However, it so happened that after the First World War, in a Germany struck by revolutionary fever, a new kind of art school was opened with the precise ambition of bringing together the arts, the trades of art (crafts) and industry: the Bauhaus. Its founder, Gropius, was an architect who favored collaboration with industry – an orientation he particularly made his own following an apprenticeship in the workshop of Peter Behrens, pioneer of industrial design. Behrens was co-founder in 1907 with Hermann Muthesius of the *Deutscher Werkbund* (German Association of Craft), an association which brought together artists and representatives of industry and whose goal was to promote, in a functionalist vein, an aesthetic of industrial objects and constructions. Initially won over by William Morris's ideas concerning the rapprochement between arts and crafts, Behrens subsequently turned away and resolutely engaged in industrial construction. His interest was in materials such as glass and steel, which he believed capable of embodying an aesthetic dimension specific to industrial constructions.

It was between 1908 and 1910 that two of the future directors of the Bauhaus, Gropius and the architect Mies van der Rohe, as well as a young Swiss architect destined to become a major representative of modern architecture, Le Corbusier, met in Behrens's workshop. The fundamental idea of Gropius was that the field of competence of architecture goes far beyond the construction of buildings: being fundamentally an activity of creating form, it is competent over everything in the human environment that has a form (buildings, of course, but also everyday objects and even the city as a whole). As French professor of literature Lionel Richard writes about Gropius, for him, the invention of forms is essential. Gropius always strived, through the consideration of all modes of visual expression, towards a science of form. But only an artist is able to invent forms. On the eve of the First World War, he was aware ... of the difficulties in finding morally satisfactory situations in a society where the power of technology and industry became predominant (Richard 2012, p. 39).

This is the background against which the Bauhaus project took shape, nourished by the experiences of Morris, Behrens and the *Deutscher Werkbund*. It should be noted that at the time, Gropius did not fully share Muthesius's ideas concerning the relationship between art and industry. Muthesius undeniably helped disseminate Morris's views in Germany; nevertheless, unlike him, he did not much believe in the possibility of a renewal of craft in the industrial age. His diagnosis was rather that industrial machinism has definitely triumphed, and that the artist/artisan must step aside in favor of the engineer. It is therefore more relevant to look for the conditions of an aesthetic of the object in industrial manufacture itself, by orienting the latter towards objects with purified forms, adapted to their function, rather than to preserve the craft ideal as the guiding principle of industry.

Gropius, on the other hand, was closer to the conceptions of the architect Van de Velde, also a follower of Morris and an important player in the *Deutscher Werkbund*. Van de Velde recommends, unsurprisingly, abandonment of the ornamental deviation of Art Nouveau, and defends a functionalist approach reconciling the beautiful and the useful. That said, unlike Muthesius, he does not militate in the direction of submission of the creators of forms (the artists) to the demands of industrial production. Artists must maintain their independence from industry. Their mission is not to provide it with standardized shapes designed from the outset for mass production.

Like Muthesius, Gropius believes that traditional craft is no longer able to meet the challenge of an industrial aesthetic. That said, he does not envision an outright dissolution of the aesthetic activity of creating new forms into the demands and processes of industry.

The Bauhaus was born from this conviction that industrial aesthetics must be placed in the hands of artists of a new genre. These must be masters of form concerned with the conditions of production of the latter (hence the importance of workshops devoted to working with materials). In other words, the Bauhaus was initially driven by Gropius's conviction that the masters of form, the new artists/artisans, must take precedence over the representatives of industry and the machine. The latter must be accepted as the most modern means of conceiving forms (Richard 2012, p. 132). It is therefore a question of finding an arrangement with it, but in no case should the specific requirements of industrial machinery and mass production govern the activity of creating forms. To parody a famous formula, *artists create, industry applies and conforms*.

The year 1922, however, marked a turning point in the history of the Bauhaus, with the desire (also that of Gropius himself) to give a more scientific orientation to teaching, and to initiate a much closer connection with the technology of machines and industry. Replacing the expressionist painter Johannes Itten, a new teacher arrived at the Bauhaus in 1923: a Hungarian constructivist artist, specifically responsible for initiating this rapprochement. This was László Moholy-Nagy, passionate about the use of science and technology in the arts (he notably created a teaching geared towards exploiting the artistic potential of photography) and enthusiastic about industrial machines. Responsible for the metal workshop at the Bauhaus, he moved it from artisanal execution of unique pieces to the manufacture of prototypes for industry (Richard 2012, p. 128).

In short, it was Muthesius's point of view that ended up triumphant at the Bauhaus; except that for Gropius, as Richard points out, in 1923 as in 1919, artisanal work remains the source of the best learning … because it requires that a single hand master the entire manufacturing process, while industrial production is based on the division of labor (Richard 2012, p. 132). The point is decisive: Gropius's vision always seems to be tied to the certainty that a rapprochement of the arts and industry can only be achieved under the auspices of a somewhat organicist conception of production. Even if this is now implemented by industrial machines within the framework of a scientific organization of work, it must retain as an ideal term of reference the representation of a tooled body ensuring the coordination of all manufacturing tasks.

The Bauhaus was undermined by a very strong internal tension, which Gropius himself did not really overcome, between, on the one hand, a training resolutely turned towards industry and the possibilities it offers for creating new forms, and, on the other hand, the perpetuation of the artisanal ideal.

4.1.5.2. *Design against the anti-biological tendency of industrial capitalism*

In this context, the arrival of Moholy-Nagy at the Bauhaus is of particular importance. As he explains in his testament book published posthumously in 1947[5] (Moholy-Nagy 1947), his teaching at the Bauhaus and then, after his emigration to the United States in 1937, at the New Bauhaus[6], was constantly driven by the intention of overcoming the mismatch between, on the one hand, the possibilities opened up by the industrial revolution and, on the other hand, the old ideologies accused of preventing the translation of these possibilities into an emotional language and cultural reality. Above all, Moholy-Nagy intends to free industrial design from the ideal craft model – a task which, according to him, means above all rethinking the conditions for a unified life in industrial society and in the process of production.

In its time, craft implied not only a way of producing, but also and above all a way of feeling and perceiving. However, like Morris before him, Moholy-Nagy deplores the deeply anti-biological dynamic of the industrial revolution. This led to a dissociation between technology on one side and biological impulses on the other. The potentialities of the machine have been diverted and oriented towards profit, which has resulted in the appearance of an increasingly impassable gulf between the biological and technological conditions of human existence.

Considering that the human being in industrial societies has become a human machine, Moholy-Nagy ultimately only reiterates a diagnosis shared by others, in particular by Morris and Mumford, concerning the divorce of technology and life. By the slogan *design for life*, Moholy-Nagy thus intends to promote a creative activity of objects that can serve as a fulcrum for the development, and not the exhaustion, of human life, understood above all as life defined by biological rhythms, the basis for the processes of psychic and collective individuation.

To ensure that the requirements of the organism can be satisfied within the framework of a harmonious social existence: this is the purpose of design as a creative activity (Anker 2010). The designer must maintain a point of view on objects that is not that of the search for optimal performance, but that of the life of living human beings and their own requirements. Nothing is more important, says Moholy-Nagy, than the search for *better ways*, from a biological point of view and not from the point of view of economic calculations.

5 Moholy-Nagy died in 1946.

6 A new design school, founded in Chicago in 1937, closed after six months for financial reasons. Moholy-Nagy reopened a school in 1939, the School of Design, subsequently the Chicago Institute of Design.

Humans have the biological potential to understand the world using all of their faculties, and to express themselves using the full spectrum of media at their disposal (words, colors, etc.). Unfortunately, they have voluntarily allowed themselves to be cut off from these faculties. Their biological interest in all things has been stifled by a seemingly easy life (Kostelanetz 1971, p. 20). Design must make it possible to overcome this state of affairs, and to restore to individuals the free use of their aesthetic and technical means.

More precisely, Moholy-Nagy emphasizes that the function of design education is to create favorable conditions for the expression of the aesthetic potential of the products of industry. Moholy-Nagy does not mean by this that these products must be evaluated according to their own criteria of beauty (the concept of technological beauty is abandoned); "aesthetic" should be understood here in the sense of *aisthesis*, of sensory experience. The objects issuing from industrial production can be the ferment of a new structuring of sensory and perceptual-motor experience; on condition, however, that they are freed from the shackles imposed by the search for profit, and on condition also that our sensitivity and our perception are educated.

In short, Moholy-Nagy already introduces, if not the term, at least the idea of bioaesthetics defended by Montani, although he offers a very different reading of it. As we have seen, the concept of bioaesthetics beckons towards the idea that *aisthesis* is organologically constituted; that is to say, that it is always supplemented by technological prostheses, but in such a way that today, in industrial societies, the orientation taken by this constitution makes individuals incapable of entering into the processes of psychic and collective emancipatory individuation.

Against the backdrop of the same organological conception of sensitivity and perception, Moholy-Nagy, for his part, defends a perspective that is, so to speak, opposed to that of Montani: our daily experience is totally shaped by surrounding objects which only transmit the imperative of profit. Their conception and our perception of them are at the service of market capitalism. However, it is possible to recreate territories freed from this consumerist and mercantile grasp – the Bauhaus was such a territory. But this implies above all returning the needs of the organism to the foreground.

The organic body is not entirely soluble in the devices of technological intervention which take it as an object – it is not only a prosthetic body, a body entirely shaped from the outside by its prostheses, but also a body-instrument. It bears values which must provide criteria for the evaluation of industrial machines and products. Moholy-Nagy formulates this with a striking metaphor: we need a biological *Bill of Rights*, a bill of rights for the organism.

4.1.5.3. *Vision in Motion*

According to Moholy-Nagy, the education of sensitivity and perception must also be done in the sense of an ability to grasp the relationships between things. "Vision in Motion" – the title of Moholy-Nagy's work – means seeing things not in a static and substantialist way, but in the processes and relations that constitute them and in which they are engaged. "Seeing everything in relationships": this is the goal of training for designers. Objects are "diagrams of forces": we need a new vision of materials, the energies, the tensions that constitute them.

Moholy-Nagy thus makes an essential contribution to reflections on the relationship between technology and life.

On the one hand, he remarkably echoes Bergson and Dewey in pointing out that the most urgent task is to make for ourselves a new perception. According to Moholy-Nagy, the aim of design is to actively contribute to "visually constituting a better world".

On the other hand, this new perception must enable us to see things according to processes and relations. In other words, industrial machinery has very deep metaphysical foundations, it is based on a substantialist metaphysics leading us to see things in a truncated way, seeing them locked into their own limits, cut off from their process of production and the relationships through which they affect the world. The critique of industrial machinery and the promotion of a new *aisthesis* therefore involve criticism of this metaphysics.

Finally, the guiding idea behind Moholy-Nagy's conception of design is that biological, sensorimotor life is a source of properly vital values and that these values must serve as terms of reference for judging industrial production.

From Morris to Moholy-Nagy, a whole set of reflections on the relationships between life, technology and aesthetics has developed, with the ultimate goal of a radical change in the practices of technological conception – design and engineering. Care as such is apparently not central to this reflection on design and life; nevertheless, all the elements are gathered here to identify the guiding principles of technological conception, and training for engineers, oriented by the global aim of care.

4.1.6. *Opening*

The human of pre-industrial civilization, writes French philosopher Georges Friedmann, did not seem to observe the world with the same natural organs as we do

(Friedmann 1966, p. 26). Man was not the same, he did not feel, he did not act, he did not think in the same way according to the epochs of his history, according to the environment in which he lived: according to the technologies at his disposal (Friedmann 1966, p. 15). This observation makes it necessary to seek as much as possible the precise knowledge of the mechanisms according to which these thousands of solicitations, excitations, stimulants which were unknown not long ago (Friedmann 1966, p. 12) operate today on the organism and reconfigure its sensori-motor assemblies.

To be content to maintain, in a very general way, that technology contributes to configuring the forms and the contents of experience, consequently falls a little short in the eyes of Friedmann. It is essential to go further and to analyze as finely as possible the real mechanisms of this reconfiguration, and especially to identify very exactly the orders of magnitude at which they operate.

It should be emphasized that the most valuable contribution of the fourth stream of the biological philosophy of technology lies undoubtedly in the importance that every author associated with it attaches to the very concept of the technical object. Among those authors more classically considered to be the representatives of a biological philosophy of technology (examined in the previous chapter), the concept of the technical object does not appear to be central. It is, so to speak, very secondary or even completely inessential in Plessner. This is undoubtedly less the case in Gehlen or in Ortega y Gasset, but their analyses do not really go beyond the opposition between the tool and the industrial machine. The technical object – tool or machine – is evoked above all to symbolize two great figures of the technological human, the artisan and the worker in large-scale industry. It is the transformation of the modes of confrontation between humans and matter, that of the modes of organization of humans among themselves and, finally, that of the fundamental structures of human experience, which primarily interest these authors. The technical object as such is not analyzed closely.

This remark is undeniably less relevant to Leroi-Gourhan, so great is the importance that he gives to the materiality and operation of technical objects. However, his analyses focus on pre-industrial technologies. He has beautifully showed how the way in which a human group fits into space and time can be read very precisely in its objects. However, he stopped at the threshold of the industrial age and, far from continuing to seeking in these new technical objects a key to understanding new modes of confrontation between humans and matter, he rather associated them with the loss of this confrontation.

Morris, Souriau, Dewey, Mumford and Moholy-Nagy, on the other hand, have closely associated, on the one hand, the design of technical objects, and on the other hand the type of experience – more or less fragmentary, more or less alienated – of which human beings surrounded by said objects are capable, at a given time and in a given society.

All of these authors, reflecting on the moral and political significance of design in the industrial capitalist societies of the past century (as well as the previous one for Morris), may not have approached the problem in terms of care. What emerges from the analyses of this chapter is that they nevertheless did find, in their reflections on contemporary technological fact, the four conditions of care in technology identified in the previous chapter.

The question they leave for us is therefore the following: can we, on the basis of their reflections on design, offer current engineers some advice on orientation, so that they may have a more concrete understanding of this? What can it mean in practice to make technological design depend on the four conditions of care in technology?

It turns out that the ancient Chinese and Greeks put a lot of thought into the question of how to go about it, very concretely, so that their technical activities were carried out with the greatest care. They even identified very precise practical principles for this. Over 2,000 years later, can these principles continue to inspire our engineers?

4.2. Towards responsible and caring innovation

4.2.1. *Technical activities and care: practical lessons from ancient China and Greece*

4.2.1.1. *A story of a Chinese cook*

Zhuangzi, a Chinese philosopher of the 4th century BCE and a major figure of Taoism, gives the following account: the cook was carving up an ox for King Hui of Liang. Wherever his hand smacked it, wherever his shoulder leaned into it, wherever his foot braced it, wherever his knee pressed it, the thwacking tones of flesh falling from bone would echo, the knife would whiz through with its resonant thwing, each stroke ringing out the perfect note, attuned to the "Dance of the Mulberry Grove" or the "Jingshou Chorus" of the ancient sage-kings.

The king said, "Ah! It is wonderful that skill can reach such heights!"

The cook put down his knife and said, "What I love is the Course, something that advances beyond mere skill. When I first started cutting up oxen, all I looked at for three years was oxen, and yet still I was unable to see all there was to see in an ox. But now I encounter it with the spirit rather than scrutinizing it with the eyes. My understanding consciousness, beholden to its specific purposes, comes to a halt, and thus the promptings of the spirit begin to flow. I depend on Heaven's unwrought perforations and strike the larger gaps, following along with the broader hollows. I go by how they already are, playing them as they lay. So my knife has never had to cut through the knotted nodes where the warp hits the weave, much less the gnarled joints of bones. A good cook changes his blade once a year: he slices. An ordinary cook changes his blade once a month: he hacks. I have been using this same blade for nineteen years, cutting up thousands of oxen, and yet it is still as sharp as the day it came off the whetstone. For the joints have spaces within them, and the very edge of the blade has no thickness at all. When what has no thickness enters into an empty space, it is vast and open, with more than enough room for the play of the blade. That is why my knife is still as sharp as if it had just come off the whetstone, even after nineteen years.

Nonetheless, whenever I come to a clustered tangle, realizing that it is difficult to do anything about it, I instead restrain myself as if terrified, until my seeing comes to a complete halt. My activity slows and the blade moves ever so slightly. Then all at once, I find the ox already dismembered at my feet like clumps of soil scattered on the ground. I retract the blade and stand there gazing at my work arrayed all around me, dawdling over it with satisfaction. Then I wipe off the blade and put it away" (Zuanghzi 2009, pp. 22–23).

This account seems at first sight to deal with the problem of acquisition by the artisan (a butcher in this case) of manual dexterity and skillful know-how. There does not seem to be any question here of an ethics centered on care: King Hui does not praise Ding for his high moral value and for his caring attitude, but for the perfection of his technique.

In addition, improvement of the cook's technique does not appear to be made necessary by the desire to avoid failure, as in the case of the man of Song: at the start of his career, the cook did not yet have this certainty of the gesture, nevertheless nothing says that he failed purely and simply in his activity. He undoubtedly succeeded in cutting up his beef carcasses, and he succeeded perhaps even very satisfactorily in spite of a coarser technique.

Consequently, Ding's progression has been oriented neither by the aim of greater care as a careful and cautious being, nor by that of care as an intrinsic attachment to a supposed moral virtue of cooks. Ding's explanations take place on a level which is neither technological nor moral; nor are they situated on an intermediate level which would seek to integrate care as a technological imperative and care as a moral requirement.

Ding's explanations are, rather, on an aesthetic level. The cook performs his cutting gestures in a way which, in fact, delights the eyes of the prince: despite the violence of the gesture which consists of hitting, pushing, hammering, cracking bones, everything is flexible, fluid and is accomplished in a rhythm of dance and song. The aesthetic dimension is not added here from the outside, and afterwards, to a technological gesture which would otherwise have reached its full effectiveness; nor is it added, from the outside and after the fact as well, to soften the rigor of a process of moral improvement of the subject engaged upon the Dao.

Cook Ding does not seek to make his technological gesture more careful in order to improve its efficiency, a condition for success; nor does he seek to develop an ethic specific to his profession. His path consists above all in surpassing any separation between himself and things. This is the Dao: to be at one with the natural world and the whole process that governs it.

At the end of his apprenticeship, the cook Ding no longer stands in front of the carcass of the ox; he no longer sees it with his eyes in the sense of this "seeing" where the observer and the observed are different, at a distance from each other. He sees with the mind, that is to say, with a "seeing" which does not come from the subject but which seems to proceed from the things themselves. Ding has completely renounced any kind of mastery; his gesture is almost somnambulistic, guided by the intrinsic organization of what he applies himself to. The technical perfection admired by King Hui is achieved when the technical gesture no longer exists as a separate activity: it merges with the organizing principle of the anatomy of the ox.

His gesture is however *almost* somnambulist, we must be precise here: in any case, Ding's technical skill does not derive from any automatism of the gesture. Each time the cook comes to a complex joint, he must prepare himself carefully – a bit like the painter Henri Matisse filmed in slow motion of whom Merleau-Ponty speaks[7]: before setting brush to canvas, in a gesture that only just touches as if by a

7 "A camera has registered the work of Matisse in slow motion … The same brush which, to the eye, did not jump from one movement to another, could be seen meditating, in a suspended and solemn time, in an imminence, like the beginning of the world, beginning ten possible movements, performing in front of the canvas a sort of propitiatory dance, coming to close several times as almost to touch it, and finally coming down like lightning in the only stroke necessary" (Merleau-Ponty 1973, p. 44).

miracle, the painter hesitates, gropes, prepares himself. There is no automatism in this, but a co-belonging never definitively acquired, which must therefore be permanently reassured, of the technical gesture and the intrinsic organization of that to which it relates[8].

The way in which Ding operates is therefore not aesthetic in the sense of a purely external beauty of the gesture: it is aesthetic in the already Kantian sense of an almost miraculous alliance – and yet one achieved through experience, work and effort – between a nature (the anatomical organization of the ox) and an end (butchering the carcass).

The technical efficiency of the gesture on the one hand, the moral improvement of the Chinese cook engaged with the Dao on the other hand, are not two dimensions that are, through learning and experience, to be better articulated – quite simply because these two dimensions are not originally given as isolated. They both derive from a more fundamental attitude, therefore of an aesthetic nature, which consists in ensuring that technical activity does not arise from the subject separated from things, but from things themselves – as if the subject itself, endowed with its technical skill, were immanent in them.

Before taking on a technical or moral significance, Ding's learning consequently takes on an anthropological significance: it consists, for the acting subject, in apprehending itself as immanent in the process which governs all things, as forming with them a single system of reality. We should no longer even speak of action, nor even of an acting subject: to act, for the subject, is indeed to intervene from the outside in the course of things in order to impose upon it a plan, a direction which is not its own. The result can only be a disturbance, a derangement. Having reached the peak of his art, the cook Ding no longer acts in the sense of imposing a butchering on the bovine material: he allows the desired effect to occur (Jullien 2004).

Having arrived at this point, Ding feels the most complete happiness, he is "satisfied": it is not a question here of a vague feeling arising to give color to technical success, or the certainty of doing well morally, but of the fundamental experience of becoming one with the process of reality. From this experience emerges the ultimate meaning of care, a Kantian-style formulation of which might be as follows: act technically upon nature in such a way that nature itself appears to act through you.

8 In the case of the painter, moreover, this organization does not pre-exist the gesture, although, as Etienne Souriau said, it imposes itself on the painter with the greatest exigency, as a "work to be done".

Ding cleans and puts away his knife, which is in impeccable condition, never worn, as if it had just come out of the whetstone even though it has been used for nearly 20 years. The technical object is not reduced to its instrumental function; it itself becomes an object of care, the depositary of this precious alliance of a nature and a finalized gesture.

Zhuangzi's account therefore allows us to rediscover, in the very practice of the cook Ding, all the elements previously identified as conditions of care in technology: experiencing a belonging of the human being to nature that has nothing to do with the theme of *deficient being*; the instrumental conception of the body – the body becomes, so to speak, the instrument of nature, everything happens as if nature were operating on itself through the intermediary of the technician; the refusal of the fixed point – on the contrary, it is a question, through technology, of becoming one with the process which governs everything; and finally, the conversion of perception, introducing a more unified experience of the world and of oneself.

The cook Ding has developed his technique at the same time as he has gradually learned to live in harmony with nature: what can be the relevance of this message today? How can we rediscover this conception of technology as a transformative activity of things which seems to proceed from the organization of things themselves, and not from an intelligence cut off from the world – whereas industrialization has very often meant, precisely, almost complete indifference with regard to the organizing principles of the beings of the world, in particular of living beings? Should we consider that there is an incompatibility, in principle so to speak, between on the one hand, the very nature of the work of the engineer today, and on the other hand, the commanding maxim of acting technologically upon nature in such a way that the technical action appears to be the act of nature itself? The account of the cook Ding remains dependent on an enhancement of the cautious artisanal gesture: can we conceive of a care passing through technology detached from this reference to craft?

4.2.1.2. *Some Greek precepts for caring technical action*

The Greeks also left us with precious considerations of the way in which they thought they could, very practically, make care prevail in technical action. Medicine was for them a privileged area for this. Not that it is ultimately a question of returning to the starting point and the bilayer model of care: but precisely, the Greeks did not at all conceive the relationship between medicine and care in terms of this model and of its pair of presuppositions, concerning respectively the human being (the body/mind compound) and the intellectualist conception of technology.

The Greeks indeed defined medicine as a very special *techně*. As we have seen, *techně* designates above all a form of knowledge. The technician's activity does not proceed from a groping experience deprived of any theoretical basis: it is based on knowledge of what it is applied to. "Men of experience (empeiria) know that the thing is so, but do not know why, while the others [i.e. artists, men who possess an art] know the 'why' and the cause" (Aristotle 1924, A1).

However, *techně* is not, however, like science, a knowledge of the general only. Still according to Aristotle, "actions and productions are all concerned with the individual" (Aristotle 1924, A1).

Techně therefore depends on a knowledge of essences and causes, but also on the ability to apply this knowledge to contexts of intervention which are always unique – which requires experience, *empeiria*. Medicine is from this point of view a *techně* like any other; it must also combine knowledge and experience: "the physician does not cure man, except in an incidental way, but Callias or Socrates or some other called by some such individual name, who happens to be a man. If, then, a man has the theory without the experience, and recognizes the universal but does not know the individual included in this, he will often fail to cure; for it is the individual that is to be cured" (Aristotle 1924, A1).

Medicine is, however, a very singular *techně* which requires of the doctor not only to take account of the essential nature of the thing to which he applies his activity (the bodies that must be treated, *therapeia*), but also to take care (*epimeleia*) of something that is beyond that to which his activity relates. This absolutely strange situation constitutes the specificity of the medical *techně*: it is a *techně* in whose definition must be understood a non-technical relation to the thing to which this *techně* relates.

Besides medicine, animal husbandry and the art of politics, there is another activity that the Greeks associated with care: agriculture. The definition of agriculture as a care is found in particular in Xenophon (Xenophon 2006). It is worth noting that Xenophon seems to refuse to consider agriculture as a true *techně*, on the pretext that it does not precisely require knowledge of the thing to which it applies. Agriculture is an easy activity; you just have to watch what your neighbor is doing, apply the same methods in a perfectly empirical and mimetic way, and wait for the crop to grow. Xenophon's argument is clear: if agriculture above all relates to an attitude of *epimeleia,* it is because in this area humans are not the artisans of success: nature does everything. It is the same in medicine and in animal husbandry: there too, the one who possesses the *techně* only succeeds if he manages to reconcile himself to a process which does not depend on him, which operates without him and over which he has no real control.

In the field of manufacturing in general, in construction, any technician must of course also reckon with the context, with the contingency of the situation, but in the final analysis, it is the technician himself who operates. He operates by arranging given elements; he does not produce anything *ex nihilo*, nevertheless from the materials which he finds he creates something new: he is *demiourgos*. On the other hand, in agriculture, animal husbandry or medicine, it is not the technician who operates, but the *physis*. To be successful in healing, the physician must begin by taking care of the power which is really active, creating the conditions for that active power to actually produce the desired result: healing. As Ambroise Paré said in the 16th century, "I dressed him, God healed him[9]". And Panckoucke again, in the 19th century, in his dictionary of medical sciences: "Whatever the character of a disease, the doctor's functions are always reduced to directing or exciting the efforts of nature and letting it act" (Panckoucke 1812).

The doctor does not do nothing, of course, and Panckoucke insists heavily on it, comparing him to the pilot of the ship. The doctor is very active, but his *techne* does not make him, unlike the artisan, the *demiourgos*, that is, the creator of what he has in view (health). The good doctor is that one whose *techne* consists mainly of extending care to that within the sick person which can really bring about healing. *Vis medicatrix naturae*.

That said, Plato did not place medicine on the same level as agriculture and especially animal husbandry. The *epimeleia* which is in question, in either case, may not have quite the same meaning. An absolutely essential point characterizes medicine and the type of care that constitutes it, unlike animal husbandry.

In *The Laws* (Plato 1997, p. 1406), Plato distinguishes two ways of administering medical care: there are on the one hand those that the slave lavishes on the slave, and on the other hand those that the free man lavishes on the free man. In the well-governed city, the slave will not heal the free man, and conversely the free man will not heal the slave. The slave heals the slave in the sense that he prescribes treatments, but he does not accompany his prescription with any explanation. In other words, he cures but he does not bother to take care; he is not bound by the need to maintain this non-technical relationship with the patient which precisely defines medical care. With the free man, on the other hand, it is not possible to act in this way: the treatment must always be explained, the cure is inseparable from this eminently moral faculty which, in medicine, consists in caring for the concrete and singular person beyond the organic thing.

9 Cited in (Donghi and Preta 1995, p. 48).

The essential point of this analysis is as follows: if the dimension of taking care is constitutive of medical practice, it is not because without it the treatment itself, the cure, would be ineffective. On this account, we do not see how Plato could justify the existence of a medicine for slaves which is less effective (surely, the free man had an interest in his slaves being in good health!). If the dimension of taking care is constitutive of medical practice, it is because it is above all a moral relationship. We understand that the slave is not held, with regard to the other slave, by the same obligation: moral improvement does not concern any of them; purely technical treatment is sufficient.

It follows from this conception that medicine must in principle resist what may happen to artisanal *technai*. Starting from the superiority of those who have a *technè*, compared to those who are dedicated to *empeiria* alone, Aristotle explains that "the masterworkers in each craft are more honourable and know in a truer sense and are wiser than the manual workers, because they know the causes of the things that are done (we think the manual workers are like certain lifeless things which act indeed, but act without knowing what they do, as fire burns)" (Aristotle 1924, A1). Any constructive, fabricating *technè* can be exercised by separating knowledge from simple mechanical execution, which socially translates into a sort of division of labor, between the *architecton* (the one who has the knowledge of the thing to be produced) and the *cheirotechnon* (the manpower, the simple performer who acts mechanically).

It goes without saying for Aristotle that medicine cannot be divided according to the duality of *architecton* and *cheirotechnon*. If it were to experience this division, indeed, medicine would be limited to treatment and would lose all sense of taking care, which however is essential to it, to the extent that the latter can only be implemented by an activity which is, precisely, something other than a mechanical application of knowledge and rules.

Let us summarize. Letting nature's processes take place rather than systematically seeking to control them; seeking knowledge of causes; refusing the vision of the caregiver as a simple button-pusher, a performer who acts mechanically like a fire burns; ensuring that the techniques used and their operations are always subject to rational explanation to patients: these are therefore the four practical principles that guide the medical *technè*, in that it must inseparably treat and take care.

Can these four practical principles serve much more generally to guide engineers today? Let us take a closer look at each of these four practical principles and their implications.

4.2.2. *The square of care in technological design*

4.2.2.1. *Letting nature do its thing. The example of alternative agriculture*

The first principle – letting nature's processes take place rather than systematically seeking to control them – can lead to two very different types of attitudes.

The first type is illustrated by the farmer in Xenophon's sense. It is only a question here of imitating one who succeeds, and not strictly speaking of deploying a *technè*, that is, a form of knowledge. The farmer only intervenes locally in the productivity of nature. To be successful it is not necessary to know the organization and the functioning of the reality to which he applies himself. Strictly speaking, he does almost nothing, except to seize the opportune moment, the *kairos*, to perform the few actions required. There is no regularity in the relationship between human and nature. These relationships are subject to an irreducible variability, to the contingency of places and times.

The second type of possible attitude is illustrated by medicine. Nature is the healing power, yet it requires the active intervention of the physician. The productivity of the *physis* does not depend on human action, and it does not proceed from the latter's own constructive effort; however, it requires its assistance in order to be able to operate in the direction of healing. There is a conjunction of the productivity of nature and human labor. The first operates on its own and provides the effect, but under the constant control of the second.

This control consists in establishing a regularity in the relations of human action and natural productivity. For Xenophon, the newcomer farmer cannot count on any general inference; his action consists only in grasping what makes the singularity of place and time. The doctor must also focus his efforts on the singular; however, he can rely on an increase in generality, on a knowledge of causes and effects. He may be led to treat Callias rather than Socrates, but in either case he may rely on knowledge of certain regularities concerning both the manifestations of the pathological and the effects of human action.

The farmer's *laissez-faire* according to Xenophon ultimately requires a relatively weak sense of *epimeleia*. In fact, nature does not need the concern of humans to produce results. Strictly speaking, it has no need at all; it is pure self-sufficient productivity in its deployment as well as in its effects. What Xenophon calls *epimeleia* is here reduced to the strict minimum; care seems to be limited to seizing the *kairos*.

At the other end of the spectrum, in a quest for mastery, human technical action is no longer enveloped by a productivity that does not depend on it. Humans exert their action upon passive, homogeneous and constant matter; there is no longer any symbiosis between them and nature: they alone make the law. Matter thus understood has no needs; it does not live, explains Simondon.

Between these two extremes, however, there is a middle way defined as a non-simplifying human act, attentive to the specific needs of the reality to which it applies.

Laissez-faire is not at all equivalent to advocating non-action, in the sense that it would be appropriate as much as possible to do nothing. The principle of *laissez-faire* does not imply any false glorification of nature. Nature has no incentive to operate in place of the engineer, but with and within engineering practices, which is very different. The caring act is not that which minimizes its perimeter, but on the contrary that which incorporates as much as possible into its axioms the constructive and organizing processes of reality. It is a question of extending technical action to all the natural processes involved in the reality to which this action relates. To technical gesture must thus correspond, explains Simondon, "an element of reality that, in order to satisfy this gesture and correspond to it, becomes the basis of this gesture, to which this gesture applies itself and in which it accomplishes itself" (Simondon 2017, p. 202). This is, moreover, very well illustrated by alternative agriculture, indicative of the prerequisites for a caring and careful technical action.

These agricultures defend a low tech approach to cultural practices, contrary to agro-industry which seeks on the contrary to promote innovations in the fields of machinery, electronics, chemistry, and biotechnologies. They therefore seem to opt for a frankly deflationary posture in technological matters. However, alternative farmers are in reality driven by a demanding technological thinking, at least if one defines technology like Simondon as a general theory of operations in the Being.

Simondon indeed opposes two ways of conceiving technology. The first, abstract and weak, consists in seeing it only from the angle of its various functionalities, in a purely utilitarian perspective; the second, concrete and ambitious, consists in seeing the device or the technological process above all from the angle of the physico-chemical processes which it embodies or which underlie its operation.

Because they are based upon the most in-depth and most contextualized analysis possible of the natural operations characterizing each particular agricultural environment, with a view to understanding how technological operations might be harmoniously articulated with them, alternative agriculture appears to be resolutely concrete. It is rather intensive agriculture which, by saturating the soils with mechanization and chemical inputs, that is to say by doing nothing but mechanically applying a general law and acting upon the real from a point of view totally indifferent to its own organization, ultimately appears very impoverished as regards the technological thought guiding it.

Alternative agriculture ultimately turns out to be much more elaborate from the point of view of understanding natural and artificial operations, as well as their integration into an overall operational schematism. Succeeding in this field consists in relying on in-depth knowledge of soils and the very complex microbiological life they harbor, as well as on that of the mutual services provided by plant varieties, giving up the attempt to identify a general law that is valid everywhere. We are here very far from the purely imitative virtues of the farmer promoted by Xenophon.

Thus, if these practices are indeed deflationary in relation to a certain conception of technology, they do not yet invite us to reject all technical thought (we would even more willingly say: technological), in fact it is quite the contrary. Alternative farmers do not recommend reducing technical action to the strict minimum so as not to disrupt natural processes; on the contrary, they considerably extend their field of action, incorporating within it all the living processes which take place within the soil. Caring technical action consists, not in adding new elements to reality (for example, an overabundance of nitrogen through massive input of fertilizers), but in thinking of natural processes as technological operations.

4.2.2.2. *Technology is in nature*

Care results in short from a double movement, respectively of incorporation and insertion: by incorporating natural processes into the operational system of their action, humans testify to their deep integration into nature. To the human cut off from nature corresponds an activity that applies to reality as to something completely foreign to it, which is imposed upon it with violence. On the contrary, the inserted human corresponds to an activity which does not need to add anything to the real, quite simply because the real is from the outset conceived as a set of operations which are coordinated, or at least able to be coordinated by the technician. Technology does not have to be added from the outside quite simply because it is already there, in the real, as in the state of envelopment or potential: it is rather a question of freeing it, of moving from the potential to the actual. The microbial life of soils is a set of very complex natural processes. To make them technological, it is not necessary to genetically modify the micro-organisms: it suffices to incorporate these processes into an operational thought, of a complexity equivalent to that of the processes which have been recruited. Taking care is first of all knowing how to welcome the complexity of the real, and not to seek to reduce it in order to force the real to correspond to predetermined categories.

Laissez-faire therefore does not mean doing nothing. It is even quite the opposite: letting things happen is much more complex and demanding, in terms of operational thinking, than imposing the same general rule everywhere, according to a simplifying approach consisting in reducing the diversity of reality to a common denominator.

4.2.2.3. *The human technician, organizer of nature*

The first practical principle, which therefore consists in designing schemes for technical action by incorporation and operational coordination of existing natural processes, is directly linked to the second, which requires the search for knowledge. Seeking knowledge as systematically and as widely as possible indeed appears to be an absolute requirement, in counterpoint to the ignorance that characterizes a growing number of technological devices and activities – geo-engineering is an example of this, but many other fields are concerned. Dagognet himself points out with disapproval the extension of more-or-less blind tinkering in the domain of living things: one cannot carry out overly adventurous manipulations on organisms (Dagognet 2003, p. 6) and one must avoid assimilating the seriousness of an improving and synthetic work on the organic to fragmented, groping, domineering, even frivolous activities on living beings (Dagognet 1988, p. 12).

It may be pointed out that these two guiding principles do not in themselves lead to meeting the requirements of care. The concretizing activity in Simondon's sense seems to proceed by way of selection and closure: certain physical processes, and not others, attract attention in that they are integrated into the operative schematism. However, on this account, an object or a technological practice tending towards concretization may very well turn out to lack serious care with regard to human health and the environment. The case of the diesel engine, mentioned above, is an emblematic illustration of this.

This objection is strong and fully justified, and Simondon himself has taken it into consideration. He thus evokes a surgical act, which nothing prevents us thinking that it may seek the greatest possible realization, for example by integrating elements or biological processes into the whole of its operating chain. However, this technical act does not appear to be any less antinomic of care.

The situation is as follows (Simondon 2014, p. 124): a young farmer suffers from a serious hormonal imbalance which has made him obese. To heal and be able to continue to practice his profession, the young man must undergo expensive surgery. To finance it, he offers to sell one of his eyes. According to Simondon, the very fact of considering such a surgical act shows that a purely operative gesture … is infra-technical: it possesses no normativity adequate to the whole of reality upon which it operates.

For the operative gesture, including by extension the operation of any machine, to be part of an authentic technological approach, it must necessarily (if we understand correctly) be evaluated reflexively with regard to the complete reality to which it relates. Technical action is to be evaluated from an axiological viewpoint: concretization indicates an ideal to be realized, which is not very often realized in the actual history of technology. Simondon thus multiplies the examples of technological choices which, for various reasons, have historically deviated from a real process of concretization (he is in particular never short of criticisms of the motor vehicle). However, this axiological judgement is not self-referential, it implies taking account of the concrete whole to which the technical act applies. It may be an organic whole, as in the case of a surgical act, but also a totality mixing the organic and the inorganic – like a territory which inherently possesses an organization whose lines of force can be expressed by technical action, thus fulfilling the true meaning of technology as relating human needs to natural processes.

Simondon gives the pre-industrial example of the windmill which connects the energy that moves the inaccessible clouds, to the millstones under which the wheat harvested by the hand of man is crushed. Technological concatenation guarantees the homogeneity of the real. Winds and lightning are of the same species as the realities which populate the immediate surroundings of man. The technical object creates a communication between previously separate, qualitatively distinct orders of reality (Simondon 2014, p. 105). This communication is not submission, the expression of a desire for mastery and domination of nature. Technology is undoubtedly devoted to allowing humans to satisfy their needs. There is initially no relation between the winds and the need for flour; the mill makes these immeasurable orders of reality compatible. However, in its essence, technology is not limited to this utilitarian dimension: it is more deeply of the order of a manifestation, of a revelation (Simondon speaks of "technophany"). Nature is rich in potential, says Simondon; it needs the organizer that it itself has produced (the human being, of course).

In a compact and fairly elliptical formula, Simondon writes: we are natural beings who have a debt of *technè* to repay the *physis* within us; the germ of *physis* which is in us must dilate in *technè* around us. We cannot accomplish our essence without making the organizers within us radiate (Simondon 2014, p. 24). Simondon therefore thinks like Canguilhem that the human being is a living being whose life expands in technicity. However, he adds three details.

First, this dilation is a fulfilment; the human fulfils its nature as a living being (the germ of *physis* within it) by becoming a technician.

Second, the life of the living human being expands into *techné* since this living being makes the "organizers" within it "radiate". What does Simondon mean? If the living human being has organizers within it, it is because there is something to organize somewhere. It is not clear what this could be, except the *physis* which is not within it but around it – external nature, in short. *Physis* is therefore not only a power of generation of all things; it is also form, structure and organization, but an organization still in the state of envelopment, of germ or of potential to be actualized.

This idea that *techné* carries nature further, by doing what nature has not done itself, is already found in Aristotle, but in the sense that this accomplishment is oriented by strictly human goals. Simondon seems to say something else here. To make the organizers within one radiate is to introduce into nature organizing principles which are not governed by human goals alone, but first of all by the lines of force contained in nature itself. More precisely, the organizing activity is inseparably the fulfilment of human goals, and the actualization of the potentials of nature. A non-human living being which constitutes a territory for itself by tracing a path in its environment already makes the organizers within it radiate. The path, says Simondon, is the dimension according to which the life of the subject is integrated into the territory by individualizing it and by individuating the subject: the path is both world and subject (Simondon 2020). Vital activity organizes nature by operating from within nature itself. It fulfils nature by fulfilling the living being which fulfils nature.

But it is exactly in the same categories that Simondon grasps human *techné*. The fact that human *techné* presents new characteristics in the living world does not change anything in what must remain its essential meaning. Like any vital organizing activity of the environment, it is an organization of nature operated from within nature, through the intermediary of a living individual whose life has precisely found itself to expand in *techné*.

Dagognet is undoubtedly the one who, after Simondon, most clearly assumed this conviction that the human task is to carry nature further, to fulfil it according to its own logic, which of course presupposes having a good knowledge of this logic. Its own logic being clarified, asks Dagognet, why not extend it and above all correct it? (Dagognet 1988, p. 12). Extending nature can only consist in promoting its deployment (Dagognet 1988, p. 69). According to Dagognet, this is not a manifestation of technical *hybris*, but on the contrary an authentic ethical posture, much more respectful of nature than the conservationist attitude which considers it as immutable and untouchable. Radically, Dagognet maintains that nature invites us, not to conservation, but to artificiality (Dagognet 1988, p. 49). This is not an unprecedented understanding of the relationship between humans and nature: as Dagognet specifies, instead of going into ecstasies over it [nature] or contemplating it, we are already implicitly (and for a very long time) committed to entering into its own combination, in order to broaden its own spectrum and promote its exuberance (Dagognet 1988, p. 83). Nature is in its essence, not a harmonious and unchanging totality, but a power generative of new beings, producing variety; its logic is that of combinations and compositions. "Intensifying nature" means manipulating and transforming when nature itself fails, says Dagognet. In other words, it is not a question of being out of control and concluding that in the end, all combinations of nature and artificializing human intervention are equal. Dagognet castigates cloning, eugenics, the commercialization of gametes – so many artificializing enterprises that do not fit into the logic of nature and of life, since they do not consist in producing variety, always something new, but on the contrary, reproducing the existing by freezing it.

The human being is therefore a living being whose nature is to actualize the potentials of nature, along the lines of force drawn by these potentials themselves. This recognition of natural potentials may thus serve to define a standard for technical action, but this is not yet sufficient to confer moral significance upon it. As we have seen, this meaning can only be posited by reference to the existence of beings to whom we grant the status of valuators. Allowing the organizers within one radiate by means of technical action, as Simondon says, only has moral value in the eyes of valuators.

This, moreover, is the third precision that Simondon adds to his definition of the human: the indebted essence of our nature. The dilation of the *physis* within us through *techné* outside of us is not in the order of a right of appropriation and exploitation; it proceeds from a "debt". This means the debt that the human being, this living being who always already has ceased to be a "simple living being" to become a technician, has therefore incurred with regard to the fact that, despite everything, it remains a living being among living beings. We are in debt because we too are living beings, and *techné* must repay this debt of life by letting ourselves be guided, and commanded, by the aim of a fulfilment of life, that is to say of valuating activity – ours but not only ours, as we recognize valuation by non-human living beings.

It is therefore advisable to be careful. To be satisfied with setting up *laissez-faire* as the guiding principle of technological design, in the aim of being careful, may indeed lead to serious misunderstandings. As we have recalled, many research and innovation sectors today highlight the virtues of piloting natural processes, rather than manufacturing controlled and executed according to a pre-established plan. Living beings are tinkered with and modified in all directions to produce what we fail to produce: this certainly involves letting things happen, more precisely, making them do what we ourselves do not do. Where is the care, though?

We should therefore specify that *laissez-faire* only takes care if technical action is commanded by the recognition of valuations where they exist. If, on the contrary, it consists in an appropriation of living beings by a being which considers itself apart and imagines itself to be outside nature, which authorizes itself to no longer consider itself obliged to repay the *physis* within it and which, correlatively, arrogates to itself the right to extenuate without examination the valuations of all living beings in order to put them at the service of its current interests: then such an act appears to be the opposite of care.

4.2.2.4. *Techno-aesthetic*

Simondon designates the alliance of the technician and *physis* by the expression of "aesthetic feeling", then, later in his writings, as that of "techno-aesthetic".

The technical object only acquires an aesthetic dimension when it goes beyond its character as an object indifferent to the world and when it "marries" with Nature, as Simondon says. When it is no longer a tool, detachable and transportable everywhere, but a reality inserted into both the natural and human world, then it takes on an aesthetic character and arouses a feeling of "a beauty proper to technical objects".

The technical object becomes aesthetic when it is no longer only that which locally ensures the effectiveness of the gesture, but also includes a reference to the totality of the world. "It is not only the technical object that is beautiful: it is the singular point of the world that the technical object concretizes. It is not only the line of pylons that is beautiful, it is the coupling of the lines, the rocks, and the valley, it is the tension and flexion of the cables: herein resides a mute, silent and ever continued operation of technicity applying itself to the world" (Simondon 2017, p. 197).

Aesthetic feeling is thus born when technicity operates upon reality by adding something to it, "but according to lines that already exist in given reality" (Simondon 2017, p. 200). From this perspective, technical action does not distort nature since it is, on the contrary, governed by the requirement of insertion; but this does not imply that it should be modest and ultimately very little active. Etienne Souriau would rather say that such a technical act is an *"instauration"* of nature as a work to be done.

The task is immense and not modest, but it demands from the technician that their action not be animated by the spirit of mastery and domination – rather by a form of openness, of welcoming that which imposes, as Souriau says, its right to exist. The Garabit viaduct that Simondon particularly liked, or the more recently built Millau viaduct, both establish a line of communication between the heights which constitute the key points of the natural environment. They establish a resonance of the environment with itself by actualizing the potentials which called for the constructive gesture.

4.2.2.5. *The imperative of knowledge*

However, this alliance of humans and nature calls for knowledge, even makes it essential, as we have seen forcefully underlined by Dagognet and Simondon: "a technical education is needed if the beauty of technical objects is to appear as an integration of technical schemas into a universe, within the key-points of this universe" (Simondon 2017, p. 197). Simondon gives the example of a radio relay placed on a mountain: if we stick to what we perceive, we only see a "tower of mediocre height, with a parabolic grid". To feel the beauty of this technical object, one must know that a wave beam is emitted by this tower, a beam which propagates from one tower to another and gives the impression that the relay is participating in a large-span, majestic structure – from which arises the aesthetic feeling.

Understood in this way, the technical act seems to essentially relate to care. From the Simondonian perspective, the implicit forms of nature, organic or inorganic, cannot be grasped according to the category of the exploitable resource. They constitute for the technical act a term of reference with normative, and even axiological, value, since they define an ideal of accomplishment.

The essential point is that our debt of *technè* to repay the *physis* within us has considerably increased over the past three or four centuries. Paleolithic humans also had a debt of *technè*, but their *technè* allowed them to make the organizers within them radiate only very weakly. The situation has changed profoundly and, as Latour observes with his usual verve, "We can no longer pose the question of morality in the same way as we would have done at a time when human beings had hardly started to scratch the earth on which they passed from life to death without anyone else noticing" (Latour 2002, p. 256).

A debt obliges us; a heavy debt obliges us heavily. The service of this debt goes through the imperative of knowledge in particular. Care associates technology with science, not in the sense that technology is applied science, but in the sense that science is now an integral part the moral debt contracted by *technè*.

In a context of finalized research, scientists are today summoned by their funders to justify the economic and societal utility of their work. To what applications does the research being performed lead? It may be appropriate to reverse the question and ask: is the technical action sufficiently supported by scientific knowledge? Not only, therefore: to what extent does the development of useful technology pay the debt of science to the benefit of society, but also: to what extent does science actually contribute to paying the debt of *technè* with regard to nature? A particularly thorny question when technical action is faced with an uncertainty which science cannot completely remove.

Might the debt of *technè* have become so massive that today we have become insolvent? Let us simply say that the effort to know causes and effects, to which the ancient Greeks had already drawn attention, is clearly not sufficient to render our *technè* more caring.

Bergson, we said, indicated the way: to transform even our perception, to learn to see what we do not see – to reconfigure the distribution of the sensible, in short. For the Kappian pair of projection and reflection, we thus substitute the pair of insertion and perception of processes.

It is not impossible, we also said, for technology to also participate in this great remaking of visibility. Is its potential for reflexivity, that is to say for bringing to light the processes which underlie it, enough to take care and to repay our debt of *technè*, at least in part? In any case, it would have the merit of bringing to our attention, on a daily basis, one essential thing: the real, frightening level that this debt has attained. Such instrumented knowledge could be the first step in an in-depth reorientation of technical action towards care.

4.2.2.6. *Refusing the view of the user as pure performer*

The third principle of the "square of care", consisting in rejecting the vision of the caregiver as a simple button-pusher, carrying out directives and instructions given by the engineer, does not seem to be oriented to the requirement of taking care of nature. This principle would rather be a reminder of the imperative of human care, well beyond the medical field. Thinking of technological design in terms of integration, rather than mastery, on the basis of the first two principles (let nature do its thing and seek knowledge), in no way leads to an approach separating technologies of interventions upon the natural world, though the latter be very strongly shaped by human activity like cultivated soils, and technologies of direct intervention upon human activity and on human environments (professional environments, housing, etc.). The latter in fact benefit from being themselves considered according to these same two practical principles, and consequently according to the concept of insertion. Indeed, a work environment, for example, is also a matter of attention, whose specific organization must be known well before seeking to transform it technologically, in such a way that design, here too, seems to marry the norms which structure ways of doing things and the real activity of the actors.

This basic refusal of a split between design and pure execution may be seen as a very useful precept, particularly where pressure exerted upon the activity of work by the injunctions of management based on very standardized procedures, for purpose of control, is extremely strong.

Automation of work processes has often gone hand in hand with a loss of workers' autonomy: by the very fact of subjecting them to procedures that increasingly restrict activity, they no longer appear to be subjects of their own norms, in Canguilhem's words. Rather than leaving them to it, that is to say leaving them significant latitude to organize their own activity, it is considered the best method to force them as much as possible to scrupulously respect ways of doing things that are held to be more rational, less impressionistic.

In reality, it often means making workers more controllable, because of the ease in tracing their activity when it is proceduralized to the extreme. This form of rationalization of activities of work is totally devoid of care, as shown by the suffering which it engenders.

However, as Pascale Molinier shows in a striking case (Molinier 2012), lack of care for humans, when it consists in reducing activity to the pure execution of strictly defined procedures, makes care for nature and living beings impossible. This case concerns slaughterhouses. We know of the terrifying videos secretly filmed by the L214 group, bringing to public knowledge the horrible suffering inflicted on animals in these places of shame. What, however, of the employees who work there? Are they monsters? No, replies Molinier, they are rather suffering beings themselves. This unrecognized suffering, kept invisible, without words, inevitably results in ways of acting at the antipodes of care.

The idea of involving the operators of technological systems in the process of designing these systems, of co-conception or co-design as it is usually expressed, has become quite commonplace and widely disseminated. It would no doubt also be desirable for, as systematically as possible, the engineers who design technological devices to have a very good understanding of the professional worlds in which, or more generally of the use for which, these devices will be implemented. This condition is far from being always respected. Yet it is one thing to involve future users in technological design; it is another to bring to the activity of design a deep knowledge of the professional values and cultures with which the technology, when it is deployed and produces all its effects, risks colliding head-on.

Processes of automation often provoke opposition from those who will have to embrace them in their work; not because "resistance to change" is inherent in "human nature" as engineers sometimes like to say, but because automation is very often carried out in order to bring about profound changes in the way of doing and in the organization of work (in a sense that can be experienced as a loss of autonomy by the users).

A technological design process commanded by the wish to take care therefore requires designers to make the effort to understand the values that future users of their systems wish to uphold, and more specifically that they take the time to properly understand conflicts of values. Preliminary conditions for this are, on the one hand, that they not hide behind the worn-out argument of the neutrality of technology, and on the other hand, that they assume that their design activity is not limited to meeting specifications (in short, performing Appraisal, in Dewey's terms), but also that it may have the effect of reducing the ability of users to define for themselves the guiding values of their activity, to which they attach a value (Prizing).

In short, this third practical principle of care in technology requires individuals to take precedence over group values, or rather, to be more precise and to avoid misunderstanding; it means preventing the strategic goals of organizations – which give engineers their instructions – from stifling the ability of individuals to be subject to their own norms in work collectives.

The value of insertion of a technological device is ultimately not only in the coupling of humans and nature, but also in the ability of engineers to think about this insertion of the device, in a professional activity for example, in a way which is not destructive to this activity, but on the contrary potentiates it.

Melanie Wilson thus relates in the early 2000s (Wilson 2002) a case illustrating resistance by nurses to deployment, in a hospital in the north of England, of a new computer system. This rejection can be explained, according to her, by the fact that the designers of said system did not take into account at the time the guiding values of nursing – that is to say, they had not fully grasped what nurses considered to be the very essence of their profession, and which absolutely had to be preserved. This system, called the *Zenith Nurse Management System*, aimed at computerizing the recording of all acts of care carried out by nurses for each patient – a task previously accomplished manually, on paper. As the terminal was located in a separate room, the nurses using it ended up feeling an incompatibility between the obligation to use the new system and care. The latter in fact involves touch, through the hands in contact with the patients, by proximity with them; the distance imposed by the use of the system thus conflicted with this fundamental value of contact. It is therefore an exemplary case of technological innovation ignoring the guiding values of the activity it was supposed to supplement.

This case concerns the medical field, but the lesson that should be drawn from it goes far beyond this field. A genuinely caring and careful technological device certainly reconfigures practices, but in such a way that the internal logic of these is extended, by the benefit of technological supplementation, rather than exhausted by it. This requirement requires that the engineer, in one way or another, have a good understanding of said logic and of what must absolutely be preserved in the environment of work, or more generally of use, into which the device must fit.

4.2.2.7. *Stimulating conflictual pluralist debate*

The fourth and final principle – to ensure that technology be subject to rational explanation to its users – obliges engineers to consider their activity and their design choices as subjects which must be able to be discussed by all whom this activity and these choices are likely to affect. This implies, on the part of engineers, an effort to justify but also to evaluate technical developments with regard to their intended purposes. Once again, the question here is the ability of individuals to retain the full range of their intellectual, technical, aesthetic (that is to say, sensitive) means, in order to be able to actively contribute to the decisions that concern them.

4.2.2.8. *Concluding remarks about the square of care*

Implementing care in technology ultimately boils down to two basic requirements.

The first is understanding the technological relationship of humans to the world according to the category of insertion, and not according to that of mastery. The exorbitant amount that our debt of *technè* has reached today forces us to couple this understanding of the human being as a living technician inserted in a struggle to reorganize the distribution of the sensible, by instrumenting our perception so as to render accessible the constituent processes of this debt.

The second requirement for embedding care into technology is to never forget that the capacity for self-normativity of individuals must take precedence over any constituted group. To insist on the importance of individuals, without yet conceiving them as isolated atoms, but by situating them in a set of relations, is this not care? (Larrère 2012, p. 240).

Drawn from Greek analyses of medicine, the "square of care" therefore aims to be not only the set of guiding principles for the practice of doctors; it also aims to impose itself as the set of guiding principles for the practice of engineers and designers.

The question finally arises whether all these practical principles must necessarily be satisfied at the same time. The answer cannot be unambiguous and theoretical: it can only be contextual.

Thus, let us consider designers concerned with finely understanding the values structuring a profession, by carrying out an ethnographic survey for this purpose: are they inexorably lacking in care when, for example, they leave in the background the question of the ecological footprint of the materials they will use to build their devices? The judgement would be somewhat harsh.

A physician who prescribes targeted therapy to one of his patients is not *de facto* in lack of care, while the underlying principle of these therapies is much more the aim of mastery than of *laissez-faire*[10]. On the other hand, if this doctor omits to tell this patient that the clinical efficacy of these therapies generally remains to be demonstrated, if they do not explain the principle of action of these therapies, if they do not make the effort to seek the most updated knowledge – then it is in the most complete breach of the other practical principles of care.

These practical principles do not make it possible to determine in advance what constitutes care in each situation, but are only guidelines – we would gladly say regulators, to use Kantian terminology. Thinking about care in technology is a matter of the faculty of reflective judgement, and not of the determining judgement. The "square of care" in no way constitutes a system of general rules to be applied unequivocally to any situation, thereby making it possible to automatically take care. Such an understanding of the meaning and scope of these principles would be contradictory to care.

10 These innovative therapies "target" molecular alterations affecting the metabolic pathways of the cell, which are known to be involved in pathological processes: they aim to inhibit enzymes, growth factor receptors, etc.

Conclusion

It is often admitted that there is, on the one hand, technology with a positive value, oriented by care, and on the other hand technology with a negative value, animated by the aim of mastery, characterized by predation exercised over any form of life (biological as well as social and subjective), and therefore antithetical to care. There would be, in summary, good technology and bad technology.

This bad technology, or at least this technology which is always suspect in principle – to put it quickly: to some extent, everything that is not artisanal-type know-how – appears to be the vessel of overall strategies aimed at giving priority, overall life, to capitalist interests. Technology and so-called neoliberal capitalism are thus often associated in a global and massive critique.

It is certainly not wrong to argue that research and technological innovation in the many sectors considered to be "disruptive", supported by a very dynamic economy of promises and by very substantial public funding (nanotechnologies, Artificial Intelligence, synthetic biology, very high-throughput genomics), often appears magnetized by the aims of consolidation and extension of a capitalism that is ceaselessly transforming itself. In this context, adopting the point of view of care seems capable of taking the place of a critical and denunciatory posture towards new forms of capitalist exploitation of life, considerably reinforced and extended thanks to the development of today's main productive forces (which Marx, moreover, had already diagnosed very well): science and technology. The posture of care would thus be a posture of defense of life in all its forms against technology, or rather against these bad technologies identified as culprits of carelessness.

However, this posture may very quickly turn out to be counter-productive and far short of what a truly transformative critique of reality would require. This is what Antonio Negri explains in the interview about his intellectual and political career in April 2018, already mentioned in the introduction (Negri 2018).

Negri here evokes the situation of Southern Italians in the first half of the 20th century and until the 1960s and 1970s, who, fleeing agricultural poverty, migrated to the North of the country, and found employment in construction and factories. In the region where Negri is from, Veneto, the process of training workers in particular led through the shoe industry. This industry, he says, employed around 10,000 people who worked in very small factories and in terrible conditions as they cut their hands and were exposed to chemicals for processing leather. In this sector, they were killing you. And so, when they got to big chemical industry, they felt that they finally became real workers! Except that they forgot that the risk of dying was even greater than before because the chemistry was particularly toxic. So, what could we do? We had to play on the contradictions, to reorganize the production cycle, isolate dangerous activities or automate them: we started the struggles with that (Negri 2018, p. 39). And again: you are faced with the fact that workers are dying due to the work cycle in the factory. If you have the capacity to intervene in the discourse and show that it is the cycle that kills, you begin to pose a general political problem for all workers (Negri 2018, p. 40).

Negri emphasizes here two things which are of the utmost importance.

Firstly, life, in the capitalist system of production, is not only the living labor force that the worker sells in exchange for wages and that Marx had theorized; it is also the vulnerability of an organism dying from the toxic environments to which it is exposed. In Marx, this vulnerability is bare; it is pure exposure, and when the living labor-power is exhausted by wear and tear on the body, disease or death, the capitalist just has to replace the worker.

Negri, on the other hand, says something fundamentally different: the vulnerability of the exposed life is also a favorable ground for struggles, it even incites them. Better still, it is a fulcrum for transforming sectoral and local demands (reorganizing the production cycle and promoting automation in a particular factory) into a general political problem for all workers.

In Negri's terms, recognition and taking charge of the vulnerability of life, in the workplace, is therefore a driving force behind the subjectivation of living work. It introduces the possibility of rationally subjectifying subversive action in the class struggle, of organizing the passion for the emancipation of living labor "from below", from within a class autonomy (Negri 2018, p. 85).

Without saying as much, Negri therefore almost surreptitiously introduces the question of concern and care (Tronto's *caring about*, the number one stage of care work), seen not as private compassion and concern, but as a strategic element in the political struggle, within a framework of Marxist thought.

Second, if the technological processes in chemical factories are responsible for the worst misconduct, since they kill the workers, technology is also the remedy since the struggles pass through the very material, that is to say technological (with the possible use of automation), reorganization of production cycles. Technologies are not foreign to the requirements of care, they make them effective and even, if we stick to the letter of what Negri says, they alone can make them effective. There is indeed no care for life if production techniques, in the factory, are not oriented by values of care. Better: there is no possible care for life, in the factory, if class struggles do not relate above all to technology. Extremely well-paid workers, controlling their rhythm of work, listened to by management, would not necessarily develop in a caring world if they continued to die on the grounds that they are daily exposed to inherently toxic production techniques.

Negri's experience of political struggles (we started struggles with that, he confesses) therefore does not at all agree with the idealist posture of invoking care from a position of oversight, and to reject effective technological processes *en bloc* without seeking to transform them, to reconfigure them. Providing care, in the factory but also elsewhere – at the hospital, at school, in administrations, etc. – is on the contrary, argues Negri, to analyze technological processes very concretely, and to organize the struggle where it is possible to change them.

It may be objected that conceived in this way, the political struggle is very weak: the revolution is far away, it is simply a matter of improving the lot of the workers a little so that they die a little less in the factories. Capitalism emerges consolidated, in the end.

However, the essential point of Negri's argument is that the objection does not bear. To fight over production cycles, over technological processes, is to hatch in factories new processes of subjectivation – which means, already, organizing resistance to capitalism. Since technology configures and makes effective relations of power and domination, it also appears as the privileged target (the only real one?) of struggles to modify these relations.

The question of care in technology takes on its full significance here: taking care in technology is not to serve the interests of capitalism. On the contrary, it is to make manifest the relations of power and domination that technology configures and materializes, and which have been made invisible, in order to renegotiate them.

Rendering visible all the production processes of electronic devices, from metal extraction to the accumulation of waste in gigantic open-air landfills in Africa or Asia, is not just encouraging Apple, Samsung and others to improve their production processes a little and, in so doing, slavishly contribute to their greenwashing policy.

It is to carry the struggles to where the capitalist relations of domination can truly be destabilized.

Another area also illustrates this point: animal welfare. In a few years, animal suffering in agro-industrial sectors has become a social issue, even in a country like France which remains very attached to the consumption of meat. Several large retail groups have recently taken up the issue by changing the organization of their supply chains, and by adopting new techniques for the traceability of animals, from breeding to slaughter. The technological cycle of production is therefore at the heart of the struggle, also in this case.

So, when Negri explains that class struggle precedes technological transformation, we will not follow him. At a pinch, we might understand this formula if classes existed independently of production systems – if they pre-existed as *sui generis* entities, already constituted when they enter into struggle through the production systems, trying to apply all their weight to give these systems, particularly on the technological level, certain determinations rather than others. However, Negri himself rejects the idea that classes can exist *sub specie aeternitatis.* They are rather to be constructed according to the characteristics of capitalism at a given time. Negri is a pragmatic Marxist: that's Marxism, the capacity that you have to read the reality that you have in front of your eyes and to elaborate constructions adequate to this reality in order to transform it (Negri 2018, p. 65)[1]. Theoretical constructions, in particular the concept of class, are only valid because of their power to truly transform reality.

Therefore, there is no sense in speaking of class independently of the technological structures of capitalist production, here and now. On the basis of analysis of these structures, it is possible to identify who exactly are the exploited – they are undeniably very different depending on whether we are dealing with the capitalism that Marx described, or with the cognitive capitalism that Amazon's *Mechanical Turk* illustrates in its most brutal aspects.

The processes of primitive accumulation of capital have changed profoundly and theoretical analysis must imperatively account for this. It is a question of rebuilding a working class in the new context of cognitive capitalism, characterized by the extension of exploitation well beyond the workplaces proper, to all social activities and all subjective forms of life. Negri calls this new class "multitude" and asks who makes it up, or rather: who should be brought into it, given the reality of capitalist relations of domination today. The multiplicities of subjects and the singularity of

---

1 This definition is in no way specific to Marxism.

the processes of subjectivation are literally composed in the experience of the class struggle (Negri 2018, pp. 95–96).

Technology does not therefore come after struggle between classes, it defines their framework and determines their target. It provides a foothold for "multitudinary" subjectivation processes.

Raising the question of care in technology therefore appears to be an essential strategic element in building this multitude, in such a way that the processes of subjectivation, that is to say forms of resistance, are set in motion.

Negri does not say so in his interview, but we have to admit that today, the multitude brings together humans but also non-humans. Farm animals slaughtered in atrocious conditions are also the multitude. Animals subject to scientific experimentation, these are again the multitude. The countless living species whose habitats have been devastated by human activity: a multitude there too.

Yet it is indeed in the struggles to develop technologies, to change production cycles, that these beings dominated and exploited by capitalism constitute multitudes and impose themselves as recalcitrant subjectivities.

Thus, research in the currently booming field of organoids (reconstructed, miniaturized, and simplified versions of human organs) is part of the desire to develop the methods of pharmacological tests, still largely dependent on animal experimentation. In the long term, the development of these organoids should make it possible to carry out toxicity tests on new drug molecules, before switching to humans, without it being necessary to use animals.

In its label on animal welfare, Carrefour, a famous French retail group, writes in black and white, talking about animals: the individual must be able to express the natural behaviors of its species, at all points of the production chain. Quite an incredible formulation, which amounts to recognizing that animals, on the one hand, are individuals and not machines for supplying animal protein to human consumers, and on the other hand, that they are valuating beings.

The new class that must be built, the multitude dear to Negri, now includes all living valuators who are diverted from their valuating activities by human technological action, or because they are literally put to work by the exploitation of their metabolism, or because the environmental conditions required for them to carry out their valuations are degraded by human activity.

Ultimately, seeking to make care exist as a guiding and structuring requirement of multitudinary relationships, which must now also include non-human living beings, is inevitably to attack the way in which these relationships are technologically configured and organized. Care in technology is therefore quite another thing than a modest, tactical concession of capitalism to ecological claims which have won over the populations. It is a strategic element in new processes of subjectivation.

Claiming to take care without bringing the struggle into the field of the design of technology and the technological organization of production chains is inevitably to leave as it is the distribution of the sensible which defines the social and political order of the moment. It is, in short, letting lack of care flourish, where production processes are kept invisible – that is to say almost everywhere – if we do not take care.

References

Achterhuis, H. (ed.) (2001). *American Philosophy of Technology: The Empirical Turn.* Indiana University Press, Bloomington.

Alder, K. (1998). Making things the same: Representation, tolerance and the end of the Ancien Regime in France. *Social Studies of Science*, 28(4), 499–545.

Alsberg, P. (1970). *In Quest of Man. A Biological Approach to the Problem of Man's Place in Nature*. Pergamon Press, Oxford.

Andrès, R. (2013). État de l'art. Design et médecine. Enjeux et pratiques du soin aujourd'hui. Cité du Design, Pôle des Technologies Médicales, Saint-Etienne.

Anker, P. (2010). *From Bauhaus to Ecohouse. A History of Ecological Design.* Louisiana State University Press, Baton Rouge.

Arendt, H. (1958). *The Human Condition.* The University of Chicago Press, Chicago.

Arendt, H. (2006). *Between Past and Future. Eight Exercises in Political Thought.* Penguin Books, New York.

Aristotle (1924). *Metaphysics.* Translated by W.D. Ross. Clarendon Press, Oxford.

Aristotle (2009). *Nicomachean Ethics.* Translated by W.D. Ross, revised by L. Brown. Oxford University Press, Oxford.

Bach-y-Rita, P. (1972). *Brain Mechanisms in Sensory Substitution.* Academic Press, New York.

Baird Callicott, J. (1985). Intrinsic value, quantum theory, and environmental ethics. *Environmental Ethics*, 7(3), 257–275.

Basalla, G. (1988). *The Evolution of Technology*. Cambridge University Press, Cambridge.

Bates, C., Imrie, R., Kullman, K. (eds) (2017). *Care and Design. Bodies, Buildings, Cities.* John Wiley & Sons, Oxford.

Baudelaire, C. (1868). *Curiosités esthétiques.* Michel Lévy Frères, Libraires Éditeurs, Paris.

Beau, R. (2013). Ethique de la nature ordinaire. PhD thesis, Université Paris 1 Panthéon-Sorbonne, Paris.

Beaubois, V. (2019). La Zone obscure du design. Une pensée des pratiques de conception (d')après Gilbert Simondon. PhD thesis, Université Paris-Nanterre, Nanterre.

Beaubois, V. and Ferrari, F.-X. (2020). L'écodesign ou l'épreuve de l'invisible écologique. *Sciences du design*, 11, 46–53.

Beaune, J.C. (1994). L'ordre et la matière. In *Ordre biologique, ordre technologique*, Tinland, F. (ed.). Champ Vallon, Coll. Milieux, Seyssel.

Benaroyo, L., Lefève, C., Mino, J.-C., Worms, F. (2010). *La Philosophie du soin. Éthique, médecine et société*. Presses Universitaires de France, Paris.

Benjamin, W. (2002). *Selected Writings, Volume 3, 1935–1938.* Translated by E. Jephcott and others. The Belknap Press of Harvard University Press, Cambridge and London.

Benjamin, W. (2016). *Technique et expérience*. Eterotopia/Rhizome, Paris.

Bensaude-Vincent, B. and Benoit-Browaeys, D. (2011). *Fabriquer la vie. Où va la biologie de synthèse ?* Editions du Seuil, Paris.

Bergson, H. (1911). *Matter and Memory.* Translated by N.M. Paul and W.S. Palmer. George Allen & Unwin Ltd, London, and The MacMillan Company, New York.

Bergson, H. (1944). *Creative Evolution.* Translated by A. Mitchell. The Modern Library, New York.

Bergson, H. (2002). *The Two Sources of Morality and Religion.* Translated by R.A. Audra and C. Brereton. University of Notre Dame Press, Notre Dame, Indiana.

Bertrand, A. (2013a). Peut-on étendre l'éthique du soin à l'environnement ? *Éthique, politique, religions*, 2(3), 45–58.

Bertrand, A. (2013b). Prendre soin de la nature ? In *Prendre soin : savoirs, pratiques, nouvelles perspectives*. Chagnon, V., Dallaire, C., Espinasse, C., Heurgon, É. (eds). Hermann, Paris.

Bihouix, P. (2014). *L'Âge des low tech. Vers une civilisation techniquement soutenable.* Éditions du Seuil, Paris.

Bihouix, P. (2019). La transition énergétique peut-elle être *low tech* ? *Revue internationale et stratégie*, 1(113), 97–106.

Boëda, E. (2013). *Technologique & technologie : une paléohistoire des objets lithiques tranchants*. Archéo-éditions, Paris.

Bonneuil, C. and Fressoz, J.-B. (2017). *The Shock of the Anthropocene. The Earth, History and Us.* Translated by D. Fernbach. Verso, London and New York.

Boorse, C. (1975). On the distinction between disease and illness. *Philosophy & Public Affairs*, 5(1), 49–68.

Brey, P. (2010). Philosophy of technology after the empirical turn. *Techné*, 14(1), 36–48.

Butler, S. (1970). *Erewhon or Over the Range.* Penguin Classics, London.

Byl, S. (1971). Note sur la polysémie d'"Όργανον et les origines du finalisme. In *L'Antiquité classique*, T.40, fasc., 1, 121–133.

Cambiano, G. (2013). Platon et les rapports entre théorie et praxis dans la médecine hippocratique. *Études platoniciennes*, 10 [Online]. Available at: https://journals.openedition.org/etudesplatoniciennes/351 [Accessed 8 November 2019].

Canguilhem, G. (1947). Milieu et normes de l'homme au travail. *Les cahiers internationaux de sociologie*. Editions du Seuil, Vol. III, Paris, 120–136.

Canguilhem, G. (1975). Le tout et la partie dans la pensée biologique. In *Études d'histoire et de philosophie des sciences*, Canguilhem, G. (ed.). Vrin, Paris.

Canguilhem, G. (1978). *On the Normal and the Pathological.* Translated by C.R. Fawcett. D. Reidel Publishing Company, Dordrecht.

Canguilhem, G. (2002). The problem of regulation in the organism and in society. In *Writings on Medicine*, Canguilhem, G. (ed.). Translated by T. Meyers. Fordham University Press, New York.

Canguilhem, G. (2008). Machine and organism. In *Knowledge of Life*, Canguilhem, G. (ed.). Translated by S. Geroulanos and D. Ginzburg. Fordham University Press, New York.

Canguilhem, G. (2011). Descartes et la technique. In *Œuvres complètes, vol. I. Écrits philosophiques et politiques, 1926–1939*, Canguilhem, G. (ed.). Vrin, Paris.

Carnino, G. and Marquet, C. (2018). Les *datacenters* enfoncent le *cloud* : enjeux politiques et impacts environnementaux d'Internet. *Zilsel*, Éditions du Croquant, 3, 19–62.

Casilli, A. (2019). *En attendant les robots. Enquête sur le travail du clic*. Editions du Seuil, Paris.

Chamayou, G. (2019). Eh bien, recyclez maintenant ! *Le monde diplomatique*, 779.

Charles, D. (1997). *La pensée technique dans l'œuvre de Victor Hugo*. Presses Universitaires de France, Paris.

Cheverst, K., Clarke, K., Cobb, S., Hemmings, T., Kember, S., Mitchell, K., Phillips, P., Procter, R., Rodden, T., Rouncefield, M. (2001). Design with care. *Journal of New Technology in the Human Services*, 14(1), 39–47.

Cheverst K., Clarke, K., Dewsbury, G., Hemmings, T. (2003). Design with care: Technology, disability and the home, In *Inside the Smart Home*. Harper, R. (ed.). Springer Verlag, London.

Clarizio, E. (2019). Une archéologie du dispositif. In *Vie, anthropologie, politique. Perspectives italiennes contemporaines en philosophie des techniques*. Clarizio, E., Guchet, X. (eds). Editions Mimésis, Paris.

Coccia, E. (2019). *Le bien dans les choses*. Editions Payot & Rivages, Paris.

Confucius (2005). *The Analects of Confucius*. Translated by A. Waley. Routledge, London and New York.

Comte, A. (2009). *The Positive Philosophy of Auguste Comte Vol II*. Translated by H. Martineau. Cosimo Classics, New York.

Coulomb, C.A. (1821). *Théorie des machines simples en ayant eu égard aux frottements de leurs parties et de la roideur des cordages*. Bachelier, Libraire, Quai des Augustins, Paris (nouvelle édition), 255–297 [Online]. Available at: https://gallica.bnf.fr/ark:/12148/bpt6k1095299/f377.image [Accessed November 9, 2019].

Crawford, M. (2009). *Shop Class as Soulcraft. An Inquiry into the Value of Work*. Penguin Press, New York.

Dagognet, F. (1988). *La Maîtrise du vivant*. Hachette, Paris.

Dagognet, F. (2003). *Penser le vivant. L'Homme maître de la vie ?* Bordas, Paris.

Deforge, Y. (1985). *Technologie et génétique de l'objet industriel*. Maloine, Paris.

Deneys-Tunney, A. (2010). *Un autre Jean-Jacques Rousseau. Le paradoxe de la technique*. Presses Universitaires de France, Paris.

Derksen, M.-H.G. (2008). *Engineering Flesh. Towards Professional Responsibility for Lived Bodies in Tissue Engineering*. Technische Universiteït Eindhoven, Eindhoven.

Descartes, R. (1999). *Meditations and Other Metaphysical Writings*. Translated by D.M. Clarke. Penguin Classics, London.

Descola, P. (2014). *Beyond Nature and Culture*. The University of Chicago Press, Chicago.

Dewey, J. (1929). *Experience and Nature*. George Allen & Unwin, London.

Dewey, J. (1939). Theory of valuation. *International Encyclopedia of Unified Science*, 2, 4, vii + 67.

Dewey, J (1980). *Art as Experience*. Perigee Trade, New York.

Didier, C. (2008). *Penser l'éthique des ingénieurs*. Presses Universitaires de France, Paris.

Donghi, P. and Preta, L. (eds) (1995). *In principio era la cura*. Gius Laterza & Figli Spa, Roma-Bari.

Du Camp, M. (1855). *Les champs modernes*. Michel Lévy Frères, Libraires Éditeurs, Paris.

Dupuis, M. (2013). *Le soin, une philosophie*. Seli Arslan, Paris.

Dupuy, J.-P. (2004). Pour une évaluation normative du programme nanotechnologique. *Annales des mines*, February, 27–33.

Ehrard, J. (1984). La main du travailleur. La plume du philosophe. *Milieux*, 19–20, Oct. 1984–Jan. 1985, 47–53.

Eriksen, M., Lebreton, L.C.M., Carson, H.S., Thiel, M., Moore, C.J., Borerro, J.C., Galgani, F., Ryan, P.G.m Reisser, J. (2014). Plastic pollution in the world's oceans: More than 5 trillion plastic pieces weighing over 250,000 tons afloat at sea. *PLoS ONE* 9(12), e111913.

Ettinger, R.C.W. (2005). *The Prospect of Immortality*. Ria University Press, Waterloo.

Evans, J. (1864). The coinage of the Ancient Britons and natural selection. *Journal of the Royal Institution of Great Britain*,7, 476–487.

Feenberg, A. (2010). *Between Reason and Experience. Essays in Technology and Modernity*. The MIT Press, Cambridge.

Ferguson, A. (1996). *An Essay on the History of Civil Society*. Cambridge University Press, New York.

Fleury, C. and Fenoglio, A. (2019). Le *design* peut-il aider à mieux soigner ? Le concept de *proof of care*. *Soins*, 834, 58–61.

Flusser, W. (1999). *The Shape of Things. A Philosophy of Design*. Translated by A. Mathews. Reaktion Books Ltd, London.

Fonteneau, Y. (2009). Les antécédents du concept de travail mécanique chez Amontons, Parent et Daniel Bernoulli : de la qualité à la quantité (1699–1738). *Dix-huitième siècle*, 1(41), 339–368.

Foucault, M. (1995). *Discipline & Punish. The Birth of the Prison*. Translated by A. Sheridan. Vintage Books, New York.

Foucault, M. (2003). *Society Must be Defended. Lectures at the Collège de France, 1975–1976*. Translated by D. Macey. Picador, New York.

Francastel, P. (1996). Technique et esthétique. *Cahiers internationaux de sociologie*, 101, 95–109.

de Fréminville, C. (1927). Perronet, précurseur de Taylor ? *Annales des ponts et chaussées*. January–February 1, 95–98.

Frémont, C. (1913). *Origine et évolution des outils*. Société d'Encouragement pour l'Industrie Nationale, Paris.

Friedmann, G. (1953). L'*Encyclopédie* et le travail humain. *Annales. Économies, sociétés, civilisations*. 8ème année, 1, 53–61.

Friedmann, G. (1966). *Sept études sur l'homme et la technique*. Denoël/Gonthier, Paris.

Friedmann, G. (1970). *La puissance et la sagesse*. Gallimard, Paris.

Fumagalli, M. (2015). *La vie mise au travail. Nouvelles formes du capitalisme cognitif*. Eterotopia, coll. Rhizome, Les Lilas.

Gaille, M. (2013). Que veut dire prendre soin de l'environnement ? Une analyse philosophique du lien entre écologie et médecine. In *Éthique, politique, religions*, 2(3), 81–101.

Garçon, A.-F. (2014). Des modes d'existence du geste technique. *e-Phaïstos*, III(1), 84–92.

Gehlen, A. (1980). *Man in the Age of Technology*. Translated by P. Lipscomb. Columbia University Press, New York.

Gehlen, A. (1988). *Man. His Nature and Place in the World.* Translated by C. MacMillan and K. Pillemer. Columbia University Press, New York.

Gens, J.-C. (2013). Prendre soin et prodiguer des soins. Le souci de soi, des autres et de la nature. In *Éthique, politique, religions*, 2(3), 31–44.

Gille, B. (1952). L'Encyclopédie. Dictionnaire technique. In *L'Encyclopédie et le progrès des sciences et des techniques*. Presses Universitaires de France, Paris.

Gilligan, C. (2003). *In a Different Voice. Psychological Theory and Women's Development.* Harvard University Press, Cambridge.

Girard, B. (2015). *Histoire des théories du management en France du début de la révolution industrielle au lendemain de la première guerre mondiale*. L'Harmattan, Paris.

Gleyse, J. (2000). *L'Instrumentalisation du corps. Une archéologie de la rationalisation instrumentale du corps, de l'âge classique à l'époque hypermoderne*. L'Harmattan, Paris.

Godard, O. (1997). Principe de précaution et responsabilité. Une révision des relations entre science, décision et société. In *Qu'est-ce qu'être responsable ?* Carré Seita, Sciences humaines, Paris.

Goffi, J.-Y. (1988). *Philosophie de la technique*. Presses Universitaires de France, Paris.

Greisch, J. (2002). *Ontologie et temporalité. Esquisse d'une interprétation intégrale de Sein und Zeit*. Presses Universitaires de France, Paris.

Guchet, X. (2005). *Les sens de l'évolution technique*. Léo Scheer, Paris.

Guchet, X. (2014). *Philosophie des nanotechnologies*. Hermann, Paris.

Guchet, X. (2016). *La médecine personnalisée. Un essai philosophique*. Les Belles Lettres, Paris.

Guchet, X. (2017). Objet versus artefact. Pour une philosophie des techniques orientée-objet. *Cahiers Costech*, 1 [Online]. Available at: http://www.costech.utc.fr/CahiersCOSTECH/spip.php?article8 [Accessed November 26, 2020].

Guerpillon, P. (2015). Marx lecteur de Darwin. Lecture, Séminaire Lectures de Marx, ENS, February 9th.

Guignard, L., Raggi, P., Thévenin, É. (eds) (2011). *Corps et machines à l'âge industriel*. Presses Universitaires de Rennes, Rennes.

Hacking, I. (2006). The Cartesian body. *Biosocieties*, 1(1),13–15.

Haraway, D. (1991). *A Cyborg Manifesto: Science, Technology and Socialist Feminism in the Late Twentieth Century*. In *Simians, Cyborgs and Women: The Reinvention of Nature*, Haraway, D. (ed.). Routledge, New York.

Haudricourt, A.-G. (1987). Domestication des animaux, culture des plantes et traitement d'autrui. In *La Technologie science humaine. Recherches d'histoire et d'ethnologie des techniques*, Haudricourt, A.-G. (ed.). Fondation de la Maison des Sciences de l'Homme, Paris.

Havelange, V., Lenay, C., Stewart, J. (2002). Les représentations : mémoires externes et objets techniques. *Intellectica*, 2(35), 115–129.

Heidegger, M. (1962). *Being and Time*. Translated by J. Macquarrie and E. Robinson. Blackwell Publishers Ltd, Oxford.

Heidegger, M. (ed.) (1971). The origin of the work of art. In *Poetry, Language, Thought*. Translated by A. Hofstader. Harper & Row, New York.

Heidegger, M. (1977). *The Question Concerning Technology and Other Essays*. Translated by W. Lovitt. Garland Publishing Inc., New York and London.

Hermès, E. and Sifer, L. (2005). Le processus identitaire des patients en hémodialyse : l'impact de la technique sur la trajectoire de vie. *Recherche en soins infirmiers*, 2(81), 56–77.

Hervé-Gruyer, P. (2018). Récit professionnel : la ferme biologique de Bec Hellouin. Une ferme expérimentale en agro-écologie, selon les principes de la permaculture. In *Evolution agrotechnique contemporaine II. Transformations de l'agro-machinisme : fonction, puissance, information, invention*. Caroux D., Dubois, M.J.-F., Sauvée, L. (eds). Université de technologie de Belfort-Montbéliard, Belfort.

Hess, G. (2013). L'éthique environnementale et le *care*. De la sobriété volontaire à la vigilance. In *Éthique, politique, religions*, 2(3), 59–80.

Hilaire-Perez, L. (2010). Technique, économie et politique entre la France et l'Angleterre (XVIIème–XIXème siècle). *Documents pour l'histoire des techniques*, 19, 9–29.

Hilaire-Perez, L. (2013). *La pièce et le geste : artisans, marchands et savoir technique à Londres au XVIIIème siècle*. Albin Michel, Paris.

Hoquet, T. (2011). *Cyborg philosophie. Penser contre les dualismes*. Editions du Seuil, Paris.

Horkheimer, M. and Adorno, T. (2002). *Dialectic of Enlightenment. Philosophical Fragments*. Stanford University Press, Stanford.

Hottois, G. (2004). *Philosophie des sciences, philosophie des techniques*. Editions Odile Jacob, Paris.

Huneman, P. (2008). *Métaphysique et biologie. Kant et la constitution du concept d'organisme*. Editions Kimé, Paris.

Huyghe, P.-D. (2015). *Art et industrie : philosophie du Bauhaus*. Circé, Paris.

Imrie, R. and Kullman, K. (2017). Designing with care and caring with design. In *Care and Design. Bodies, Buildings, Cities*. Bates, C., Imrie, R., Kullman, K. (eds). John Wiley & Sons, Oxford.

Ingold, T. (2000). *The Perception of the Environment. Essays on Livelihood, Dwelling and Skill*. Routledge, London and New York.

Ingold, T. (2007). Materials against materiality. *Archaeological Dialogues*, 14(1), 1–16.

Jarrige, F. (2014). *Techno-critiques. Histoire des résistances au* progrès *technique*. Editions La Découverte, Paris.

Jonas, H. (1982). *The Phenomenon of Life. Toward a Philosophical Biology*. The University of Chicago Press, Chicago.

Jonas, H. (1985). *The Imperative of Responsibility. In Search for an Ethics for the Technological Age*. The Chicago University Press, Chicago.

Jonas, H. (1996). Tool, image and grave: On what is beyond the animal in man. In *Mortality and Morality. A Search for the Good after Auschwitz*, Jonas, H. (ed.). Northwestern University Press, Evanston.

Jonas, H. (2012). *L'Art médical et la responsabilité humaine*. Les Editions du Cerf, Paris.

Jullien, F. (2004). *A Treatise on Efficacy. Between Western and Chinese Thinking*. Translated by J. Lloyd. University of Hawai'i Press, Honolulu.

Kant, I. (2006). *Anthropology from a Pragmatic Point of View*. Translated by R.B. Louden. Cambridge University Press, Cambridge.

Kant, I. (2007). *Critique of Judgement*. Translated by J. Creed Meredith. Oxford University Press, New York.

Kant, I. (2009). *Idea for a Universal History with a Cosmopolitan Purpose*. Translated by H.B. Nisbet. Cambridge University Press, Cambridge.

Kapp, E. (2018). *Elements of a Philosophy of Technology. On the Evolutionary History of Culture*. Translated by L.K. Wolfe. University of Minnesota Press, Minneapolis.

Kitcher, P. (2004). On the autonomy of the sciences. *Philosophy Today*, 48, 51–57.

Koestler, A. (1967). *The Ghost in the Machine*. Hutchinson & Co, London.

Kostelanetz, R. (ed.) (1971). *Moholy-Nagy*. Penguin, London.

Kroes, P. and Meijers, A. (eds.) (2000). *The Empirical Turn in the Philosophy of Technology*. Elsevier Science Ltd, Oxford.

Kroes, P. and Meijers, A. (2006). The dual nature of technical artifacts. *Stud. Hist. Phil. Sci.*, 37.

La Mettrie, J.O. (1996). *Machine Man and Other Writings*. Translated by A. Thomson. Cambridge University Press, Cambridge.

de Laborde, L. (1856). *De l'union des arts et de l'industrie, 2 volumes*. Imprimerie impériale, Paris.

Lafitte, J. (1980). *Reflections on the Science of Machines*. Translated by L. Grant. Mechanology Press, London, Canada.

Laforgue, P. (2003). Machine et industrialisme, ou romantisme, modernité et mélancolie. Quelques jalons (1840–1870). *Revue d'histoire littéraire de la France*, 103(1), 63–92.

Laplane L. and Solary E. (2017). Identité des cellules souches normales et cancéreuses. *Médecine/science*, 33, 899–904.

Larrère, R. (1994). Histoires de parcs : les ambitions et les contextes [Online]. Available at: http://www.or2d.org/or2d/Accueil_files/or2d-10-Larrere-nov2014.pdf [Accessed November 15, 2019].

Larrère, C. (2002). Avons-nous besoin d'une éthique environnementale ? *Cosmopolitiques*, 1, 69–85.

Larrère, C. (2006). Développement durable : quelques points litigieux. *Les Ateliers de l'éthique*, 1(2), 9–18.

Larrère, C. (2010). Les éthiques environnementales. *Natures, sciences, sociétés*, 4(18), 405–413.

Larrère, C. (2012). Care et environnement : la montagne ou le jardin ? In *Tous vulnérables ? Le care, l'environnement et les animaux*, Laugier, S. (ed.). Editions Payot & Rivages, Paris.

Larrère, C. and Larrère, R. (2015). *Penser et agir avec la nature. Une enquête philosophique*. Editions La Découverte, Paris.

Larrère, R., Lizet, B., Berlan-Darqué, M. (eds) (2009). *Histoire des parcs nationaux. Comment prendre soin de la nature ?* Editions Quae, Paris.

Latour, B. (1996). *Aramis or the Love of Technology*. Translated by C. Porter. Harvard University Press, Cambridge.

Latour, B. (2000). The Berlin key or how to do words with things. In *Matter, Materiality and Modern Culture*, Grave-Brown, P.M. (ed.). Routledge, London.

Latour, B. (2002). Morality and technology. The end of the means. Translated by C. Venn. *Theory, Culture & Society*, 19(5/6), 247–260.

Laugier, S. (2012). Frontières du care. In *Tous vulnérables ? Le care, l'environnement et les animaux*, Laugier, S. (ed.). Editions Payot & Rivages, Paris.

Laugier, S. (2015). Care, environnement et éthique globale. *Cahiers du genre*, 2(59), 127–152.

Le Roy, E. (1931). *Les origines humaines et l'évolution de l'intelligence*. Boivin, Paris.

Leikas J. (2009). *Life-Based Design. A Holistic Approach to Designing Human-Technology Interaction*. Edita Prima Oy., Helsinki [Online]. Available at: http://www.vtt.fi/Documents/2010_P726.pdf [Accessed November 17, 2019].

Lemonnier, P. (2012). Des objets pour penser l'indicible. In *La Préhistoire des autres. Perspectives archéologiques et anthropologiques*, Schlanger, N., Taylor, A.-C. (eds). Editions La Découverte, Paris.

Lenay, C. (2006). Enaction, externalisme et suppléance perceptive. *Intellectica*, 1(43), 27–52.

Leroi-Gourhan, A. (1943). *Evolution et techniques. Tome 1 : L'homme et la matière*. Albin Michel, Paris.

Leroi-Gourhan, A. (1945). *Evolution et techniques. Tome 2 : Milieu et techniques*. Albin Michel, Paris.

Leroi-Gourhan, A. (1993). *Gesture and Speech*. Translated by A. Bostock Berger. The MIT Press, Cambridge.

Levi, P. (1995). *The Monkey's Wrench*. Translated by W. Weaver. Penguin Classics, London.

Lévi-Strauss, C. (1976). *Structural Anthropology, Vol. 2*. Translated by M. Layton. The University of Chicago Press, Chicago.

Linhart, R. (1981). *The Assembly Line*. Translated by M. Crosland. The University of Massachusetts Press, Amherst.

Lock, M. and Nguyen V.K. (2010). *An Anthropology of Biomedicine*. John Wiley & Sons, Hoboken.

Lombard, J. (2006). Aspects de la *technè* : l'art et le savoir dans l'éducation et le soin. *Le portique*, 3–2006 (I) [Online]. Available at: http://leportique.revues.org/876 [Accessed November 8, 2019].

Maag, G. (1983). "Les machines ne sont rien sans l'art". De l'union des arts de l'industrie, du comte de Laborde, et les réactions de la presse. *Romantisme*, 41, 41–56.

Mabi, C. (2014). Le débat CNDP et ses publics à l'épreuve du numérique. Entre espoirs d'inclusion et contournement de la critique sociale. PhD thesis, Université de technologie de Compiègne, Compiègne.

Macherey, P. (2016). Canguilhem et l'idée de milieu. Exposé présenté au colloque "Georges Canguilhem. Science, technique, politique : perspectives actuelles", Liège, April 22 [Online]. Available at: https://philolarge.hypotheses.org/1737 [Accessed November 9, 2019].

Madge, P. (1997). Ecological design: A new critique. *Design Issues*, 13(2), 44–54.

Marin, C. (2013). Qui prend soin de qui ? In *Prendre soin : savoirs, pratiques, nouvelles perspectives*, Chagnon, V., Dallaire, C., Espinasse, C., Heurgon, É. (eds). Hermann, Paris.

Marshall, E.G., Clarke, B., Peddle, S., Jensen, J. (2015). Le programme Care by Design. Une nouvelle façon de coordonner les soins primaires et actifs dans des établissements de soins prolongés. *Canadian Family Physician. Le médecin de famille canadien*, 61, March 129–34.

Marx, K. (n/a). Un chapitre inédit du Capital : résultats du procès de production immédiat [Online]. Available at: https://www.marxists.org/francais/marx/works/1867/Capital-inedit/index.htm [Accessed November 9, 2019].

Marx, K. (1973). *Grundrisse*: *Foundations of the Critique of Political Economy*. Translated by M. Nicolaus. Penguin Classics, London.

Marx, K. (2019). *Capital: Volume 1. A Critique of Political Economy.* Translated by S. Moore and E. Aveling. Dover Publications Inc., Mineola, New York.

Mauss, M. (1973). Techniques of the body. Translated by B. Brewster. *Economy and Society*, 2(1), 70–88.

Meacham, D. and Studley, M. (2015). Care in the air. Caring robots and environments of care. *Multitudes*, 1(58), 173–183.

Merleau-Ponty, M. (1973). Science and the experience of expression. In *The Prose of the World*, Merleau-Ponty, M. (ed.). Translated by J. O'Neill. Northwestern University Press, Evanston.

Meskus, M. (2018). *Craft in Biomedical Research. The iPS Cell Technology and the Future of Stem Cell Science*. Palgrave Macmillan, New York.

Mitcham, C. (1994). *Thinking through Technology. The Path Between Engineering and Philosophy*. The University of Chicago Press, Chicago.

Moholy-Nagy, L. (1947). *Vision in Motion*. Paul Theobald & Co, Chicago.

Mokyr, J. (1996). Evolution and technological change: a new metaphor for economic history? In *Technological Change. Methods and Themes in the History of Technology*, Fox, R. (ed.). Routledge, London and New York.

Mol, A. (2008). *The Logic of Care. Health and the Problem of Patient Choice*. Routledge, London and New York.

Mol, A., Moser, I., Pols, J. (eds) (2010). *Care in Practice. On Tinkering in Clinics, Homes and Farms*. Transcript-Verlag, Bielefeld.

Molinier, P. (2012). Cochons et humains. A propos d'une tension inhérente à nos façons de traiter les animaux. In *Tous vulnérables ? Le care, l'environnement et les animaux*, Laugier, S. (ed.). Editions Payot & Rivages, Paris.

Molinier, P. (2018). *Le care monde. Trois essais de psychologie sociale*. ENS Éditions, Paris.

Montani, P. (2014). *Bioesthétique. Sens commun, technique et art à l'âge de la globalisation*. Vrin, Paris.

Montel, N. (1992). L'organisation du travail sur les chantiers de terrassement. Le cas du percement de l'isthme de Suez. *Culture technique*, 26, 134–145.

Moor, J.H. (2006). The nature, importance and difficulty of machine ethics. *IEEE Intelligent Systems*, 18–21.

Morris, W. and Morris, M. (eds) (2012a). Art under plutocracy [1883]. In *The Collected Works of William Morris*. Cambridge University Press, Cambridge.

Morris, W. and Morris, M. (2012b). The arts and crafts of today [1889]. In *The Collected Works of William Morris*. Cambridge University Press, Cambridge.

Mumford, L. (1934). *Technics and Civilization*. Routledge and Kegan Paul Ltd, London.

Mumford, L. (1952). *Art and Technics.* Columbia University Press, New York.

Mumford, L. (1972). The human heritage. In *Technology, Power, and Social Change*, Thrall, C.A., Starr J.M. (eds). Lexington Books, Lanham.

Navel, G. (1949). *Man at Work.* Translated by G. Reavy. Dennis Dobson Ltd, London.

Nègre, V. (2016). *L'Art et la matière. Les artisans, les architectes et la technique (1770–1830).* Classiques Garnier, Paris.

Negri, A. (2018). *Travail vivant contre capital : une conférence donnée au séminaire Lectures de Marx à l'École Normale Supérieure de la rue d'Ulm, précédée d'un entretien avec Antonio Negri.* Les éditions sociales, Paris.

Nelson, R. and Winter, S. (1982). *An Evolutionary Theory of Economic Change.* Harvard University Press, Cambridge.

Nelson, N.C., Keating, P., Cambrosio, A. (2013). On being "actionable": Clinical sequencing and the emerging contours of a regime of genomic medicine in oncology. *New Genetics and Society*, 32(4), 405–428.

Nordmann, A. (2012). Object lessons: Toward an epistemology of technoscience. *Scientiae Studia. Revista Latino-Americana de Filosofia e História da Ciência*, 10, special issue, 11–31.

O'Neill, J. (1992). The varieties of intrinsic value. *The Monist*, 75(2), 119–137.

Ortega y Gasset, J. (1941). Man the technician. In *History as a System and Other Essays Toward a Philosophy of History*, Ortega y Gasset, J. (ed.). W.W. Norton & Company, New York and London.

Ortega y Gasset, J. (1961). *The Modern Theme*. Translated by J. Cleugh. Harper, New York.

Ortony, A., Norman, D.A., Revelle, W. (2005). The role of affect and proto-affect in effective functioning. In *Who Needs Emotions? The Brain Meets the Machine,* Fellous, J.M., Arbib, M.A. (eds). Oxford University Press, New York.

Panckoucke, C.F.L. (1812). *Dictionnaire des sciences médicales, par une société de médecins et de chirurgiens*, 31, article Médecin. Paris.

Pelluchon, C. (2019a). *Nourishment. A Philosophy of the Political Body.* Translated by Smith J.E.H. Bloomsbury Academic, London.

Pelluchon, C. (2019b). Promouvoir un monde habitable. *Etudes*, 1(January), 39–50.

Pelluchon, C. (2020). *Réparons le monde. Humains, animaux, nature*. Editions Payot & Rivages, Paris.

Picon, A. (1992). Gestes ouvriers, opérations et processus techniques. La vision du travail des Encyclopédistes. *Recherches sur Diderot et sur l'Encyclopédie*, 13(October), 131–147.

Pierron, J.-P. (2010). *Vulnérabilité. Pour une philosophie du soin.* Presses Universitaires de France, Paris.

Pierron, J.-P. (2013a). Présentation. In *Éthique, politique, religions*, 2(3), 11–29.

Pierron, J.-P. (2013b). Prendre soin de la nature et des hommes. In *Éthique, politique, religions*, 2(3), 11–29.

Pierron, J.-P. (2013c). Au-delà de l'anthropocentrisme : la nature comme partenaire. *Revue du Mauss*, 2(42), 41–48.

Pierron, J.-P. (2019). *Prendre soin de la nature et des humains. Médecine, travail, écologie.* Editions Les Belles Lettres, Paris.

Pierrot, N. (2010). Les images de l'industrie en France, peintures, dessins, estampes, 1760–1870. PhD thesis, Université Paris 1 Panthéon-Sorbonne, Paris.

Pitt-Rivers, A.F.L. (1906). *The Evolution of Culture and Other Essays*. Clarendon Press, Oxford.

Planche, G. (1857). L'art et l'industrie. *Revue des deux mondes*, deuxième période, 10, 185–210.

Plato (1997). *Complete Works*, Cooper J.M. (ed.). Associate Editor Hutchinson, J.S. Hackett Publishing Company, Indianapolis.

Plessner, E. (2019). *Levels of Organic Life and the Human: An Introduction to Philosophical Anthropology*. Translated by M. Hyatt. Fordham University Press, New York.

Pols, J. (2012). *Care at a Distance. On the Closeness of Technology*. Amsterdam University Press, Amsterdam.

Pontille, D. and Denis, J. (2015). Material ordering and the care of things. *Science, Technology, & Human Values*, 40(3), 338–367.

Postma, D.W. (2006). *Why Care for Nature? In Search of an Ethical Framework for Environmental Responsibility and Education*. Springer Netherlands, Amsterdam.

Puig de la Bellacasa, M. (2011). Matters of care in technoscience. Assembling neglected things. *Social Studies of Science*, 41(1), 85–106.

Puig de la Bellacasa, M. (2012). "Nothing comes without its world": Thinking with care. *The Sociological Review*, 60(2), 197–216.

Puig de la Bellacasa, M. (2015). Making time for soil: Technoscientific futurity and the pace for care. *Social Studies of Science*, 45(5), 691–716.

Pulcini, E. (2013). *Care of the World. Fear, Responsibility and Justice in the Global Age*. Springer Netherlands, Amsterdam.

Quéré L. (2015). Retour sur l'agentivité des objets. Exposé, Sciences et Technologies de l'Institut Marcel Mauss, March 9, 2015.

Rabinow, P. (1996). Artificiality and enlightenment: From sociobiology to biosociality. In *Essays on the Anthropology or Reason*, Rabinow, P. (ed.). Princeton University Press, Princeton.

Rancière, J. (2004). *The Politics of Aesthetics. The Distribution of the Sensible.* Translated by G. Rockhill. Continuum, London.

Reich, W.T. (1995). History of the notion of *care*. In *Encyclopedia of Bioethics*, Reich, W.T. (ed.). Simon & Schuster Macmillan, New York.

Reuleaux, F. (1876). *The Kinematics of Machinery. Outlines of a Theory of Machines.* Translated by A.B.W. Kennedy. MacMillan and Co, London.

Richard, L. (2012). *Comprendre le Bauhaus. Un enseignement d'avant-garde sous la République de Weimar.* Infolio Éditions, Paris.

Ricoeur, P. (2001). De la morale à l'éthique et aux éthiques. In *Le Juste 2*, Ricoeur, P. (ed.). Editions Esprit, Paris.

Robert, P. (2005). *La Logique politique des technologies de l'information et de la communication.* Presses Universitaires de Bordeaux, Pessac.

Rodotà, S. (1995). Itinerari della cura. In *Principio era la cura*, Donghi, P., Preta, L. (eds). Gius Laterza & Figli Spa, Roma-Bari.

Rolston III, H. (1994). Value in nature and the nature of value. In *Philosophy and the Natural Environment*, Attfield, R., Belsey, A. (eds). Cambridge University Press, Cambridge.

Schaer, R. (2013). *Répondre du vivant.* Editions Le Pommier, Paris.

Schaffer, S. (1999). Enlightened automata. In *The Sciences in Enlightened Europe*, Clark, W., Golinksi, J., Schaffer, S. (eds). The Chicago University Press, Chicago.

Schiefsky, M.J. (2007). Art and nature in ancient mechanics. In *The Artificial and the Natural. An Evolving Polarity*, Bensaude-Vincent, B., Newman, W.R. (eds). The MIT Press, Cambridge.

Schillmeier, M. (2017). Design and the art of care: Engaging the more than human and the less than human. In *Care and Design. Bodies, Buildings, Cities*, Bates, C, Kullman, K., Imrie, R. (eds). John Wiley & Sons, Oxford.

Schillmeier, M. and Domènech, M. (eds) (2010). *New Technologies and Emerging Spaces of Care.* Ashgate Publishing Limited, Farnham.

Schummer, J. (2006). Gestalt switch in molecular image perception: The aesthetic origin of molecular nanotechnology in supramolecular chemistry. *Foundations of Chemistry*, 8, 53–72.

Schwartz, Y. (2008). Le travail dans une perspective philosophique. *Ergologia*, 0(March), 121–154.

Schwartz, Y. (2012). *Expérience et connaissance du travail.* Les Éditions sociales, Paris.

Seed, A. and Byrne, R. (2010). Animal tool-use. *Current Biology*, 20, R1032–1039, December 7.

Sennett, R. (2008). *The Craftman.* Yale University Press, New Haven, London.

Serres, M. (1974). *Hermès III. La Traduction.* Les Éditions de Minuit, Paris.

Serres, M. (1995a). *The Natural Contract.* Translated by E. MacArthur and W. Paulson. The University of Michigan Press, Ann Arbor.

Serres, M. (1995b). *Conversations on Science, Culture and Time: Michel Serres Interviewed by Bruno Latour*. Translated by R. Lapidus. The University of Michigan Press, Ann Arbor.

Sève, L. (1994). *Pour une critique de la raison bioéthique*. Editions Odile Jacob, Paris.

Sewell, W.H. (1980). *Work & Revolution in France. The Language of Labor from the Old Regime to 1848*. Cambridge University Press, Cambridge.

Sharp, L. (2013). *The Transplant Imaginary. Mechanical Hearts, Animal Parts and Moral Thinking in Highly Experimental Science*. University of California Press, Oakland.

Simondon, G. (2014). *Sur la technique (1953–1983)*. Presses Universitaires de France, Paris.

Simondon, G. (2017). *On the Mode of Existence of Technical Objects*. Translated by C. Malaspina and J. Rogove. University of Minnesota Press, Minneapolis.

Simondon, G. (2020). *Individuations in Light of Notions of Form and Information*. Translated by T. Adkins. University of Minnesota Press, Minneapolis.

Sloterdijk, P. (2012). The time of the crime of the monstruous: on the philosophical justification of the artificial. Translated by W. Hoban. In *Sloterdijk Now*, Elden, S. (ed.). Polity Press, Cambridge.

Solée, R.V., Valverde, S., Rosas Casals, M., Kauffman, S.A., Farmer, D., Eldredge, N. (2013). The evolutionary ecology of technological innovations. *Complexity*, 18(4), 15–27.

Soler, J. and Trompette, P. (2010). Une technologie pour la santé : traces et expertises. Chercheurs, familles et médecins autour de la définition des crises d'épilepsie. *Revue d'anthropologie des connaissances*, 4(2), 323–357.

Souriau, P. (1904). *La beauté rationnelle*. Félix Alcan Éditeur, Paris.

Sparrow, R. and Sparrow, L. (2007). In the hands of machines? the future of aged care. *Mind Mach*, 16, 141–161.

Spengler, O. (1976). *Man and Technics. A Contribution to a Philosophy of Life*. Translated by C.F. Atkinson. Greenwood Press, Westport.

Steiner, P. (2010). Philosophie, technologie et cognition. État des lieux et perspectives. *Intellectica*, 1–2(53/54), 7–40.

Stevens, H. (2016). Hadooping the genome: The impact of big data tools on biology. *BioSocieties*, 11, 352–371.

Stiegler, B. (1998). *Technics and Time 1. The Fault of Epimetheus*. Translated by R. Beardsworth and G. Collins. Stanford University Press, Stanford.

Stiegler, B. (2008). *Technics and Time 2. Disorientation*. Translated by S. Barker. Stanford University Press, Stanford.

Stiegler, B. (2010). *Technics and Time 3. Cinematic Time and the Question of Malaise*. Translated by S. Barker. Stanford University Press, Stanford.

Svandra, P. (2009). Éloge du soin. Une éthique au *cœur de la vie. Sources philosophiques, pratique et conditions de l'engagement soignant*. Seli Arslan, Paris.

Tétry A. (1948). *Les Outils chez les êtres vivants*. Librairie Gallimard, Paris.

Tresch, J. (2012). *The Romantic Machine: Utopian Science and Technology after Napoleon*. Chicago University Press, Chicago.

Tronto, J.C. (1993). *Moral Boundaries. A Political Argument for an Ethic of Care*. Routledge, New York and London.

Varela, F.J. (1979). *Principles of Biological Autonomy*. North Holland, Amsterdam.

Verbeek, P.-P. (2005). *What Things Do. Philosophical Reflections on Technology, Agency and Design*. Pennsylvania State University Press, University Park.

Verbeek, P.-P. (2010). Accompanying technology: Philosophy of technology after the ethical turn. *Techné: Research in Philosophy and Technology*, 14(1), 49–54.

Verbeek, P.-P. (2011). *Moralizing Technology. Understanding and Designing the Morality of Things*. The University of Chicago Press, Chicago.

Vergez, A. (1956). Technique et morale chez Platon. *Revue philosophique de la France et de l'etranger*, 146, 9–15.

Vérin, H. (1993). *La Gloire des ingénieurs. L'Intelligence technique du XVIe au XVIIIe siècle*. Albin Michel, Paris.

Vernant, J.-P. (2006). *Myth and Thought Among the Greeks*. Translated by J. Fort. Zone Books, New York.

Vigarello, G. (1978). *Le corps redressé. Histoire d'un pouvoir pédagogique*. Jean-Pierre Delarge, Editions universitaires, Paris.

Wajcman, G. (2012). *L'Objet du siècle*. Editions Verdier, Paris.

Willems, D. (1995). Mesurer sa respiration. Reconstituer le corps avec un objet médical. *Techniques & Culture*, 25–26, 55–70.

Wilson, M. (2002). Making nursing visible? Gender, technology and the care plan as script. *Information Technology & People*, 15(2), 139–158.

Winner, L. (1980). Do artifacts have politics? *Daedalus*, 109(1), 121–136.

Winnicott, D.W. (2017). Cure [1970]. In *The Collected Works of D.W. Winnicott, Volume 9, 1969–1971*. Edited by L. Caldwell and T. Robinson. Oxford University Press, Oxford.

Worms, F. (2013). Le soin comme orientation éthique et politique dans le moment présent. In *Prendre soin : savoirs, pratiques, nouvelles perspectives*. Chagnon, V., Dallaire, C., Espinasse, C., Heurgon, É. (eds). Hermann, Paris.

Wynsberghe, A. (2013). Designing robots for care: care centered value-sensitive design. *Sci Eng Ethics*, 19, 407–433.

Wynsberghe, A. and Gastmans, C. (2008). Telesurgery: An ethical appraisal. *Journal of Medical Ethics*, 34(10), e22.

Xenophon (2006). *The Shorter Socratic Writings.* Cornell University Press, New York.

Zarka, Y.-C. (2013). De l'homme-machine à la machine post-humaine : la vision machinique du monde. *Cités*, 3(55), 3–8.

Zhuangzi (2009). *The Essential Writings with Selection from Traditional Commentaries.* Translated by B. Ziporyn. Hackett Publishing Company, Indianapolis.

Ziman, J. (ed.) (2000). *Technological Innovation as an Evolutionary Process.* Cambridge University Press, Cambridge.

Index

G, I

L, M

N, O

P, R

T, V, W

Other titles from

iSTE

in

Interdisciplinarity, Science and Humanities

2020

ATCHOUA Julien, BOGUL Jean-Jacques, DIALLO Saikou
Digital Technologies and African Societies: Challenges and Opportunities (Africa Digital Age Set – Volume 1)

CHERICI Céline
From Clouds to the Brain: The Movement of Electricity in Medical Science

GUILLAUME Astrid, KURTS-WÖSTE Lia
Making Sense, Making Science

SCHEPS Ruth, MAUREL Marie-Christine
Mathematics in the Visual Arts

PAVIE Xavier
Critical Philosophy of Innovation and the Innovator (Innovation and Responsibility Set – Volume 6)

2019

ANCORI Bernard
The Carousel of Time: Theory of Knowledge and Acceleration of Time

BOULEAU Gabrielle
Politicization of Ecological Issues: From Environmental Forms to Environmental Motives

GAYON Jean†, interviewed by PETIT Victor
Knowledge of Life Today: Conversations on Biology

KHATCHATOUROV Armen, with the collaboration of CHARDEL Pierre-Antoine, FEENBERG Andrew, PÉRIÈS Gabriel
Digital Identities in Tension: Between Autonomy and Control
(Innovation and Responsibility Set – Volume 5)

LENOIR Virgil Cristian
Ethically Structured Processes
(Innovation and Responsibility Set – Volume 4)

NIKOLOVA Blagovesta
The RRI Challenge: Responsibilization in a State of Tension with Market Regulation
(Innovation and Responsibility Set – Volume 3)

ZEMBRI-MARY Geneviève
Project Risks: Actions Around Uncertainty in Urban Planning and Infrastructure Development

2018

LOISY Catherine, COULET Jean-Claude
Competence and Program-based Approach in Training: Tools for Developing Responsible Activities
(Innovation and Responsibility Set – Volume 1)

PANSERA Mario, OWEN Richard
Innovation and Development: The Politics at the Bottom of the Pyramid
(Innovation and Responsibility Set – Volume 2)

2017

MAESSCHALCK Marc
Reflexive Governance for Research and Innovative Knowledge
(Responsible Research and Innovation Set – Volume 6)

PELLÉ Sophie
Business, Innovation and Responsibility
(Responsible Research and Innovation Set – Volume 7)

2016

GIANNI Robert
Responsibility and Freedom: The Ethical Realm of RRI
(Responsible Research and Innovation Set – Volume 2)

GRUNWALD Armin
The Hermeneutic Side of Responsible Research and Innovation
(Responsible Research and Innovation Set – Volume 5)

LENOIR Virgil Cristian
Ethical Efficiency: Responsibility and Contingency
(Responsible Research and Innovation Set – Volume 1)

PELLÉ Sophie, REBER Bernard
From Ethical Review to Responsible Research and Innovation
(Responsible Research and Innovation Set – Volume 3)

REBER Bernard
Precautionary Principle, Pluralism and Deliberation: Science and Ethics
(Responsible Research and Innovation Set – Volume 4)

www.ingramcontent.com/pod-product-compliance
Lightning Source LLC
LaVergne TN
LVHW011911200926
848484LV00002B/594
9781786305596